Boiler Dynamics and Controls

Boiler Dynamics and Controls

Course Notes

F. Paul de Mello

To order additional copies of this book, contact:
Xlibris LLC
1-888-795-4274
www.Xlibris.com
Orders@Xlibris.com
141415

CONTENTS

Text Chapters

I.	Boiler Process Dynamics and Control-Overview	11
II.	General Principles and Structures in Boiler Controls	41
III.	Drum Boiler Pressure Effects	50
IV.	Drum Boiler Feedwater Controls	70
V.	Fuel and Air Controls for Drum Type Boilers	85
VI.	Furnace Draft Controls	98
VII.	Steam Temperature Controls	128
VIII.	Miscellaneous Loop Controls	134
IX.	Controls for Once-Through Boilers	139
X.	Direct Digital Boiler Control	165
XI.	Modeling from First Principles	185

Appendix Material

A.	Dynamic Systems, Differential Equations—Transient And Steady State Solutions—Operational Impedance	227
B.	Laplace Transforms	240
C.	Transfer Functions, Block Diagrams	260
D.	Analog Computers—State Space—Numerical Methods of Differential Equation Solution	265
E.	Feedback Control System Concepts	279
F.	Notes On Process Control	303

F. P. DE MELLO

Consulting Engineer

Mr. de Mello graduated with BS and MS degrees in Electrical Engineering from MIT where he was elected to Tau Beta Pi and Sigma XI. His academic experience included several test engineering and laboratory assignments with the General Electric Company (GE) between 1945 and 1948.

In 1948, he joined the Rio Light and Power Company in Brazil and over several years held position of increasing technical responsibility in system planning and design studies concerning expansion of the Rio, Sao Paulo, and City of Santos systems.

In 1955, Mr. de Mello joined the Analytical Engineering Section of GE's Apparatus Sales Division in Schenectady. Here he undertook design and analysis studies of controls of industrial, power apparatus and aircraft power systems, making extensive use of analog computers. In 1959, he was assigned to specialized studies of dynamics of electrical machines, excitation control, prime-mover systems, and overall power systems.

From 1961 to 1969, he conducted and guided extensive research efforts on modeling of dynamics of power systems and power plants for use in advanced boiler and plant control design studies. He made major pioneering contributions in the development of digital computer methods for dynamic analysis and process control design. Of particular note were the development of computer techniques for the simulation of complex boiler dynamics and for the synthesis of multivariable boiler-turbine controls for which he was awarded GE's Managerial and Ralph Cordiner Awards. He also made significant contributions in the study of electrical machine dynamics, their voltage and governing controls and to the analysis and implementation of system load-frequency controls.

Mr. de Mello joined Power Technologies, Inc., at the time of its formation in August of 1969 as Principal Engineer, Dynamics and Control, and Secretary-Treasurer. He was appointed Vice President-Secretary in 1973. He was a Director of PTEL, PTI's affiliate in Brazil. From 1974 to 1976, he was project manager for PTI and PTEL in system and design studies for transmission from Itaipu, the world's largest 800 kV system, and served on the advisory Board of the Study Group for Itaipu transmission. Prior to his appointment as Principal Consultant in 1987 he was Manager of PTI's Consulting Services Department.

Mr. de Mello has three patents and authored more than 100 technical papers in IEEE, ISA, American Power Conference, World Power Conference, and other utility industry publications, and also lectured to professional society groups. He has served on the IEEE Systems Controls Subcommittee and the joint IEEE Working Group on Plant Response and also served as US representative on CIGRE Study Committees 38 and 39. He has taught dynamic and operational subjects in PTI's Power Technology Course and conducted one week courses given to over 2000 engineers worldwide on Power System Dynamics.

Mr. de Mello is a Life Fellow of IEEE, a Life Fellow of ISA, a member of the National Academy of Engineering, a registered Professional Engineer in New York State. He was awarded the IEEE Charles Concordia Award in 2003.

ACKNOWLEDGEMENT

This contribution to the technology of modeling and control of large steam generators, both drum boilers, and once through, critical and subcritical, was largely based on my experience gained while working at the Analytical Engineering Department of the Apparatus Sales Division of the General Electric Co., Schenectady, NY, in the mid-60s, before GE's entry into the analog operational amplifier-based process control business with the GEMAC line. While most of this information has been published before in various ISA and IEEE papers, it is presented here in a cohesive manner to serve as a text to those involved in the important field of plant modeling and control, implemented these days with distributed digital systems. The principles of control are the same, whether analog or digital, although the digital approach makes it much easier to use adaptive features, nonlinear logic, multiplication, division, function generation, square roots, etc., which were expensive to implement with analog controls. The design process with modern digital simulation tools and modeling capability can be used to greatly improve the performance of controls through the phases of start-up and wide ranges of operating conditions some of which, in the past, had to be handled by manual control subject to operator error. Other very important applications of the technology are in the field of operator training simulators.

A most important phase of the modeling effort involved field testing through extensive night work at power stations. Especial recognition for encouraging the work of model validation by tests goes to Carolina Power and Light with the test on Plant Robinson in 1961, to Georgia Power which insisted on having the design of the Hagan analog controls for Plant McDonough be done by GE, based on simulation, and for the Plant McDonough tests in 1964; to South Carolina Electric & Gas which awarded

GE the supply of a GEMAC boiler turbine control system for its sub-critical once-through unit at the Canadys plant; to Florida Power Corp for similar effort in the design of boiler controls for its Crystal River Plant; to Baltimore Gas & Electric for the tests at the Crane plant; to Consolidated Edison with their tests at Astoria related to furnace draft and implosion studies; to NY Power Authority at the Charles Poletti Plant for similar studies.

In all this work I was blessed by the collaboration of colleagues who were of immense help in the conduct of the tests (John C. Westcott, recently deceased) and in carrying out the simulation studies, and testing (D. N. Ewart, D. J. Ahner, and R. J. Mills). I am also grateful for the encouragement and support of my manager at GE, Dr. L. Kirchmayer (deceased), W. M. Stephens (deceased) of Georgia Power, and V. C. Summer (deceased) of South Carolina Electric & Gas. In the computer simulation effort I had a team of expert "Green Berets," John Undrill whose expertise in modeling dynamic systems excelled not only in electrical large scale network dynamic and steady state performance modeling, but also in the plant area, both hydro and fossil-fired. Witness his work in furnace draft problems in Chapter VI.

Lou Hannett and Jim Feltes were always there to tackle computer simulation work for any type of electrical, electro-mechanical and thermo-mechanical system. Dick Mills made enormous contributions in the large scale plant operator training simulator area, starting at GE, developing the software for the first nuclear plant simulator at Dresden II in the late 60s, and then at PTI in similar applications in Scandinavia, and extending this to the fossil plant simulator business at PTI headed by John Westcott. I mention this because a key element in this business is the efficient and accurate modeling of components in the path of water and steam flow through heat exchangers, steam generators with their economizers, waterwalls, superheaters, reheaters, as well as the dynamics of the combustion process and heat transfer through the gas path, contained in these notes.

I would also remember my deceased wife, Barbara, who bore with my frequent absence for business including prolonged trips with plant testing, while she had the responsibility of handling four children who turned out to be responsible and successful citizens, and to my wife Margaret who puts up with me in my senior years.

Finally, this publication would not have been possible without the help from my good friend Cyrus Taft who first suggested that the work merited publication, and who worked with ISA to make this possible.

CHAPTER I

BOILER PROCESS DYNAMICS AND CONTROL-OVERVIEW

INTRODUCTION

Control of the steam generation process in power plants is a vital function to assure reliability, economy, and safety of power production. It is also one of the most complex and difficult tasks representing a challenge to control system designers and plant operators. These comments are in the context of conditions in the mid 1960's when the transition from pneumatic to operational amplifier technology was under way.

The capabilities of modern control technology implemented through present day analog and computer control hardware have opened vast areas of opportunity in the automation and control of power plants. These capabilities fill a vital need as the trend to larger sizes and new designs of boiler-turbine systems, at higher pressures and temperatures, increases the importance of achieving closer, safer and more sophisticated control of plant variables.

Utility engineering know-how and dedication to the technology is no doubt the most important ingredient in capitalizing on the state of the art and applying it, in logical steps, to yield badly needed improvements in the control of power plants.

Before proceeding with the development of details of the steam generation process and its controls, it is appropriate to place in proper perspective the role of energy supply dynamics in the overall power system structure.

The schematic in Fig. 1.1 shows how various systems and subsystems of automatic controls are involved in the overall control of the power system. At a more detailed level the block labeled "Prime Mover System" is described schematically, by its main components, typically as in Fig. 1.2. The prime mover system must maintain process conditions, such as, pressures, temperatures, and flow rates at optimum and safe values while responding to the demands of automatic or manual dispatch control. This control, in turn, acts to coordinate, instant by instant, the production from each source to meet system-load requirements at minimum cost. An understanding of the reactions of the various processes and interactions among systems and subsystems is necessary to evaluate response requirements and control system design criteria.

The area addressed in these notes concerns the power plant control system. This system, often referred to as boiler control, must act to maintain process conditions at desired levels and to protect the plant by automatic override or runback action against any upset which may endanger the safety of the equipment. The well-engineered design of a dynamically integrated control system for such a complex process as that of steam power generation cannot be based entirely on intuition and costly trial and error, especially where extrapolation of past experience is not directly applicable. Fortunately, today's control technology and computation aids permit systhesis and evaluation of the overall boiler-turbine control prior to its installation. The indispensable requirement for engineering of the overall control system is the ability to simulate the dynamic characteristics of the boiler-turbine and auxiliaries system.

In addition to the fundamental requirement of understanding the process, that is, having the ability to quantitatively predict boiler dynamic performance, it is essential that one have a practical knowledge of control concepts and control analysis techniques.

The material in these notes addresses both the understanding of the process physics and control requirements.

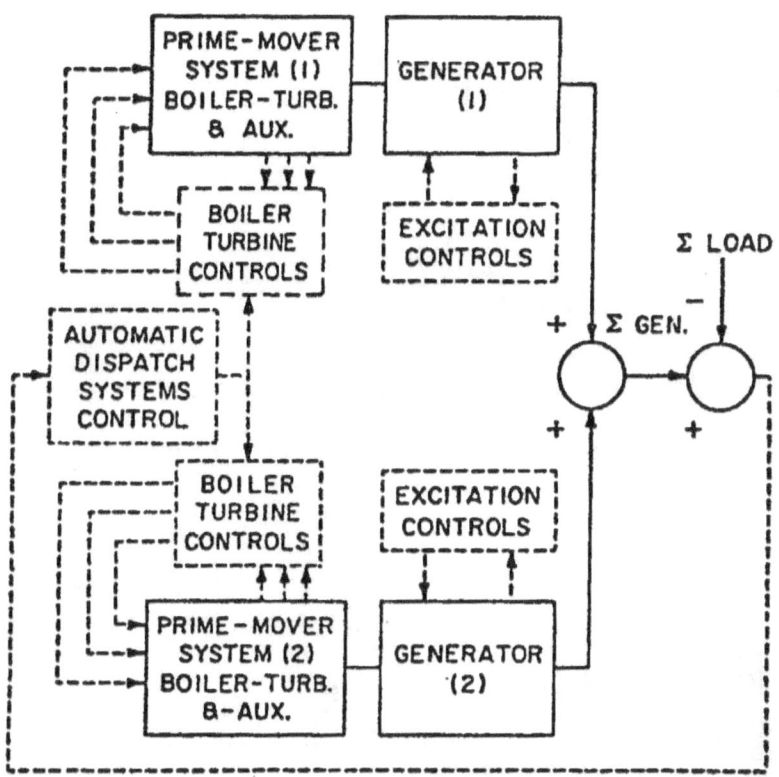

FIGURE 1.1 Power System and Power Plant Control Loops

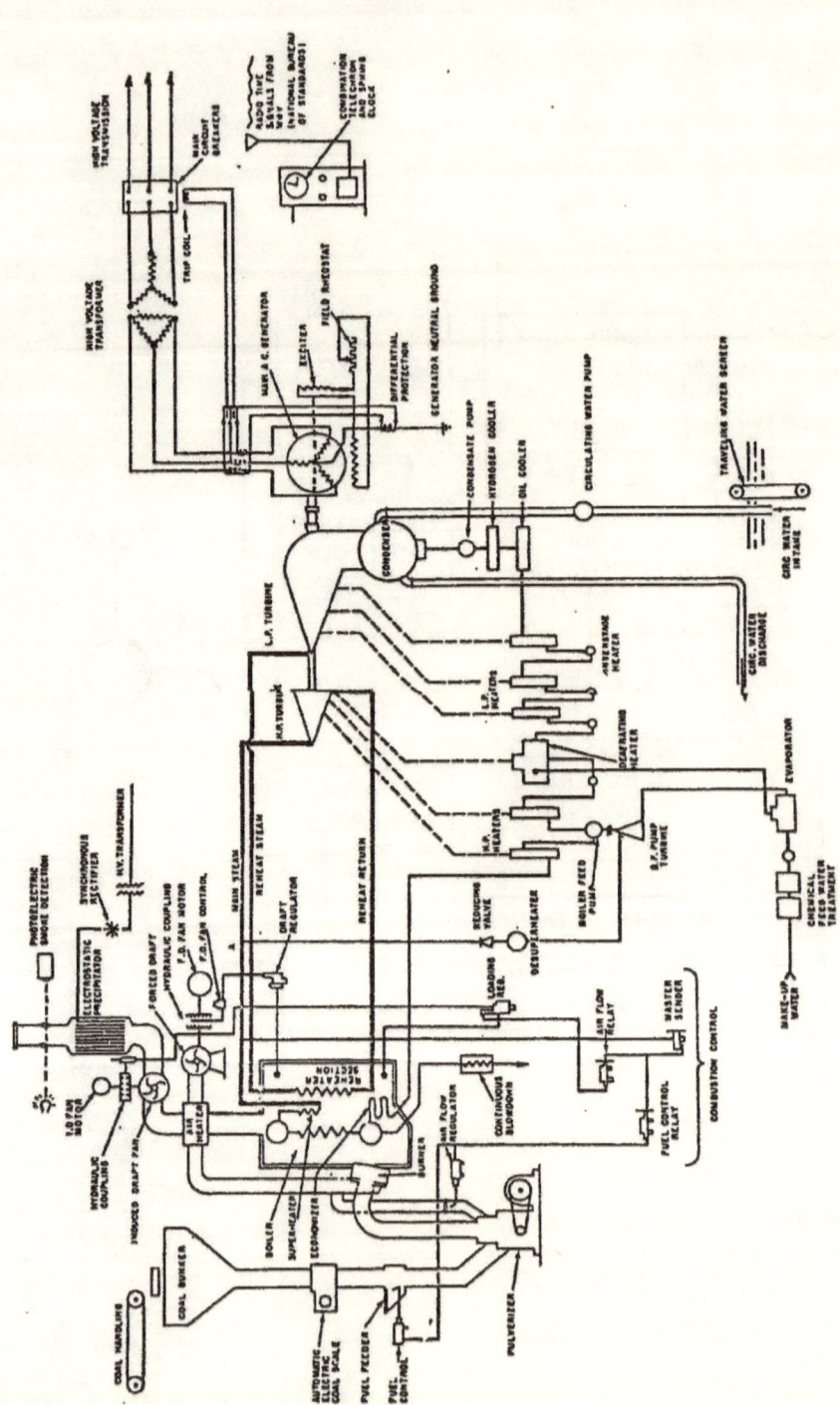

FIGURE 1.2 Elements in Typical Coal-Fired Plant

MULTIVARIABLE PROCESS

The term "multivariable process" characterizes those processes in which two or more mutually <u>coupled</u> variables are to be controlled. A boiler-turbine system is an excellent example of a multivariable process.

Fig. 1.3 describes symbolically the boiler process identifying dependent and independent variables. Independent variables are those whose values may be arbitrarily set, such as fuel, air, tilts, spray, etc. They may be considered as inputs to the process. The dependent variables describe the reaction of the process to the inputs. They are throttle pressure, main steam temperature, reheat steam temperature, drum level, etc. The important point is that these variables are mutually coupled. A change of each input variable affects some or all of the output variables. The output-input relationships are time dependent and can be described by differential equations. Although these relations are, in general, nonlinear in nature, they may often be approximated by linear differential equations where one is concerned with small excursions about a given operating point. Operationally then, a multivariable process can be described by a matrix of equations, as indicated in Fig. 1.4 which, for the sake of illustration, deals with two variables only.

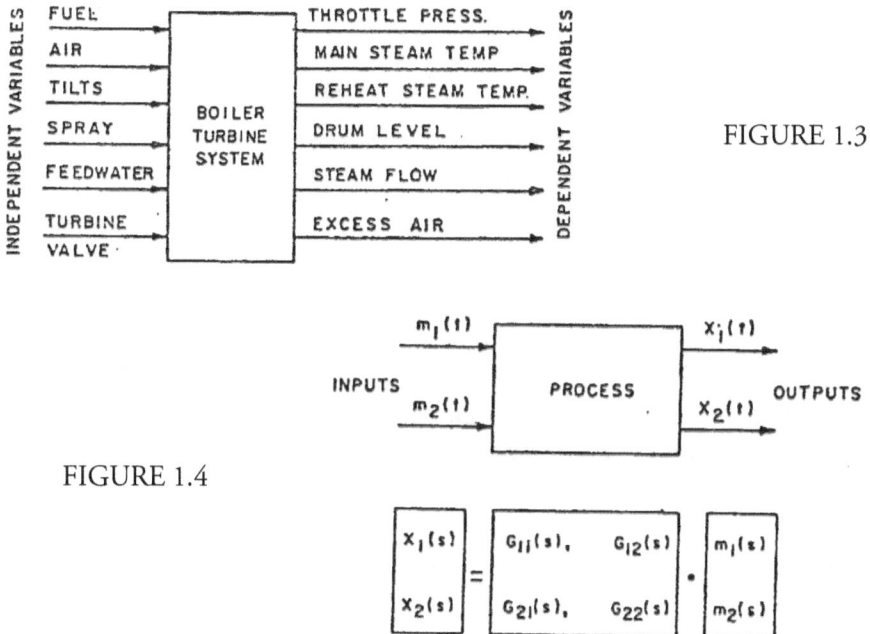

FIGURE 1.3

FIGURE 1.4

The matrix notation of Fig. 1.4 is shorthand for the following equations:

$$x_1(s) = m_1(s)G_{11}(s) + m_2(s)G_{12}(s)$$

$$x_2(s) = m_1(s)G_{21}(s) + m_2(s)G_{22}(s)$$

The transfer functions $G_{11}(s)$ and $G_{22}(s)$ represent the self terms:

$$G_{11}(s) = \frac{x_1(s)}{m_1(s)} \text{ with } m_2(s) = 0$$

$$G_{22}(s) = \frac{x_2(s)}{m_2(s)} \text{ with } m_1(s) = 0$$

$G_{12}(s)$ and $G_{21}(s)$ are the mutual coupling terms:

$$G_{12}(s) = \frac{x_1(s)}{m_2(s)} \text{ with } m_1(s) = 0$$

$$G_{21}(s) = \frac{x_2(s)}{m_1(s)} \text{ with } m_2(s) = 0$$

The strength of the terms $G_{12}(s)$ and $G_{21}(s)$ relates to the degree of coupling in the process. In the case of the boiler process, the order of the matrix is much larger, although some coupling terms are not very significant. When coupling terms are negligible, the control of each variable can be done on an independent loop basis. In this case, the process matrix reduces to a diagonal matrix, and the problem of control becomes one of several independent single-variable control loops.

Fig. 1.5 shows the concept of multivariable control where, in principle, the controller matrix should be multivariate in nature designed to compensate for the process interactions so as to effect overall responsive and stable performance. Prediction is the fundamental concept of dynamic control. The design of a control system with the correct predictive characteristics can be attempted provided that the process response characteristics are known.

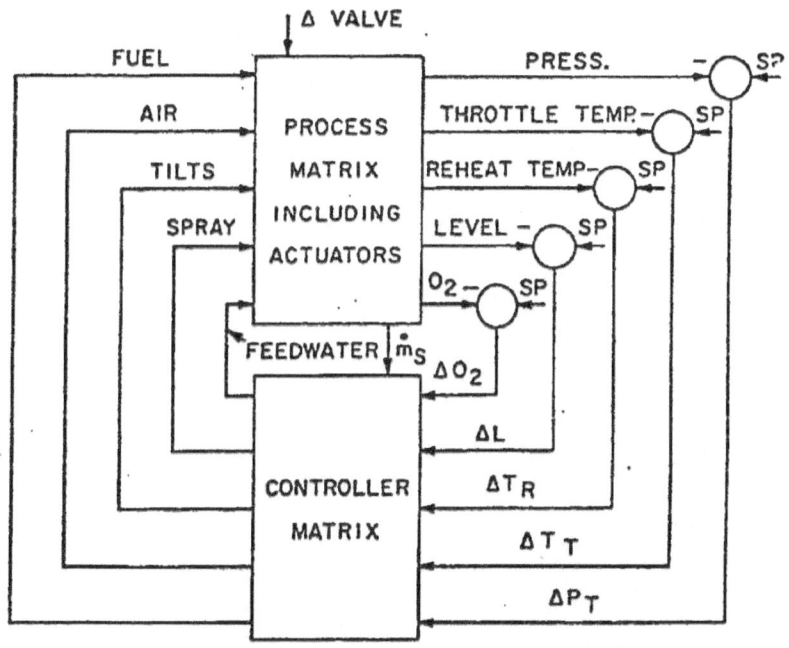

FIGURE 1.5 Multivariable Controls

PROCESS MODELING

The development of a mathematical model describing the dynamic-response characteristics of the power plant process is a most demanding and time-consuming task. This development must proceed hand in hand with the analytical being confirmed by the experimental.

There is always a strong temptation for the control engineer to try to derive a model of the process from analysis of tests only, using the input-output black-box approach, whereby hopefully he need not get involved in the analysis of the basic-process physics. However, with this approach, an empirical fit of second or third order expressions to match the tested time responses would serve little more than to model the tested unit, and would not give us the ability to predict the dynamics of other units of different designs. Further, such an approach can easily lead to error since the observed changes might have occurred as a result of some inadvertent effect which might not have been suspected without analysis. For instance, in some typical tests not involving a change in spray valve, it was discovered that a part of the temperature changes occurred because of variations of spray flow due to variations in the pressure head across the valve. This effect might not have been detected had it not shown up from comparison with results from analytical simulation where, at first, spray flow was assumed constant with constant spray valve opening. Experience has shown that analytical modeling, i.e., determining the process dynamics by solution of equations describing basic process phenomena is an indispensable step along with confirmation with tests.

THE PROCESS

Fig. 1.6 describes a typical boiler process.

Steam generated in the waterwalls leaves the drum, passes through the primary superheater and then through the secondary superheater from where it goes through the high-pressure turbine. It then re-enters the boiler through the reheater from where it passes through the low-pressure turbine and to the condenser.

The combustion products path is also indicated in Fig. 1.6. The waterwalls absorb radiant heat in the furnace. The hot gases leaving the furnace give up heat by radiation and convection to the secondary superheater, reheater, primary superheater and economizer in succession.

Constant volume circulating pumps maintain a practically constant rate of circulation through the down-comers and waterwalls.

De-superheating spray water is introduced between the primary and secondary superheater sections for the control of main—steam temperature.

Burner tilts are used to control reheat temperature, and the firing rate is governed by the pressure controls. Feedwater flow follows steam demand and is controlled to maintain drum level.

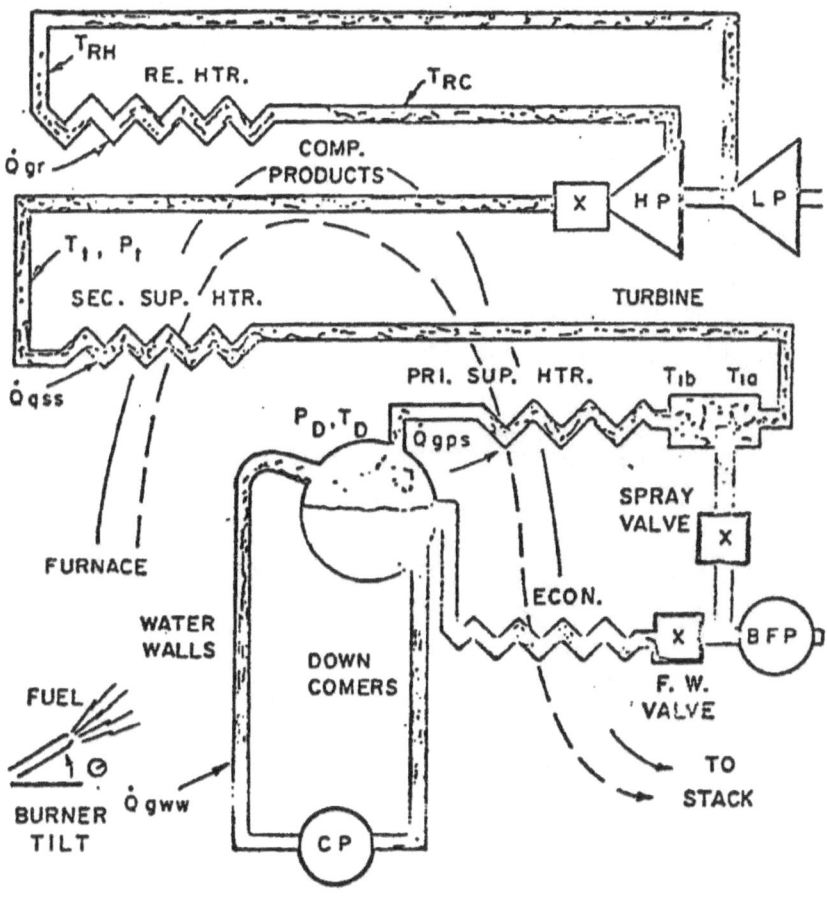

FIGURE 1.6 Schematic of Drum-type Boiler and Reheat Turbine

RESPONSE CHARACTERISTICS FROM TESTS

Typical of dynamic tests are "open loop" response tests whereby the response of pertinent process variables is recorded following an abrupt change in particular input variables. Process-variable changes are recorded with great sensitivity by showing the steady-state absolute values suppressed to zero.

Fig. 1.7 shows an example of results for an open loop test on a controlled-circulation, coal-burning drum-type boiler. An abrupt small change in coal-feeder position was effected, and deviations of pertinent variables, such as, throttle pressure, superheat, inter-stage and reheat temperature, drum level, etc., are recorded. All controls were on manual except for feedwater whose controls were left on automatic so as to hold drum level.

The dynamic-response characteristics of a boiler-turbine are recorded by many such tests at different operating load points. Typical input disturbances are changes in: turbine valve position, spray flow, air flow, burner tilts, or recirculation flow, etc.

Of significant value also are closed loop tests wherein the response of boiler variables is recorded following an abrupt change in turbine valve position with all controls on automatic. Fig. 1.8 is an example of such a test.

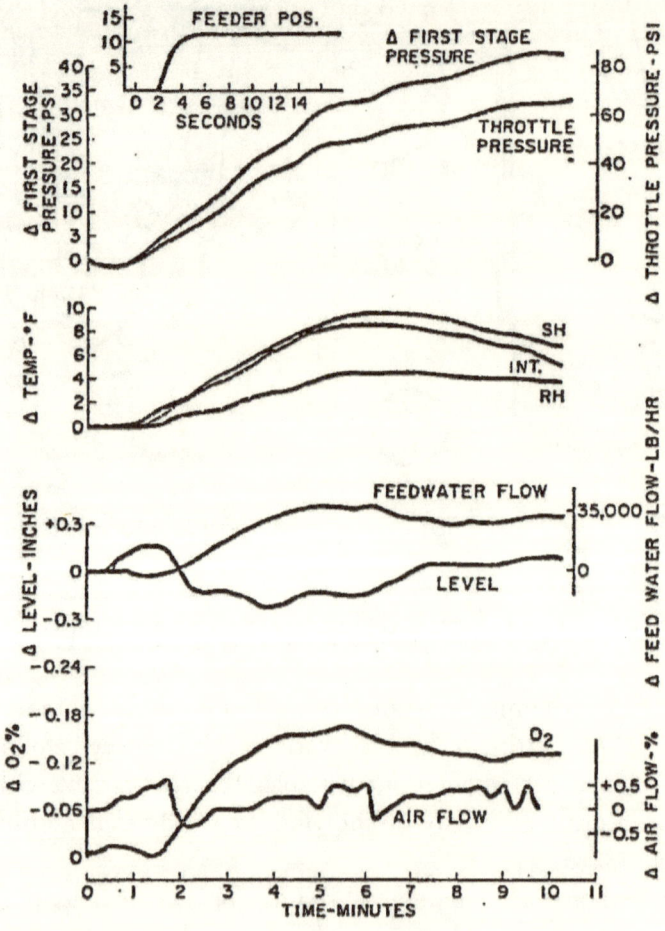

FIGURE 1.7 Open Loop Test. Change in Fuel

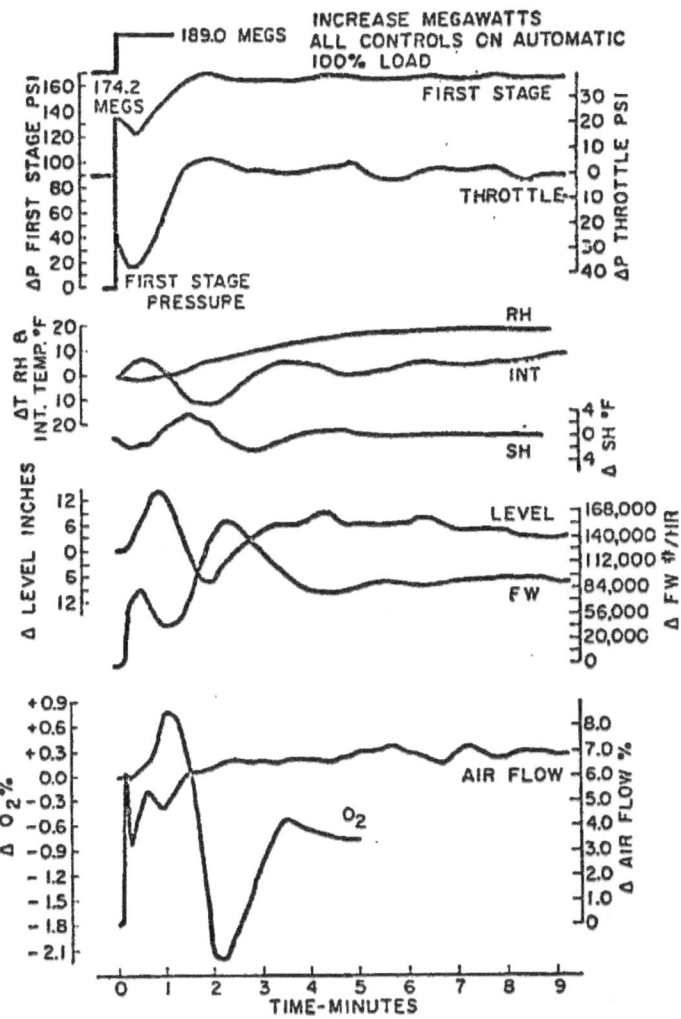

FIGURE 1.8 Closed Loop Test. Increase in MW

SIMULATION WITH LINEARIZED MODELS

Analog computer models or their digital equivalent derived from physical process considerations and constants describing the boiler-turbine are powerful tools for developing an understanding of boiler dynamics and for the study of controls. Such data as boiler section volumes, masses, steady-state values of flows, temperatures, pressures and heat balances are necessary to derive the model.

The boiler process can be described by equations relating fluid flow, heat flow and the thermodynamic properties of the substances in each section of the boiler. The true equations are nonlinear partial differential equations with respect to fluid path (distance) and with respect to time. The nonlinear nature of the equations arises from the fact that most equations involve products of variables (energy = flow × enthalpy) and from such nonlinear relations as between pressure drop and flow, the nonlinear properties of the fluids (water, water and steam and superheated steam), the nonlinear relations governing convection and radiant heat transfer as well as the fairly complex process of boiling and steam-bubble formation with its complicated effects on film coefficients.

With engineering judgment and use of linearization techniques, approximate linearized models of the boiler dynamics can be derived. A brief description of a typical process and the linearized-mathematical simulation of some process phenomena are included in order to illustrate the approach in the derivation of such models.

LINEARIZED PROCESS EQUATIONS

Pressure in the boiler is considered as a result of accumulated steam mass within the boiler storage volume. The equivalent storage factor or capacitance involves the total steam volume within the boiler including superheater, the relation between a weighted specific volume and pressure, and the mass of saturated water in waterwalls and drum. Friction drop through the superheaters is accounted for as function of steam flow rate.

Fig. 1.9 shows the equivalent system representation for changes in pressures and changes in flows.

Fig. 1.10 is a block diagram showing the linearized flow and pressure drop equations and formation of drum pressure proportional to the integral of rate of steam generation in the waterwalls less steam flow out of the drum.

Fig. 1.11 is a block diagram describing the solution of the rate of steam generated in the waterwalls. The rate of steam generated is proportional to the rate of heat absorption from waterwalls to the steam and water mixture. There is also a correction due to pressure since the enthalpy of saturated steam decreases with increasing pressure, and hence it takes less heat to generate a given mass of saturated steam at a higher pressure than at a lower pressure.

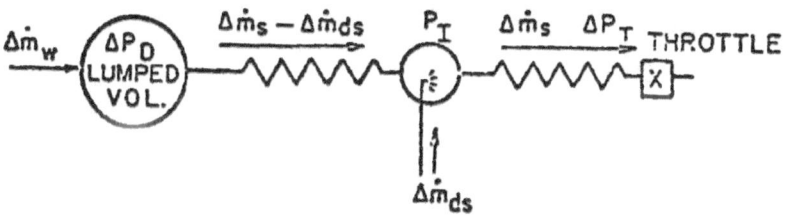

Δm_w = CHANGE OF STEAM GENERATION IN WATER WALLS

Δm_s = CHANGE OF STEAM FLOW THROUGH TURBINE

Δm_{ds} = CHANGE IN SPRAY WATER FLOW

ΔP_D = CHANGE IN DRUM PRESSURE

ΔP_I = CHANGE IN INTERMEDIATE PRESSURE

ΔP_T = CHANGE IN THROTTLE PRESSURE

FIGURE 1.9 Equivalent System Relating Pressures and Flows

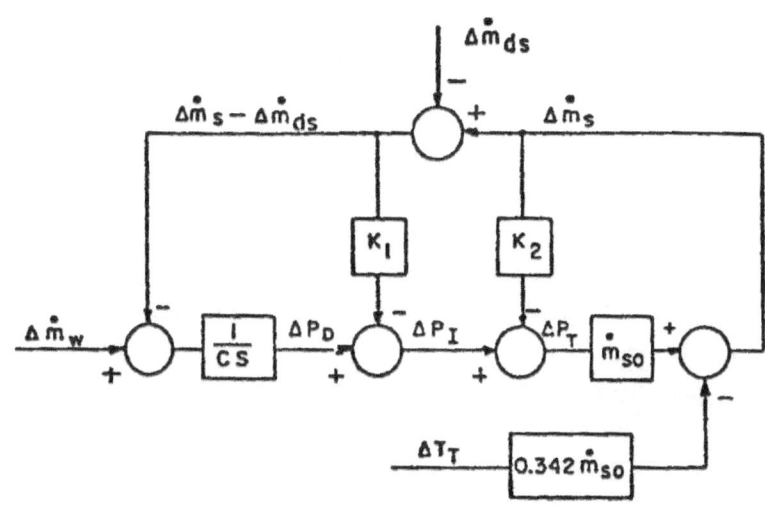

C = BOILER STORAGE FACTOR
K_1 = FRICTION DROP COEFFICIENT
K_2 = FRICTION DROP COEFFICIENT
ΔT_T = CHANGE IN MAIN STEAM TEMP
m_{so} = STEADY STATE STEAM FLOW RATE

FIGURE 1.10 Block Diagram for Solution of Pressure and Flow

K6w = WATER WALL METAL CAPACITANCE
K4w = WATER WALL HEAT TRANSFER FACTOR
hg = ENTHALPY OF SAT, STEAM
he = ENTHALPY OF WATER AT ECONOMIZER OUTLET
T_n = SAT. TEMP

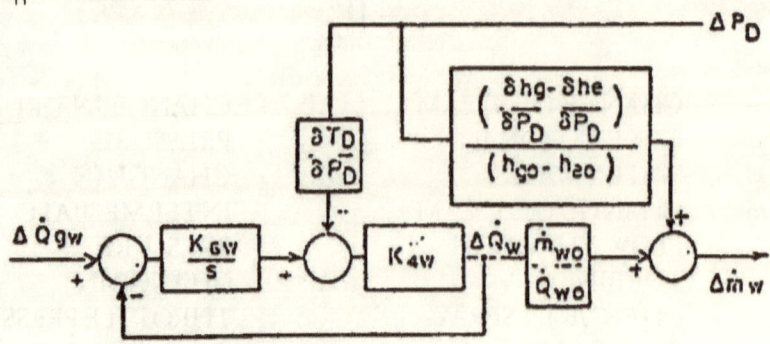

Q   = HEAT FLOW RATE ABSORBED BY WATERWALLS
P_D = DRUM PRESSURE
m_w = STEAM RATE GENERATED IN WATERWALLS
SUBSCRIPT   O = STEADY STATE VALUE
             Δ = CHANGE

FIGURE 1.11 Block Diagram for Solution of Steam Generated in Waterwalls

Rate of heat flow from waterwalls to water and steam mixture is a function of the temperature difference between waterwall metal and the water and steam mixture. The temperature of the water and steam mixture is a function of drum pressure.

The rate of change of waterwall metal temperature is proportional to the net rate of heat input to the metal.

The equations representing the process of mass flow and heat transfer in superheater and reheater tubes are nonlinear partial differential equations in space and time. The classical approach to solution of such equations is to break up the space dimension into discrete intervals and convert the partial-differential equations into a set of difference equations in the space dimension and ordinary differential equations in the time domain. The approach is analogous to converting a distributed-parameter problem into a series of lumped-parameter sections.

Fig. 1.12 shows the conversion of a length of superheater section into lumped-parameter subsections for which inlet and outlet conditions are defined. The metal mass in each subsection is assumed at an average metal temperature. The greater the number of subsections, the more faithful is the approximation, although gains in accuracy soon reach the law of diminishing returns.

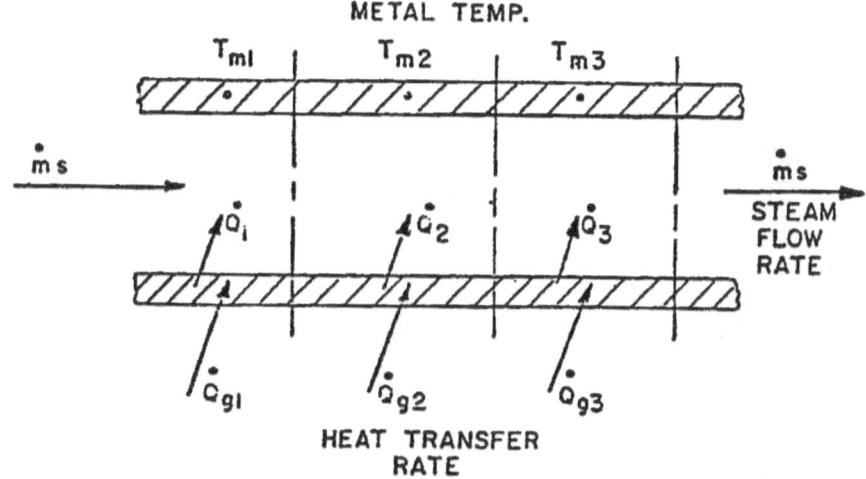

FIGURE 1.12 Lumped Parameter Subsections of a Superheater

Faithful mathematical models of superheaters have been derived by visualizing the continuous-flow process as a sequence of pulsations whereby the volume of a given subsection is filled up instantaneously with fresh steam at inlet conditions, and after residing in that subsection for a time equal to the residence time, it is instantaneously expelled to the next subsection. This cine-photographic transformation of the flow process leads to considerable simplifications in analysis permitting the solution of part of the phenomena in closed form and, therefore, avoiding the familiar temperature reversal or wringing effect characteristic of lumped parameter models. Fig. 1.13 shows the model of a superheater subsection showing outlet temperature as affected by changes in inlet temperature, mass flow rate, heat absorption from gases and pressure. Boiler heating sections would comprise several such models in cascade whereby the outlet conditions in one subsection become inlet conditions of the next.

The heat transfer from hot gases to a section of boiler metal can be related to gas temperature at the furnace exit, that section's metal temperature, rate of flow of combustion products and the heats absorbed upstream of that particular section. A set of equations are written to describe these phenomena.

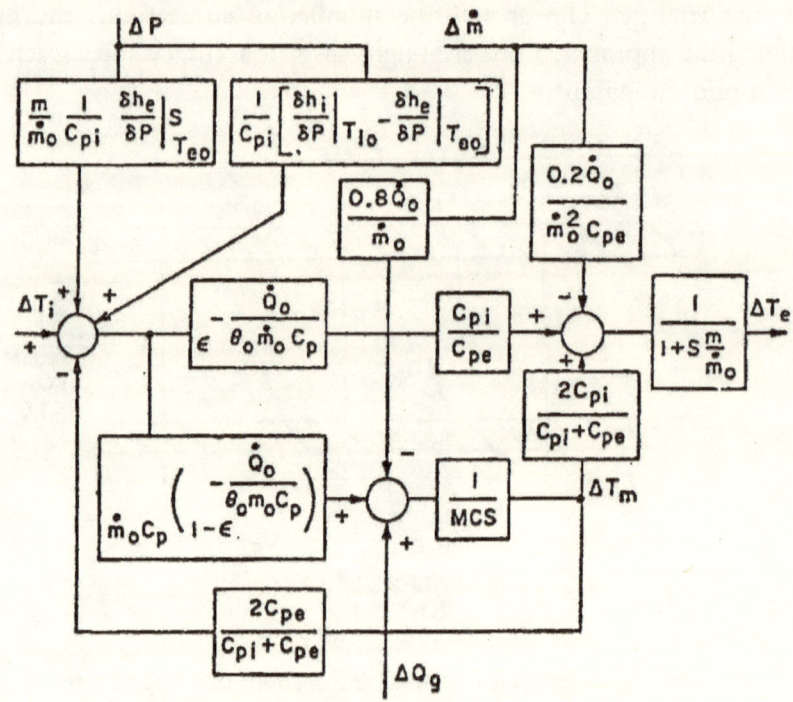

m = RESIDENT STEAM MASS

ΔT = CHANGE IN TEMP.

ΔP = CHANGE IN PRESS.

C_p = SPECIFIC HEAT

$\Delta \dot{Q}$ = CHANGE IN HEAT TRANSFER RATE

Δms = CHANGE IN STEAM FLOW RATE

θ_o = STEADY—STATE TEMP. DROP FROM METAL TO STEAM

h = ENTHALPY

MC = METAL HEAT CAPACITANCE

SUBSCRIPTS:

    i = INLET

    e = EXIT

    o = STEADY-STATE

FIGURE 1.13 Block Diagram for One Subsection Superheater Model

The entire boiler-turbine system is thus simulated with the simultaneous solution of the equations describing the phenomena on the various parts, some of which have been illustrated in Figs. 1.9 to 1.13.

Typical results confirming the validity of simulation are shown on Figs. 1.14 and 1.15. Fig. 1.14 shows the response of superheater outlet steam temperature to a change in spray flow. Fig. 1.15 shows the pressure and temperature deviations under the action of automatic combustion and temperature controls following an abrupt change in turbine MW.

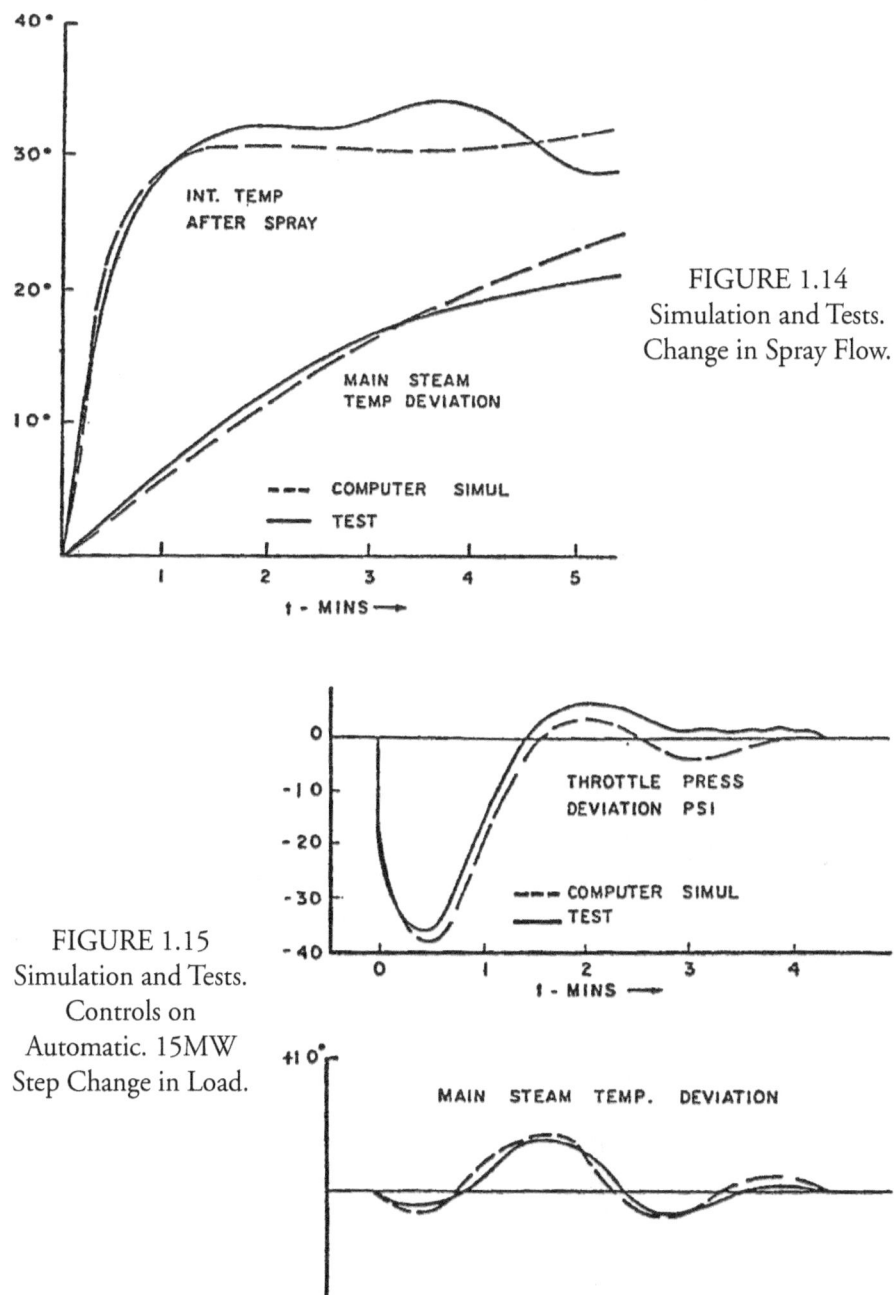

FIGURE 1.14
Simulation and Tests.
Change in Spray Flow.

FIGURE 1.15
Simulation and Tests.
Controls on
Automatic. 15MW
Step Change in Load.

NONLINEAR DIGITAL SIMULATION

Before the advent of digital computers and digital dynamic simulation programs, large and complex prime-mover system-dynamic problems were tackled on analog computers with the use of linearization techniques. To preserve the nonlinear nature of the equations would have required a large number of multipliers and function generators much beyond the capability of the largest conceivable analog facility.

Limitations of linearized solutions are the inaccuracies which must be tolerated when dealing with the extremely nonlinear character of steam properties. This particular difficulty is more critical in once-through boilers where there are no constraints as to the point where transition occurs permitting the properties of the fluid in a given section to change radically depending on whether it is liquid, mixed or superheated at any particular instant.

Digital techniques for solution of nonlinear differential equations have a logical application in the solution of the boiler nonlinear dynamics. Aside from their application to the simulation of once-through boilers, these techniques applied to the simulation of drum-type boilers have provided standards of accuracy against which the effects of various modeling simplifications, such as the linearized model approach, could be judged. The accuracy of a simplified model cannot be entirely judged by comparison with tests since correspondence between tests and simulation relies in part on the validity of certain physical and design data. This accuracy can be better judged by comparison with a rigorous nonlinear model derived from the same common basic data.

Fig. 1.16 shows the lumped parameter, nonlinear equations typically applying to a superheater subsection.

Although these equations apply specifically to superheater and reheater subsections, their general form is descriptive of the physics of flow and energy transfer from metal to inner fluid throughout the boiler. The main difference in the various sections pertains to variations in the state equations and in the relationship governing heat transfer from metal to fluid. The simultaneous solution of many dozens of such equations can be achieved using digital solution techniques.

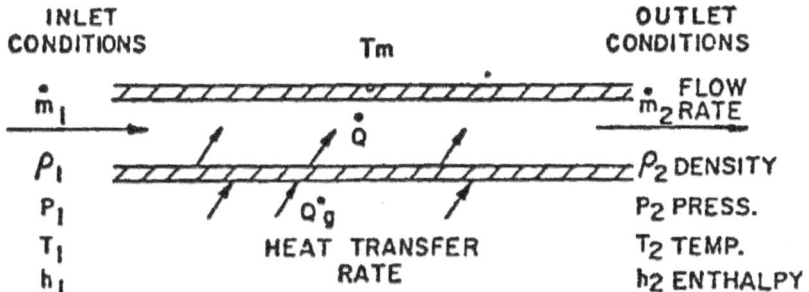

MASS BALANCE

$$\dot{m}_1 - \dot{m}_2 = V\frac{d}{dt}\left[\frac{\rho_1 + \rho_2}{2}\right]$$

ENERGY BALANCE

$$\dot{m}_2 h_2 - m_1 h_1 + V\frac{d}{dt}\left[\frac{\rho_1 h_1 + \rho_2 h_2}{2} - \frac{P_1 + P_2}{2J}\right] = \dot{Q}$$

PRESSURE DROP

$$P_1 - P_2 = \frac{2f}{(P_1 + P_2)}\left(\frac{\dot{m}_1 + \dot{m}_2}{2}\right)^2$$

HEAT TRANSFER

$$\dot{Q} = K = \left(\frac{\dot{m}_1 + \dot{m}_2}{2}\right)^{0.8}\left(T_m - \frac{T_1 + T_2}{2}\right)$$

METAL TEMP

$$T_m = \frac{1}{MC}\int(\dot{Q}_g - \dot{Q})dt$$

STEAM TABLES

$$\rho_2 = t_1\,(P_2, h_2)$$
$$T_2 = f_2\,(P_2, h_2)$$

FIGURE 1.16 Nonlinear Equations Describing Superheater
Lumped Parameter Subsection

CONTROL CONCEPTS

The foregoing description of simulation methods and techniques points out the very important fact that today's computational aids permit simulation of very complex dynamic processes. The real value of simulation lies in the fact that it permits an engineering synthesis of controls tailored for a particular plant and allows a prediction of that plant's controlled performance before it is built. Although simulation represents the major part of the problem, it nevertheless is a means to an end, namely, the means to proper design of plant controls. A brief discussion of some applicable control concepts is, therefore, appropriate.

MULTIVARIABLE AND NONINTERACTING CONTROL

In its most general meaning, a multivariable-control system is one that will yield satisfactory controlled performance of the process variables under expected operating conditions. The term "satisfactory" admits of an infinite number of interpretations. Limits can be put on acceptable transient deviations of the controlled variables and on settling times following specified disturbances such as load changes. These limits must be compatible with physical realizability within available control effort.

The problems of interaction between control loops of coupled-multivariable processes gave rise to the term "non-interacting control" which describes types of integrated multivariable-control systems in which the corrective action for deviations in one variable is coordinated with such other action as is needed to minimize consequent transient deviations in the other variables. Actually, the control system itself must be "interacting" in order to achieve so-called non-interacting process performance.

Fig. 1.17 is a block diagram of a coupled two-variable process where x_1 and x_2 are the output variables to be controlled, and m_1 and m_2 are the inputs to the process. One way of proceeding with the design problem is to effectively uncouple the process and then treat it as two independent single-variable loops, as shown on Fig. 1.18. There is much appeal to reduce the problem to the well-known single-variable problem. This non-interaction criterion, however, is unnecessarily restrictive and sometimes physically unrealizable when $\dfrac{G_{12}(s)}{G_{22}(s)}$ or $\dfrac{G_{21}(s)}{G_{11}(s)}$ turns out to be an expression with higher order numerator than denominator (which would mean that the cross-coupled transfer functions G_{12} or G_{21} have a faster response than the self terms G_{22} or G_{11}).

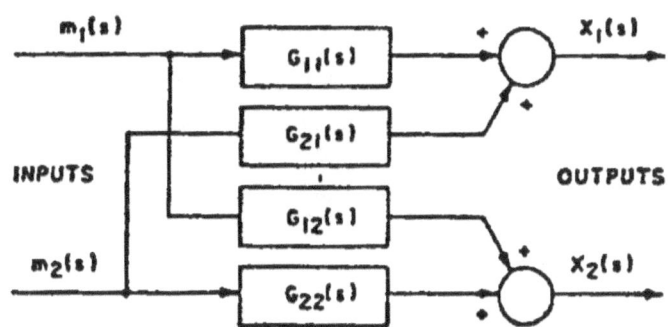

FIGURE 1.17 Block Diagram of Coupled Two-Variable Process

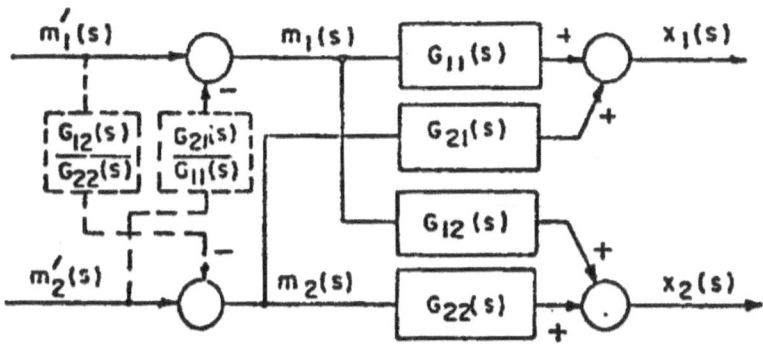

FIGURE 1.18 Cross Coupling for Non-interaction

The procedure found useful is to work out the design of each loop at a time and include the cross coupling that may be necessary between loops. The cross-coupling function when needed often turns out to be a simple transfer function designed to be effective over a fairly limited range of frequencies.

To illustrate the problem refer to Fig. 1.19, which shows the pressure control and the reheat-temperature control loops. Pressure is maintained by means of combustion control, whereas reheat temperature in this example is controlled by tilting burners. Fig. 1.20(a) shows the performance that might be expected upon a load change with a fairly loose reheat-temperature control. The frequency response of the two loops is sufficiently different that no appreciable interaction occurs. Temperature control is sluggish, however. Fig.1.20(b) shows the resulting interactions that would occur if reheat-temperature control were tightened (increase in gain). Tilting burners change the distribution of heat to the waterwalls as well as to the rest of the boiler. Hence, burner tilts will also affect steam generation and, therefore, the boiler pressure. Fig. 1.20(c) illustrates what can be accomplished with a simple rate over a time-constant network coupling the two loops.

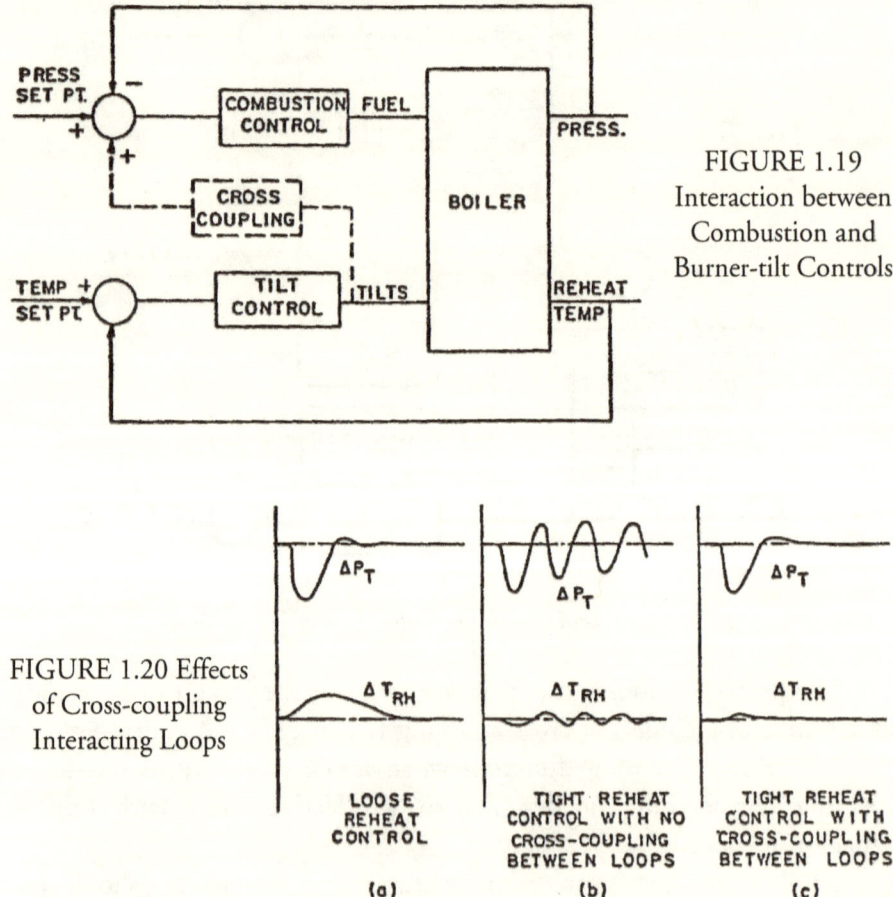

FIGURE 1.19
Interaction between Combustion and Burner-tilt Controls

FIGURE 1.20 Effects of Cross-coupling Interacting Loops

FEEDBACK CONTROL AND FEEDFORWARD CONTROL

Fig. 1.21 shows a single-variable process whose output variable q(t) is to be controlled by manipulation of the input variable m(t). The process is subject to the disturbance u(t). Define the response function $\frac{q(s)}{u(s)} = G_1(s)$ and the response function $\frac{q(s)}{m(s)} = G_2(s)$. If it is assumed that there are no errors in the estimate of $G_1(s)$ and $G_2(s)$, one might control the process in feedforward fashion, as shown in Fig. 1.22, where control action is tied to the disturbance u(s) by the anticipatory function $\frac{G_1(s)}{G_2(s)}$. Then $\Delta q(s) = \Delta u(s) G_1(s) - \frac{G_1(s)}{G_2(s)} \Delta u(s) G_2(s) = 0$.

This perfect cancellation of the effect of the disturbance is only possible if $\frac{G_1(s)}{G_2(s)}$ is physically realizable, i.e., the time response of function $G_1(s)$ cannot lead that of function $G_2(s)$. Again, any errors in estimates of $G_1(s)$ or $G_2(s)$ would directly affect q(s) as would also any other miscellaneous process disturbances other than u(s). Typically in a boiler, a mismatch of fuel to steam flow of 1% would result in a temperature deviation in the neighborhood of 20°F. Obviously, the accuracy required to maintaind temperature within 1 or 2°F precludes use of the feedforward method by itself.

If now feedback is used, the configuration would be as in Fig. 1.23 and $\Delta q(s) = \frac{\Delta u(s) G_1(s)}{1 + G_2(s) H(s)}$.

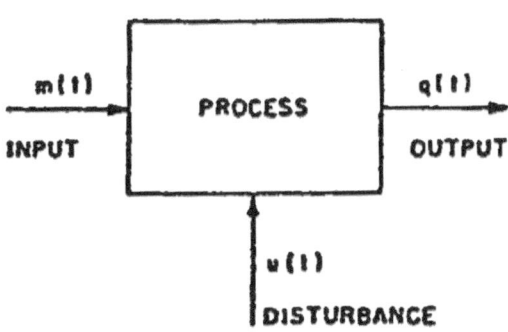

FIGURE 1.21 Single Variable Process

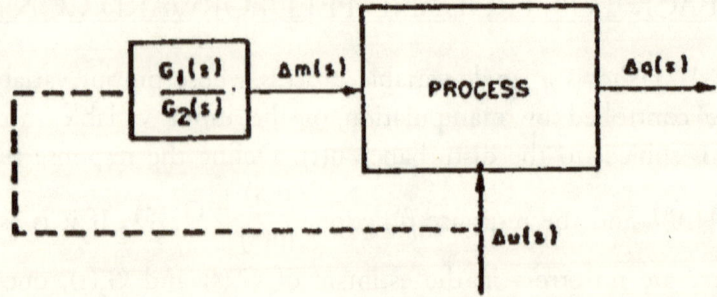

FIGURE 1.22 Feedforward Control of Single Variable Process

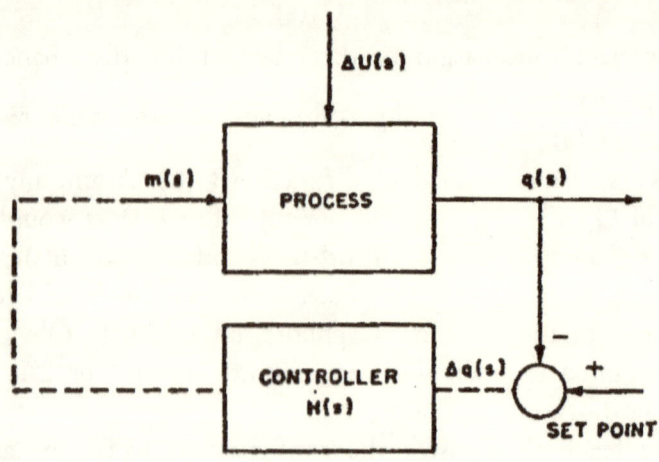

FIGURE 1.23 Feedback Control of Single Variable Process

If H(s) is a controller with reset action (an integration), the above expression will guarantee $\Delta q = 0$ in the steady state. Further, by making $G_2(s)$ H(s) large (tight control loop), transient deviations are minimized. The problem of tightening the feedback loop and preserving stability for a process with appreciable lags is one of providing proper compensation in the function H(s). Often because of improper compensation in H(s), the feedback control is loosened to preserve stability and the transient deviations Δq become excessive. In such systems where feedback is weak, feedforward or so-called anticipatory action is used to minimize the deviations. (Steam flow anticipation has been used extensively in boiler control).

The need for accurate feedforward-anticipatory functions in control is characteristic of systems where feedback can only be used in a weak sense.

This often occurs in systems with large transport lags and several time constants. The principal area of application of such methods lies in the difficult area of control of once-through boilers.

In designing multivariable controls, it is very important that proper consideration be given to the consistency of process dependent effects used as intelligence to the control. One cannot merely treat the process as a matrix of equations relating all inputs to all outputs which are then mechanically manipulated to arrive at controller characteristics and configurations. These equations describe process input-output characteristics, some of which may be dominant and, therefore, consistent, and others which may reflect relatively inconsistent and unpredictable composite effects. For instance, in a drum-type boiler, an increase in fuel will always produce an increase in pressure; an increase in air flow will always decrease boiler pressure; an increase in desuperheat spray will always decrease throttle temperature, and so on. These are the primary or dominant effects. Their magnitude may change somewhat with operating conditions, but not their sign. On the other hand, the composite effects are the net result of opposing effects. For instance, an increase in fuel will produce an increase in steam pressure, and hence of steam flow. The increased steam flow tends to decrease steam temperatures, whereas the increase in fuel will have an opposing effect, the net result being an increase or decrease in temperature depending on which effect is the stronger. The sooting, slagging and other conditions of the boiler could make this composite effect radically different from day to day. Evidently, a well-engineered control system should be insensitive to these inevitable process composite effect changes, and treatment of the process as a black box without the necessary analysis of the nature of reliability between input and output characteristics could lead to a theoretically beautiful but practically unworkable control system.

Application of the feedforward anticipatory action should beware of the pitfalls of having to rely on the consistency of composite effects.

ADAPTIVE CONTROL

Fig. 1.23 is an example of control of a single-variable process where the output $q(s)$ is to be maintained at a reference value despite the effect of disturbances $\Delta u(s)$. $G_2(s)$ is the response function of the plant to input $m(s)$. $H(s)$ is the controller function. The deviations

$\Delta q(s)$ can be expressed as a function of the disturbances $\Delta u(s)$ as $\Delta q(s) = \dfrac{u(s)G\ (s)}{1 + G\ (s)H(s)}$. The purpose of good control is to make $G_2(s)$ H(s) large for a limited bandwidth and infinite in the steady state (reset action on H(s)) so that $\Delta q(s)$ is transiently very small and zero in the steady state.

Now, stability considerations may require especial shaping of the controller function H(s) to compensate for the time-response characteristics of $G_2(s)$, and it is conceivable that as the process characteristics $G_2(s)$ change with operating conditions, that the controller characteristics H(s) may have to suffer compensating changes in order to maintain satisfactory performance. Temperature control loops do sometimes require controller-parameter changes with operating point. It should be recognized that the need for adaptive control system parameter changes often can be eliminated by proper use of inner loop feedback which essentially automatically compensates for process changes. The following examples will illustrate the point.

Fig. 1.24 shows a typical combustion control system where the signal calling for more or less fuel is directed to all the feeders in parallel. At light loads when the number of mills is reduced, the gain of the controller measured in terms of change in fuel for a given pressure error would be correspondingly reduced if no changes were to be made in the controller gain. A so-called adaptive system may be equipped to change controller gain as a function of number of mills. On the other hand, if the system be incorporated with a controller that calls for total fuel with feedback from the signals to all the feeders in operation, as shown on Fig. 1.25, the control would be equally "adaptive" without the need for controller gain changes.

The modern term for a control structure which is insensitive to variations in process characteristics is "robust".

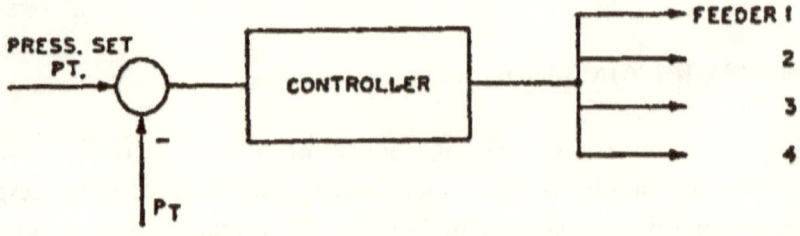

FIGURE 1.24 Pressure Control Loop

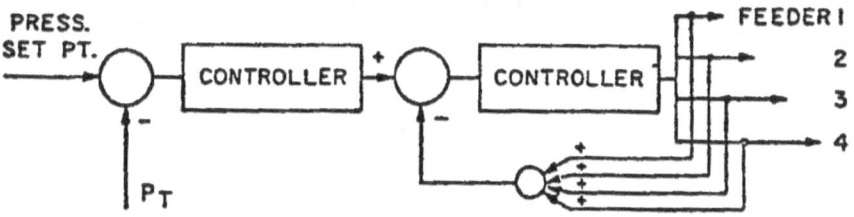

FIGURE 1.25 Pressure Loop Controlling Total Fuel

Another example is the use of feedback from temperature after spray in the main steam temperature control system. It is evident that at half load, a given change in spray water will produce twice the change in steam temperature that the same change in spray water would produce at full load. An "adaptive" system accordingly may be equipped to change gain of the temperature controller as a function of load. The control system which uses the cascade type of control with feedback from intermediate temperature after spray, and which offhand may not be recognized as "adaptive" accomplishes essentially the same thing as analysis of Fig. 1.26 will indicate.

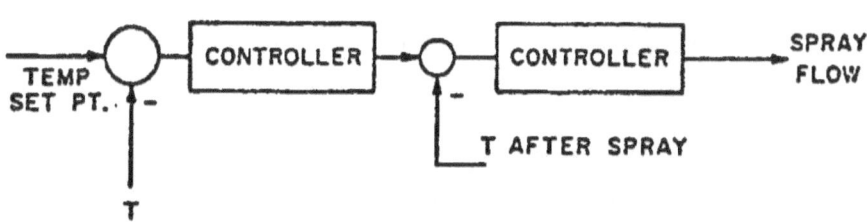

FIGURE 1.26 Temperature Control with Feedback
from Temperature After Spray

The foregoing is to stress that the need for better control sometimes can be satisfied through the use of ordinary sound feedback control techniques using analog type of hardware. It should also be evident that the need for adaptive features is greater in systems that rely heavily on feedforward action. Adaptive control-parameter changes may be indicated in cases where changes in process response occur as a result of deteriorating conditions such as, for instance, slagging in waterwalls.

PHYSICAL REALIZABILITY AND CONTROL EFFORT

Performance specifications as to limits in transient deviations of controlled variables under given load-changing conditions cannot be set arbitrarily small. Synonymous with tight specifications on transient deviations should be the acceptance of the fact that large control effort has to be exerted. In the limit, if minimization of controlled-variable deviations is to be sought with no regard to amount of control effort expended, the mathematics results in a maximum effort (bang-bang) type of control system.

In a pressure-control system, one can visualize having to over-fire to maximum and then drop the fire to minimum in rapid succession in order to hold pressure within very narrow limits. Obviously, in addition to the danger of putting out the fire and having a boiler explosion, such type of operation could be uneconomical inasmuch as the control effort measurable in terms of heat lost up the stack might not be justified.

CONTROL DESIGN

The synthesis of a control configuration for a complex process such as just described presents a real challenge to the control engineer. Attempts to automate the design process by optimization theories of modern control methodology have not been practical because the process is highly nonlinear, the control logic must recognize constraints such as valves at limits, etc., and the information available from the process generally involves much fewer states than the order of the system. The most practical approach will continue to be the use of simulation in interactive process whereby the engineer develops a design by an orderly process of trial and evaluation.

The options of accomplishing a given control function are many and the mix of feedforward, feedback, cross-coupling for non-interaction, and adaptive features that are used requires a strong element of creativity and judgment. Such difficult questions as sensitivity of the configuration to variations in process characteristics, reliability versus complexity, etc., must be carefully weighed.

COMPUTERS IN CONTROL

Typical modern boiler-turbine control systems have been configured as large operational amplifier analog type systems with hundreds of elements such as described in Fig. 1.27. Also shown in Fig. 1.27 is the natural evolutionary step in configuring the control system with digital computers rather than analog operational amplifier type systems. This step, already taken in some pioneering applications, promises major benefits in the plant control function. Digital control permits use of practically unlimited control logic, including such functions as nonlinear and adaptive control. Sophistication is often discouraged in the case of analog controls since it invariably involves more components, and complexity must be weighed against its effect on reliability. In the digital machine complexity, when needed, is easily implemented in software and has no significant effect on reliability.

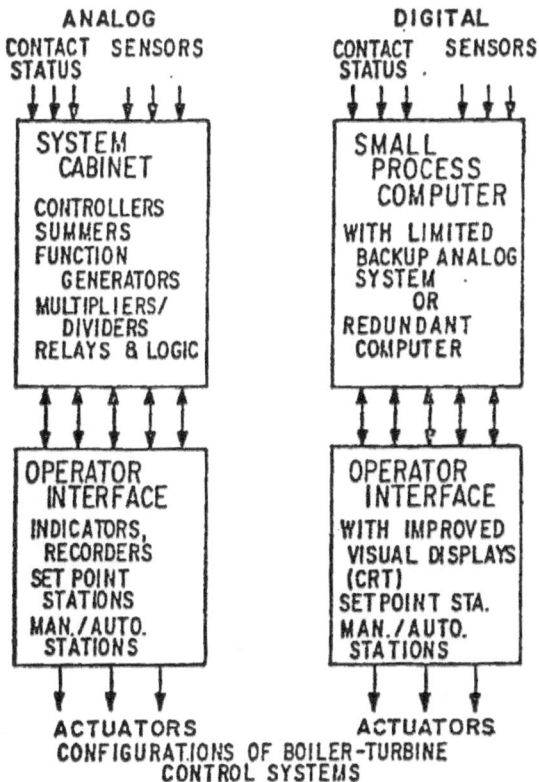

FIGURE 1.27

The digital approach opens many opportunities for the user to implement his philosophy of controls and to adapt the function through software to his particular needs. These same opportunities were not present in the case of analog systems in which new functions meant added hardware and circuitry changes.

REFERENCES

1.1   de Mello, F.P., "Plant Dynamics and Control Analysis", IEEE Transactions PAS, Vol S82, 1963, pp. 664-78.
1.2   de Mello, F.P., "Process Dynamics in Utility Systems", ISA Paper 505-70, October 26-29, 1970.

CHAPTER II

GENERAL PRINCIPLES AND STRUCTURES IN BOILER CONTROLS

The entire steam generation process may be modeled in great detail by breaking the flow path into finite volume sections, and solving for the variables at the boundaries of the sections using the mass balance, energy balance, pressure drop, heat transfer equations as listed in Fig. 1.15 of Chapter I.

A natural boundary in the process physics is the tube metal which separates the gas path from the water and steam fluid path. The coupling between the two processes is the metal temperature which integrates up or down as a function of the difference between heat transfer from gas to metal and from metal to fluid.

The techniques of model development from first principles are covered in Chapter XII.

While this approach is desirable and necessary in many situations, especially in modeling of once-through boilers, a great deal of insight may be derived from simplified models that define the response of primary output variables to those input variables which have a dominant effect. Examples of several of these simplified models are illustrated in following chapters along with single loop control configurations defining the dominant control actions.

Wherever possible we shall develop a sense of appreciation for sound control principles illustrated by examples.

The job of designing a system to perform a given control task usually involves a choice among a number of ways to accomplish the control function. The selection of the best way requires use of judgment in deciding to what extent concepts such as feedforward, feedback and adaptive control should be applied. By sound judgment we imply proper recognition and use of basic guiding principles that should influence control design.[1] Some of these principles are:

1. Simplicity

Among various systems that perform equally well, the best is usually the one that has the fewest components and need for least number of adjustments.

2. Use of Process Intelligence for Control

Strong use of actual feedback from process information wherever possible gives actual rather than inferential information to the control system.

In a situation where the control function can be accomplished well by feedback, additional use of feedforward requires that the action from feedback be reduced to prevent overshoot. Since feedforward action can only account for a few measured disturbances, this means that system performance will suffer for the cases of unmeasured upsets.

3. Need for Adaptive Features

It is inherent that the need for adaptive or self-calibrating features is much greater in systems that rely heavily on feedforward action. The use of feedback wherever process dynamics permit provides a great degree of adaptation in ways that are more subtle and less apparent than in systems with a liberal number of multipliers modifying feedforward action. Feedback by its very nature minimizes the effects of changes in process parameters.

4. Stability of Controls

Taking full advantage of process information in control configurations gives rise to closed loops and care and sophistication must be exercised in dynamic shaping of control action. Not only must the controls have the proper proportional, reset and rate action for the linear (small perturbation) mode of operation, but they must provide for smooth transition in and out of nonlinear ranges of operation when components or equipment go into limits or saturation.

Although the boiler process is multivariable, and, strictly speaking, control of inputs can be conceptually visualized as directed by a matrix from process outputs (Fig. 1.4), it is logical to compose the controls from basic primary loops relating certain process variables and the inputs that have dominant effects on these variables. Additional coordination or cross coupling between loops is then superposed as needed.

A review of some control concepts as applied in the boiler control area is appropriate prior to discussion of specific applications.

PROPORTIONAL FEEDFORWARD CONTROL

Proportional feedforward control is described in Fig. 2.1 as the most elementary form of open-loop control.

This concept has been used in large measure usually in combination with other forms of control and has been attractive because of its apparent simplicity. It is also attractive due to the fact that it does not give rise to stability problems as may arise with closed-loop feedback systems, and in days past when feedback control was a relatively mysterious field (pre-World War II) and hardware was not available to give proper dynamic compensation and to prevent such things as reset windup, it is understandable that the industry relied heavily on this concept of control.

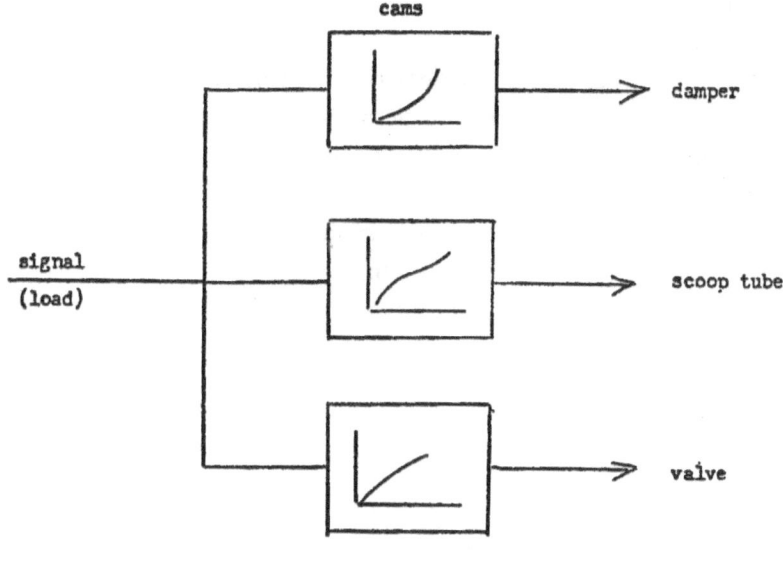

FIGURE 2.1

The basic idea is that using one fundamental signal indicative of the demand from the process, one could position all necessary inputs in feedforward open-loop fashion. Of course, this works well in elementary types of systems, but is by itself entirely inadequate in boiler control for the following reasons:

Calibration—The boiler process is complex, nonlinear and is influenced by several factors in addition to the fundamental load demand. These factors are usually not measurable, The instrumentation technician would have to spend a career calibrating the system, characterizing cams, etc., and would still find performance unsatisfactory because of accuracy and dynamics. Accuracy because, even with the best possible characterization, the control requirements are much too stringent for this type of control by itself to be effective. For instance, in a once-thru unit, a 1% error in the required amount of fuel would result in a steam temperature deviation of 17°;

The other serious shortcoming of this simple proportional feedforward system is the absence of dynamic shaping of control action. Due to the complex and high order storage lags in the process and the interacting nature of the variables, proper control action must often undergo rather involved trajectories in time when taking the process from one state to another. The oversimplification of forcing the process inputs instantaneously from one steady-state value to another in response to a step change in demand can sometimes result in large transient deviations of critical variables.

A homey analogy of the inadequacies of this control concept can be found in the example of trying to help the job of backing a trailer from one position to another by instantly positioning the car's steering wheel to its expected final value!

We can trace the next step in the evolution of control configurations used in boiler controls to that shown in Fig. 2.2.

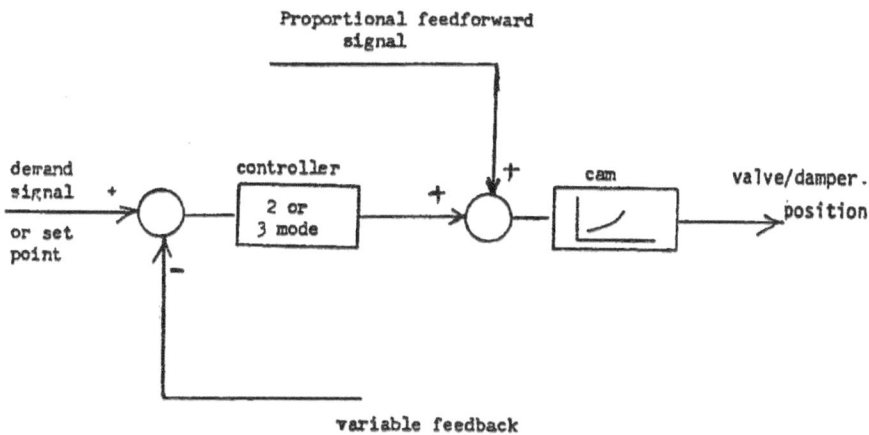

FIGURE 2.2

Here we have still kept as the primary control the proportional feedforward action previously discussed but have added a feedback trimming action from a relatively slow reset, proportional + reset, or sometimes proportional + reset + rate control operating on the error of the basic process variable.

This concept is often useful but has been misapplied. One area of misapplication has been where process dynamics are complex enough to invalidate the effectiveness of a simple proportional feedforward signal, as previously described. An example is proportional anticipation of feedwater demand from steam flow. Having this anticipation as instantaneous can add to the drum level swell effects.

The other case is where the process dynamics are fast enough making the configuration with feedback only, with the error operating on a reset, two-mode, or three-mode controller entirely adequate. In these cases where the process dynamics are fast, addition of feedforward, as in Fig. 2, requires detuning or slowing down of the feedback action to avoid over-control and consequent overshoot. Detuning of feedback control deteriorates the performance of the control system under disturbances other than the primary disturbance from which feedforward action is provided.

Fig. 2.4 exhibits examples of the open-loop step response characteristics of such processes normalized to have the same final value of unity.

Process Responses to Step Change f(t)

$$A = M_0 = \int_0^\infty (1 - f(t))dt$$

$$M_1 = \int_0^\infty t(1 - f(t))dt$$

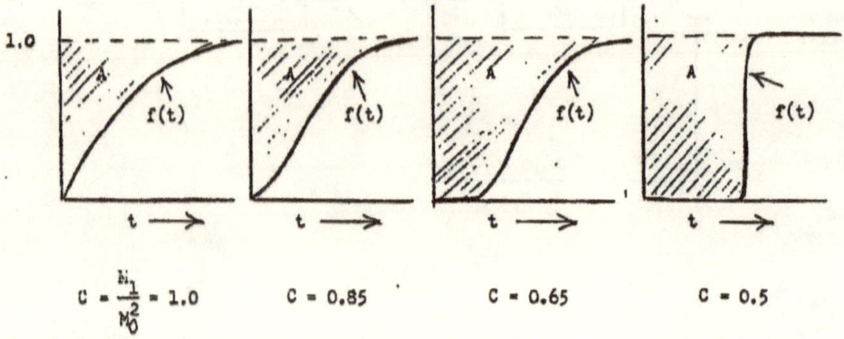

$$c = \frac{M_1}{M_0^2} = 1.0 \qquad c = 0.85 \qquad c = 0.65 \qquad c = 0.5$$

FIGURE 2.4

At the extreme left, we have a first order lag or single time constant system. Progressively higher order lag systems are shown in the successive response traces until reaching the extreme case of a pure dead time system at the extreme right. The controllability of these systems deteriorates with the increasing order of the lags.

Without resorting to traditional frequency response or S-plane analyses, one relative measure of the controllability of these processes is to characterize them by a dimensionless ratio of the moment of the area A (see Fig. 2.4) about the $t = 0$ axis to the area A squared. (Note that area A, also called loss area, is the zero order moment M_0 of the function $(1 - f(t))$ where the nth moment of that function is defined as $M_n \overset{\Delta}{=} \int_0^\infty t^n (1 - f(t))dt$).

One obvious conclusion is that if the system with feedback only is effective from a response standpoint for both the primary process disturbance as well as for other disturbances, then it is the preferable system both from a performance as well as from reliability and ease-of-adjustment points of view. One of the fundamental axioms of good control design is "simplicity".

FEEDBACK CONTROL

A most important area concerns control modes in feedback configurations, as shown in Fig. 2.3. In discussing this configuration it is appropriate to note that the response characteristics of the process dictate the combination and amount of control modes which should be used for greatest effectiveness. Some of these modes, proportional, integral, and derivative (rate), are discussed in Appendix F.

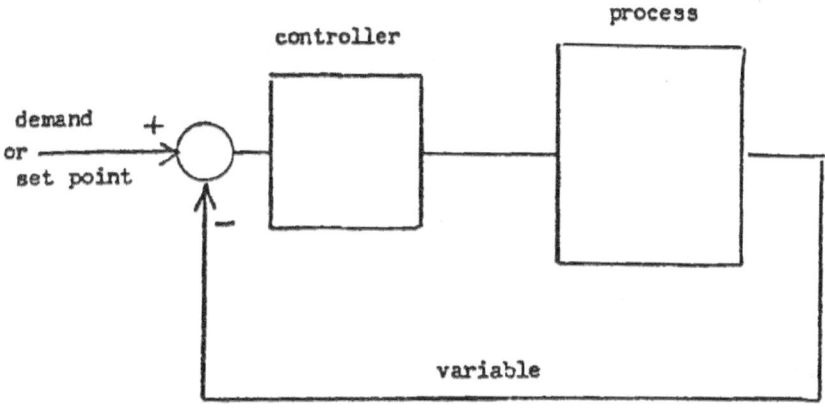

FIGURE 2.3

In examining the control needs in feedback configurations, it is helpful to relate these to the particular process dynamics. Let us examine first a wide range of processes whose open-loop responses can be characterized by various orders of lags (i.e., a process with no zeros or complex poles but with varying number of real poles).

The dimensionless figure of merit $C = \dfrac{M_1}{M_0^{\,2}}$ or $C = \dfrac{M_1}{A^2}$ has the time dimension scaled out of it and is purely a function of the shape of the response curve.

The class of processes, shown on Fig. 2.4 has this figure of merit, $C = \dfrac{M_1}{M_0^{\,2}}$, ranging from a minimum value of 0.5 for a pure dead time case at the extreme right of Fig. 2.4 to 1.0 for a single time constant system shown at the far left of the same figure.

A common misconception in this industry has been that proportional control is always easiest to apply and that integral control is harmful from a stability standpoint.

Analysis of the control characteristics required for a particular process shows that the dead time process, or the process with multiple order lags, cannot stand a high proportional gain. In fact, a gain of less than unity is required for stability in the case of a pure dead time system. In such processes integral action is necessary to produce the proper attenuation with increasing frequency so that the open-loop gain drops below unity before the phase lag of the process becomes excessive.

It is important to note that over a wide range of values of C, especially where the process has some dead time, the acceptable value of proportional gain is so low that proportional control only would be almost as good as no control at all! (A proportional gain of 1.0 would allow an error between demand and feedback of 50%—a proportional gain of 0.2 would allow an error of 83%).

High proportional gains are only applicable in the control of low order lag systems, but unfortunately these low order systems are not as universal as might first appear from a survey of academic papers on control analysis.

Beside the commonly occurring processes, whose response characteristics are described in Fig. 2.4, there are those where the system open-loop function has zeros in addition to high order poles. Fig. 2.5 shows an example of the response of one such system encountered in the control of air flow with changes of fan speed through a hydraulic coupling.

FIGURE 2.5

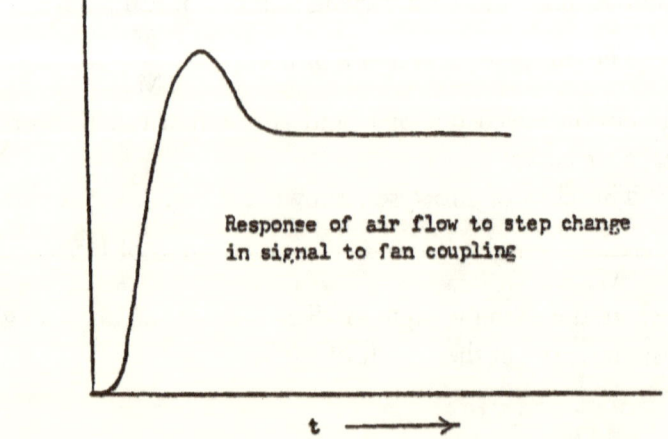

Response of air flow to step change
in signal to fan coupling

t ⟶

response trace of air flow to a change in signal to coupling exhibits a dead time, followed by an overshoot which is due to the charging of the boiler volume. An integral controller with very little proportional action is indicated for this type of process.

The control of feedwater flow, through changes in the turbine—driven feed-pump speed, exhibits similar requirements, i.e., pure reset action and zero proportional gain. The universal applicability of proportional control as the first step in control is, therefore, invalid.

CASCADE CONTROLS

The principle of cascade control, as illustrated in Fig. 2.6 is widely used in boiler controls. Some limited degree of cross coupling between loops is also used.

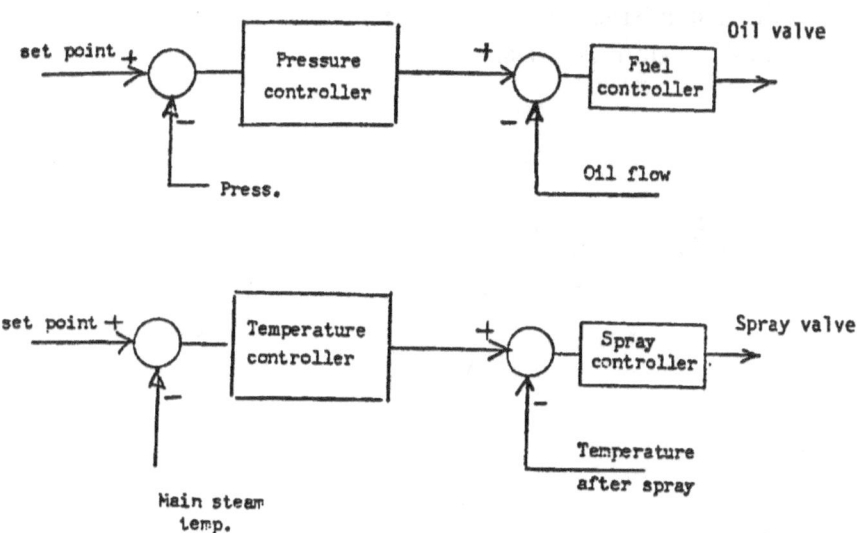

FIGURE 2.6

The advantage of cascade control arrangements is that they linearize the process through use of sub-loops. These sub-loops also automatically compensate for disturbances as they occur through intermediate stages of the process without having to wait for the effect of the disturbances felt in the main process variable.

CHAPTER III

DRUM BOILER PRESSURE EFFECTS

Although boiler pressure transient behavior is affected by many simultaneous effects requiring a high order dynamic model, for purposes of this discussion basic effects can be derived from a very much simplified model, as shown in Figs. 3.1(a) and (b).

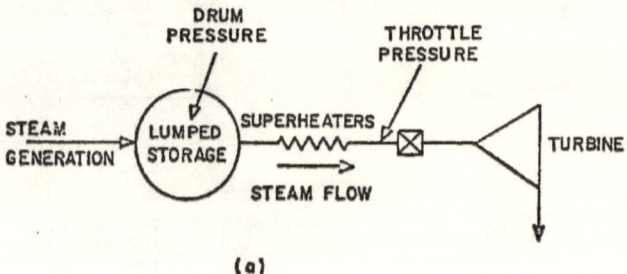

(a)

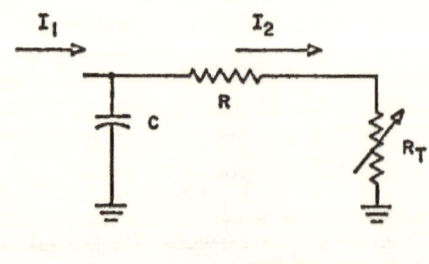

(b)

FIGURE 3.1   Boiler Turbine Representation

(a)   Schematic of boiler-turbine system
(b)   Electrical analog of boiler pressure flow phenomena

For the first few seconds following a change in R_T or turbine valve, the voltage across the capacitor (drum pressure) does not change. However, the throttle pressure will suffer a deviation due to the change in friction drop ($\Delta I_2 R$) in the superheaters and piping.

Since pressure drop is proportional to the square of flow, whereas voltage drop is linearly related to current, the above analogy is approximate and holds for small changes about an operating point. The value of resistance R varies with operating load level as can be seen from the following:

$$P_{DT} = \text{Pressure drop from} = Km_s^2 \text{drum to} \quad (3.1)$$
throttle

where

$$K = \text{friction drop coefficient}$$
$$m_s = \text{steam flow rate}$$

For small perturbations

$$\Delta P_{DT} = \text{change in pressure drop} = 2Km_{so}\,\Delta m_s$$
from drum to throttle $\quad (3.2)$

where

$$m_{so} = \text{steady state steam flow at the particular operating point}$$

and

$$\Delta m_s = \text{change in steam flow}$$

In the analog of Fig.3.2(a) $R = 2Km_{so}$.

The effect of turbine valve changes can be represented in linearized small perturbation form as follows:

$$m_s = K_v P_T \quad (3.3)$$

where

$$K_v = \text{coefficient proportional to valve opening}$$

For small perturbations, neglecting second order terms

$$\Delta \dot{m}_s = K_{vo}\Delta P_T + \frac{\Delta K_v}{K_{vo}}\dot{m}_{so} \quad (3.4)$$

Steam generation (I_1) is proportional to heat release in the furnace, and follows this heat release with a small time constant (5 to 7 seconds) due to the waterwall metal and film coefficient. The process can therefore be represented by the block diagram of Fig. 3.2(a) for the linearized treatment, and in Fig. 3.2(b) using the nonlinear flow relations of equations 3.3 and 3.1.

In these figures,

subscript o		denotes steady state value
prefix Δ	=	denotes change drom steady state value
m_w	=	steam generation
P_D	=	drum pressure
P_T	=	throttle pressure
m_s	=	steam flow out of drum
K_v	=	coefficient proportional to valve opening
T_w	=	waterwall time constant
C	=	boiler storage constant

In the above linear, small oscillation model, the parameters that change with load level are R and K_{vo}. All other parameters are essentially invariant.

Using the per unit system based on rated values, i.e., base steam flow = full load steam flow, base pressure = rated pressure, etc., typical values of parameters in the block diagram of Fig. 3.2(a) are:

K_v	=	coefficient proportional to load, 1.0 p.u. at full load, 0 at no load
R	=	friction drop coefficient
	=	2 × (p.u. load level) × (p.u. pressure drop from drum to turbine throttle at full load) = 0.2 at full load or 0.1 at 50 load
T_w	=	5 to 7 sec
C	=	100 to 300 sec

The value of pressure drop at full load is about 10% or 0.1 p.u. so that at the full load operating point a typical value of R is 0.20 p.u.

The boiler storage constant C is related to the stored mass of saturated liquid and vapor as well as the mass of superheated steam in the superheaters and steam leads. Typically C represents the number of full load flow seconds for a one per unit change in pressure assuming a linear relationship between stored mass and pressure. This storage constant varies between 120 to 300 secs in drum type boilers, and 90 to 200 secs in once-thru units.

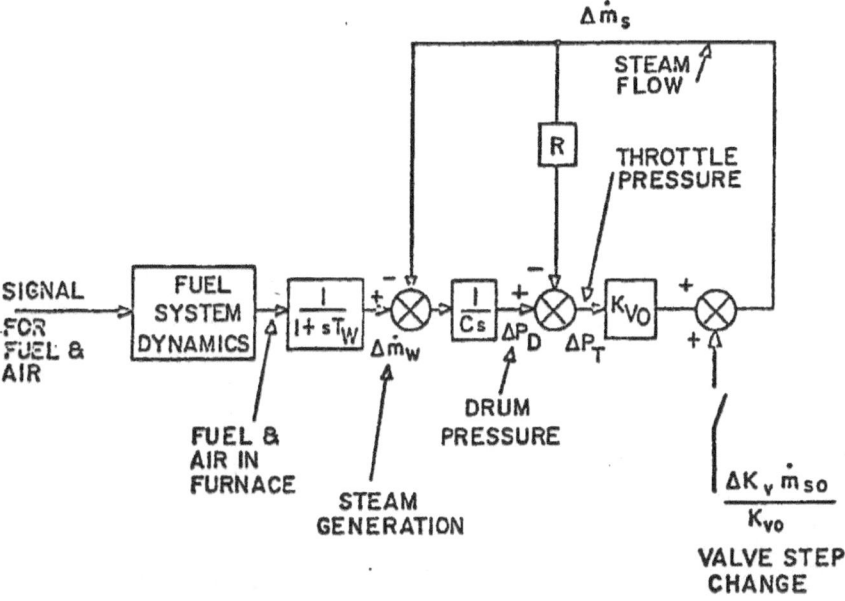

FIGURE 3.2(a) Block Diagram of Process (Linear)

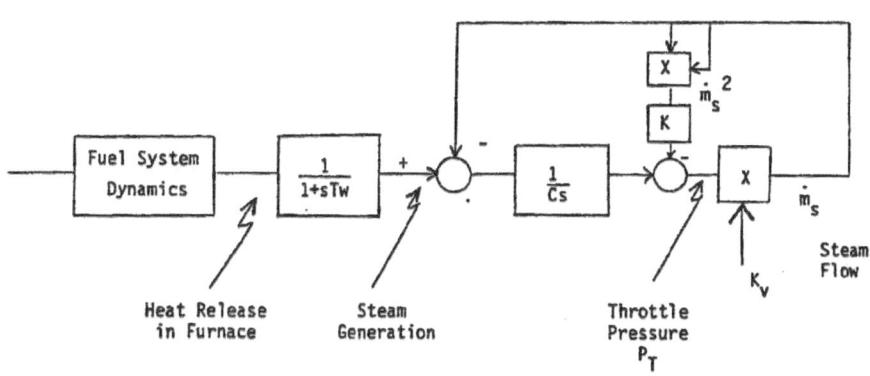

FIGURE 3.2(b) Block Diagram of Boiler Pressure Effects (Nonlinear)

The open loop responses of boiler pressure to a change in turbine valve are shown in Fig. 3.3(a) about the 100% and 50% load points. The initial instantaneous change has a great deal of significance. It is proportional to the amount by which the drum pressure will have to change to bring the throttle pressure back to set point. Since boiler storage is in large measure proportional to drum pressure, this initial change in pressure can be used to advantage as process intelligence to guide the amount of over-firing needed to change boiler storage. This point will be covered further when discussing control performance with various control configurations.

The means of controlling boiler pressure is through manipulation of the fuel and air, i.e., the heat release in the furnace which governs the amount of steam generation.

The response of boiler pressure to changes in heat release around the 100% and 50% load points is shown in Fig. 3.3(b). The response of boiler pressure to a signal calling for a change in fuel and air must, of course, also include the dynamics of the fuel system, which can be quite different depending on such factors as pulverizer dead times and other lags in the fuel subsystem. The effects of these lags are illustrated by examining cases with fuel system dynamics represented by a single time constant of 20 seconds (representative of gas, oil, and some coal-fired plants) and alternately where the fuel system is represented by a dead time of 30 seconds in series with a 20 second time constant (representative of many coal fired plants).

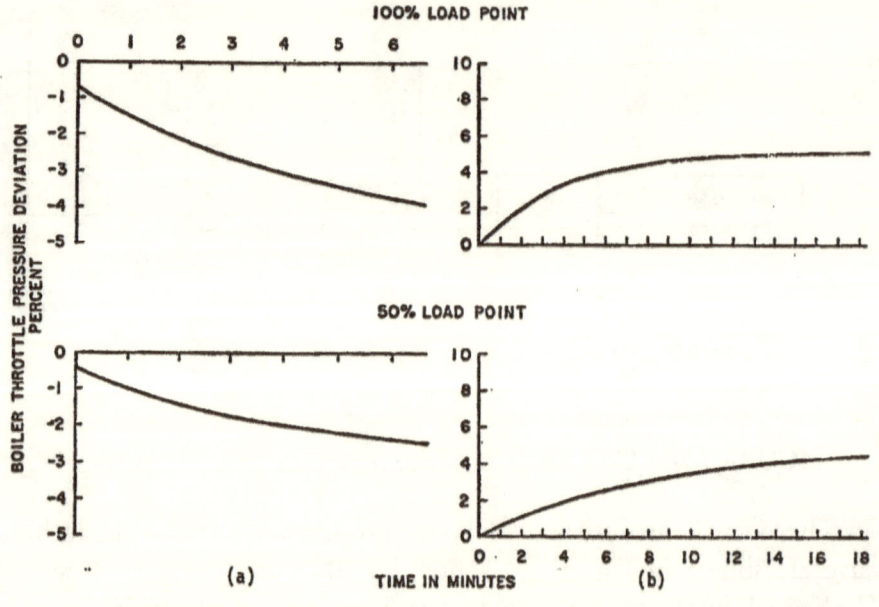

FIGURE 3.3   Process Open Loop Responses

    (a)   5% step change in turbine valve.
    (b)   5% step change in signal to fuel and air.
           Fuel system with 20 sec time constant.

Drum Type Boiler Pressure Control Configurations

With the boiler process represented as in Fig. 3.2(a), several pressure control schemes are examined for performance, both at the 100% and 50% load points. Performance is evaluated under normal load change duty, i.e., for step changes in turbine valve and also under fuel upset conditions, i.e., for a step change of fuel caused by upsets not originating in the controls.

The schematic of Fig. 3.4 identifies the overall process and controls. Details of block (C) were shown in Fig. 3.2(b). Block (B) is a sample time constant $\frac{1}{1+S20}$ or a dead time and a time constant $\frac{e^{-S30}}{1+S20}$ simulating the fuel system.

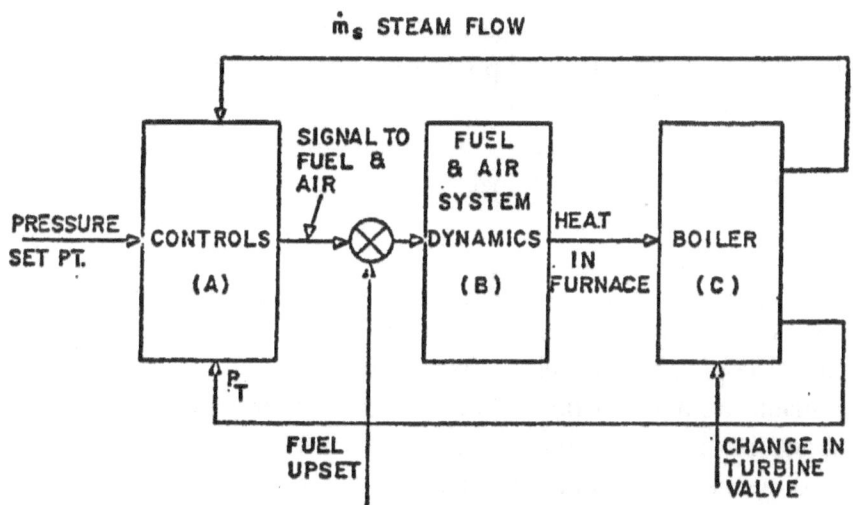

FIGURE 3.4 Schematic of Process and Controls

Block (A) contains the various control schemes under investigation which will be discussed below:

1. Pressure Control from Pressure Error Alone and 3 Mode Controller (Fig. 3.5(a))

    Examine first the case where the fuel system is represented by a single time constant (1/(1+20s)). The case of a fuel system with a dead time will be discussed under item 3 below.

Pressure control performance, with this type of control, is shown in Fig. 3.6(a) and (b). In these figures three quantities are graphed: throttle pressure, steam flow and signal to fuel system.

Fig. 3.6(a) shows the response to a step disturbance in turbine valve position about the 100% and 50% load points. It is noted that in both cases, throttle pressure is returned to normal in about 2 minutes. The oscillation about the normal point is very small and dies out completely in less than 5 minutes.

Fig. 3.6(b) shows the response to a fuel upset at full load and half load points. Here again throttle pressure was brought back to normal in about 2 minutes.

The performance at the 100% load point compared with the performance at the 50% load point shows this system to be about optimal with the same settings for both load levels.

For the same load change, the integrated amount of over-firing required around the 50% load point is only one half that required around the 100% load point. Examination of the control signal trace on Fig. 3.6(a) reveals that this desired amount of over-firing is automatically accomplished from proportional and rate action operating on pressure error alone.

2. Pressure Control with Anticipation from Steam Flow (Fig. 3.5(b))

It is noted that in this configuration two quantities are used to control the pressure, namely, pressure error and steam flow. The control obtainable from steam flow alone through proportional and rate action is shown in Fig. 3.7 for full load and half load. The disturbance is a step in turbine valve position. This steam flow anticipation may be thought as feedforward action only, although strictly speaking, since steam flow is pressure sensitive, this signal does provide an objectionable positive feedback component of pressure error. With a very slight increase in proportional gain from steam flow, in the absence of feedback from pressure, a runaway condition develops.

The results of Fig. 3.7 point out that the optimum feedforward component adjustment at 100% load is far from optimum at 50% load.

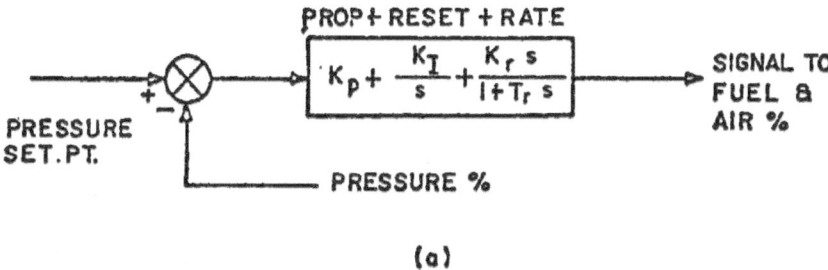

(a)

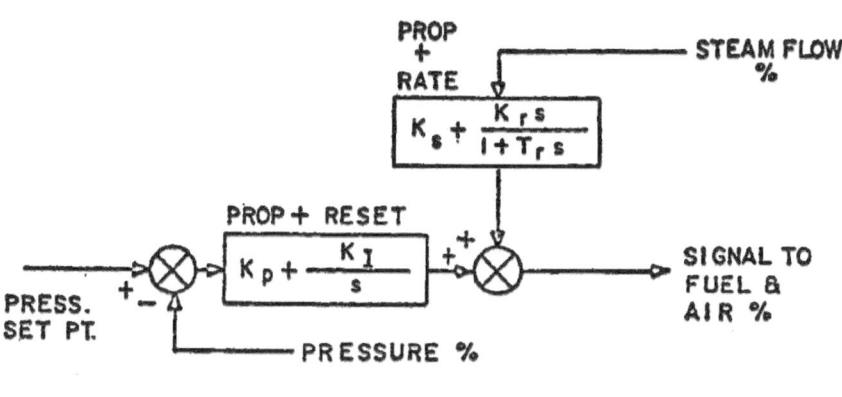

(b)

FIGURE 3.5 Pressure Control Configurations

(a) Three mode controller operating on pressure error alone
(b) Two mode controller on pressure error and feedforwardfrom steam flow

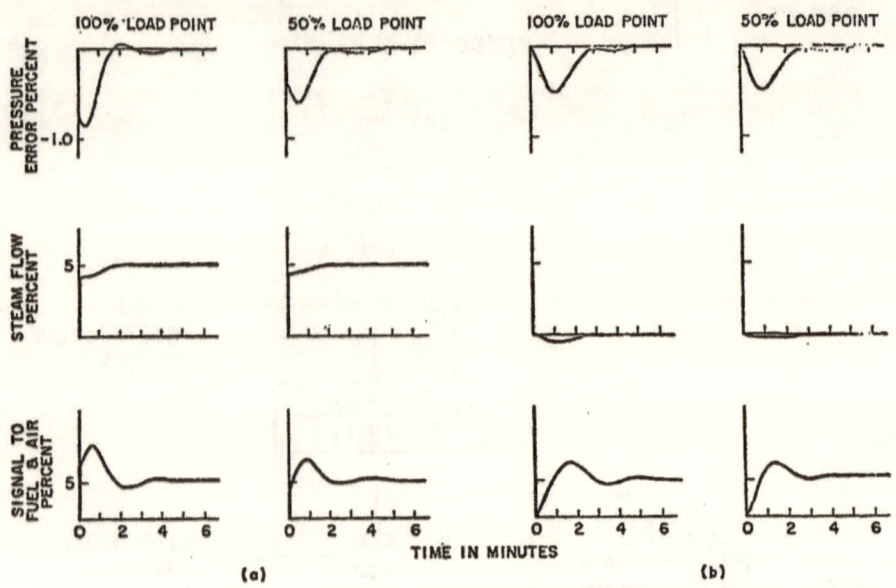

FIGURE 3.6   Performance for Control System of Fig. 3.5(a).
$K_p = 9.6$ $K_I = 0.1$ $K_r/T_r = 12$ $T_r = 10$ sec

(a)   Response to turbine valve step
(b)   Response to fuel upset

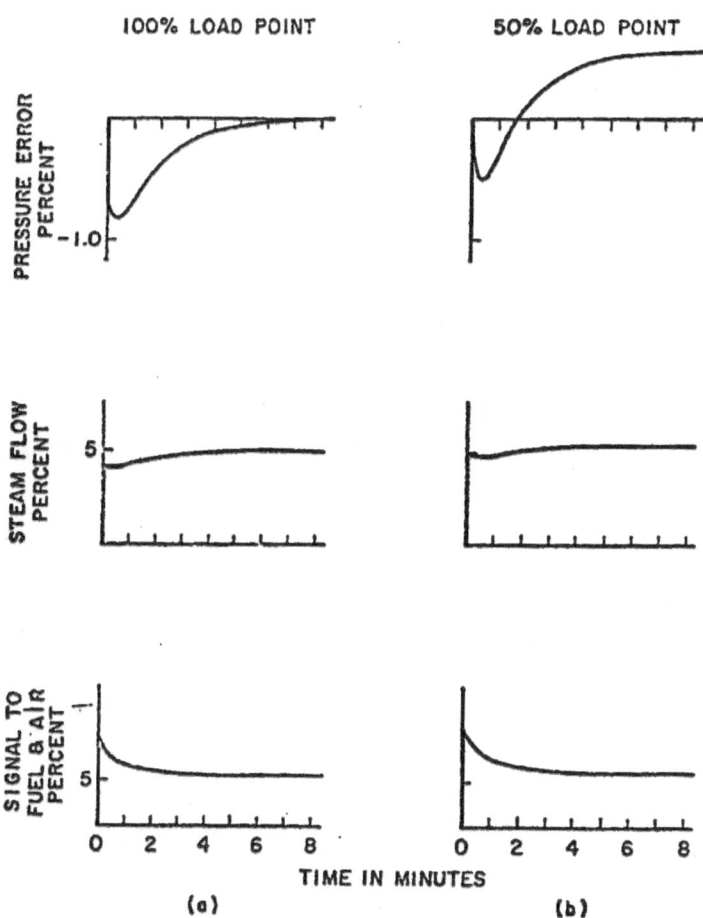

FIGURE 3.7   Response to a Turbine Valve Step with
Control System of Fig. 3.5(b) Using Steam Flow
Anticipation Only.

$K_p = 0$ $K_I = 0$ $K_S = 1$ $K_r/T_r = 1$ $T_r = 64.5$ sec

(a)   Full load point
(b)   Half load point

Next this control scheme is considered employing both pressure and
steam flow for two cases:

(a) Pressure Control with Proportional Anticipation from Steam Flow

Fig. 3.8(a) shows the response of the boiler to a step disturbance in
turbine valve position at full load and half load. The response to a fuel
upset is shown in Fig. 3.8(b) for full load and half load. It is seen that the
response to a fuel disturbance is poor. This is due to the setting of the reset
control which must be low to avoid pressure overshoots following load
changes. Increasing integral action improves the response to fuel upsets;
however, it does adversely affect the response to turbine valve position.
This is illustrated in Figs. 3.9(a) and (b). Fig. 3.9(a) shows the response
of the boiler to a step change in valve position at full load and half load,
with increased reset action. The corresponding responses to a fuel upset
are shown in Fig. 3.9(b). Note the large swings in throttle pressure.

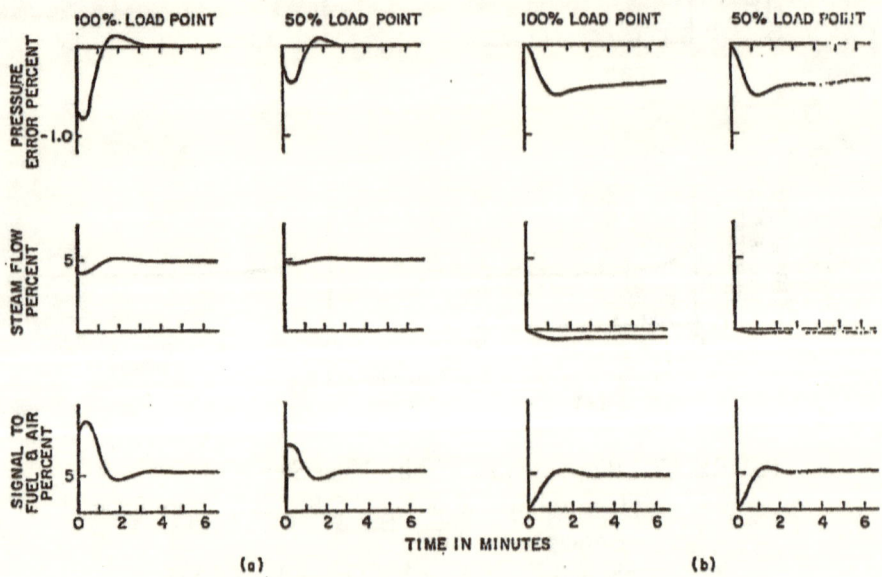

FIGURE 3.8   Performance for Control System of Fig.3.5(b)
with Proportional Anticipation from Steam Flow.
$K_p = 9.6 \; K_I = .005 \; K_s = 1 \; K_r/T_r = 0$

(a)   Response to a turbine valve step
(b)   Response to a fuel upset

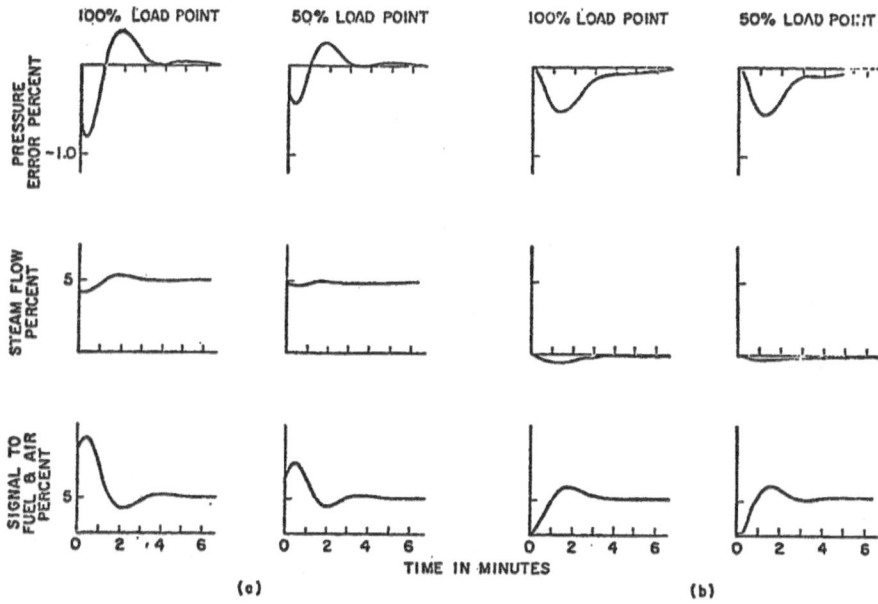

FIGURE 3.9   Performance for Control System of Fig. 3.5(b). K_p = 9.6
K_I = .075 K_s = 1 K_r/T_r = 0

    (a)   Response to a turbine valve step
    (b)   Response to a fuel upset

(b) Pressure Control with Proportional Plus Rate Anticipation from Steam Flow

Fig. 3.10 contains typical boiler responses with this type of control at the 100% load point. It is noted that the swings are excessive. The same case except that integral action is omitted is shown on Fig. 3.10(b). This configuration (with no reset) may be improved and the response looks like that in Fig. 3.10(c). Clearly, the response is still not satisfactory. The introduction of even a slight amount of reset action at this setting causes overshoots, as shown in Fig. 3.10(d). The response to a fuel upset is also oscillatory.

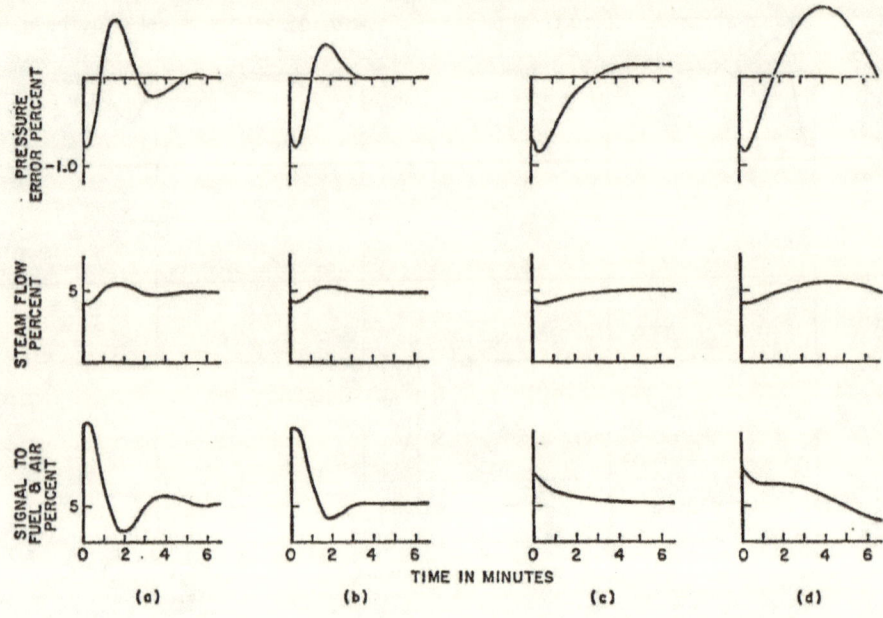

FIGURE 3.10   Typical Boiler Responses at 100% Load Point
With Control as in Fig. 3.5 (b). K_p = 9.6
$K_s = 1$ $K_r/T_r = 1$ $T_r = 64.5$.

(a)   Small amount of reset action, K_I = .01
(b)   No reset action, K_I = 0
(c)   No reset action, K_I = 0, K_p = 7.2
(d)   Same as (c) with slight reset action,
        K_I = .005 (unstable response)

3.   Fuel System With Dead Time

Now the fuel system is represented with a time constant of 20 seconds and dead time of 30 seconds: $\dfrac{e^{-S30}}{1+S20}$. Performance obtainable with (a) the control scheme shown in Fig. 3.5(a) (b) the control scheme shown in Fig. 3.5 (b); and in Fig3.5 (c), a new scheme that uses feedback compensation for dead time will be discussed below.

(a) Fuel System with Dead Time, Pressure Control from Pressure Error Alone (Fig. 3.5(a))

Figs. 3.11(a) and (b) show the response of such a system to a step change in turbine valve position, and to fuel upset, respectively, at 100% load point. Throttle pressure was brought back to normal in about 3 minutes (which compares well with 2 minutes obtained without dead time).

(b) Fuel System With Dead Time, Pressure Control from Pressure Error and Anticipation from Steam Flow (Fig. 3.5(b))

Figs. 3.12(a) and (b) show the response of such a system to a step change in turbine valve position and to fuel upset, respectively, at the 100% load point. When these curves are compared to Figs. 3.11(a) and (b), one observes excessive overshoot and slow settling to the final value. This is more pronounced in the response to a fuel upset as in Fig. 3.11(b). Just like the case without dead time, one can improve the response to a fuel upset, at the expense of the response to changes in turbine valve position. This type of control cannot be optimal for a wide range in the operating point.

(c) Fuel System with Dead Time, and Especially Compensated Controller (Fig. 3.13)

In Fig. 3.13(a) block A represents a regular 3 mode controller (like the one shown in Fig. 3.5(a)). Block B is a feedback stabilizer to compensate for the dead time in the fuel system[2,3]. The function of the stabilization block B can be understood with the help of Fig. 3.13(b). Assume the open loop response of the system without dead time to a step input is as shown by the solid curve C in Fig. 3.13(b). Clearly then, the response of a similar system with dead time τ will be as shown by the translated dashed curve D of Fig. 3.13(b). The difference between these two curves is also shown as the dotted curve E in Fig. 3.13(b). Then symbolically one may formulate: E = C − D. It follows then that C = E + D. The significance of this is that the response of a system with dead time will look like that of a system without dead time if a compensator having a response E is added to it. It is noted that all curves C, D and E are responses to a step input.

As far as the control action is concerned, it should be the same for a system with or without dead time, and this is assured by having the total feedback to the controller look like a process with the same dominant time constant but with no dead time.

The general shape of the stabilization response is as shown on the top right hand corner of Fig. 3.14. This response can be approximated with the second order transfer function of block B (Fig. 3.13(a)). The nomograph of Fig. 3.14 is self-explanatory for deriving the values of K, T_1 and T_2 of block B given the characteristics of the response curve t_1 and t_2.

With this compensator and a 2 mode controller, the response of the boiler to a step disturbance in turbine valve position is shown in Fig. 3.15(a) at the full load point. The pressure is brought back to normal in about 3.0 minutes, and the response is almost non-oscillatory. The response to a fuel upset is shown in Fig. 3.15(b) for full load. It is noted that, for the particular set of dead time and process response constants chosen in this example, the performance obtainable with this scheme is only slightly better than the conventional 3 mode controller. It does possess advantages in cases where the ratio of dead time to process dominant time constant is greater than in the example in this paper.

It should be noted that the three mode controller for a system with no dead time is tuned at quite different values than one to control a system with dead time (compare settings on Fig. 6 with those of Fig. 3.11). However, with the dead time compensator as in Fig. 3.13(a), K_p and K_I are about the same as for the process without dead time (compare settings in Fig. 3.6 with settings in Fig. 3.15).

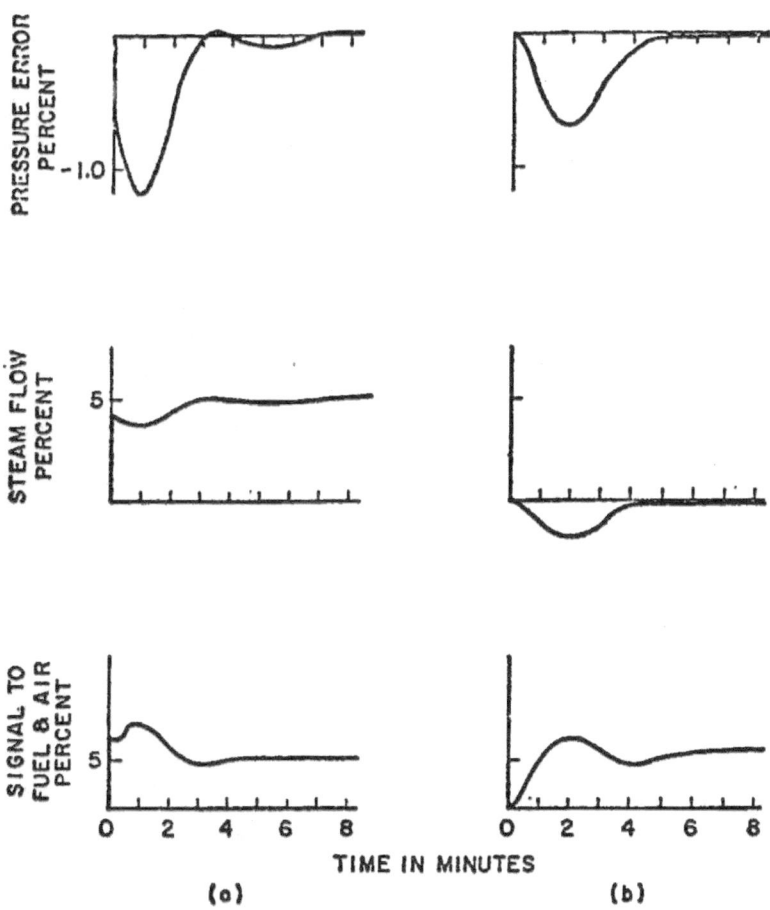

FIGURE 3.11   Performance With Dead Time in the Fuel System
and Control as in Fig. 3.5(a). K_p = 5.2
K_I = .017 K_f/T_r = 2.2
T_r = 10 sec

(a)   Response to a turbine valve step
(b)   Response to a fuel upset

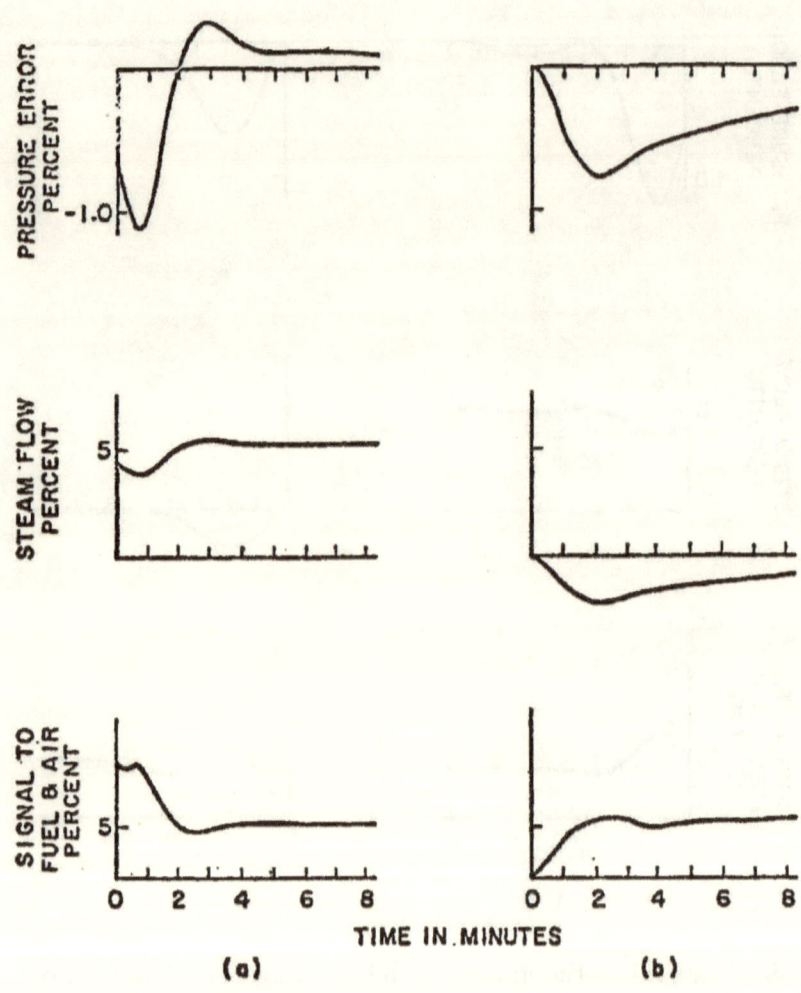

FIGURE 3.12   Performance with Dead Time in the Fuel System and Control as in Fig. 3.5(b). $K_p = 5.2$ $K_I = .005$ $K_s = 1$ $K_r/T_r = 1$ $T_r = 64.5$ sec

   (a)   Response to a turbine valve step
   (b)   Response to a fuel upset

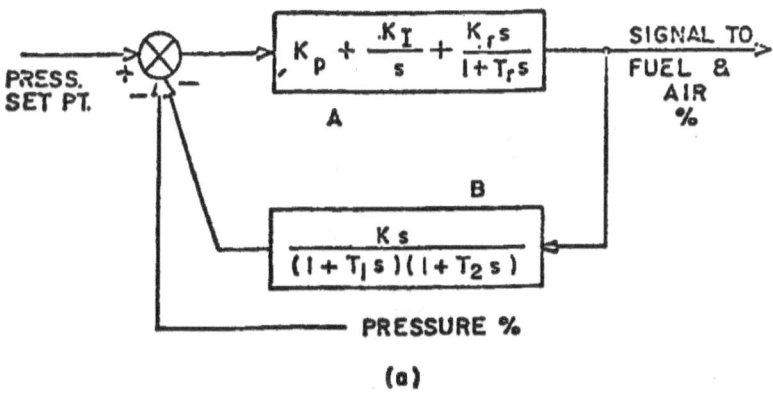

PRESS.
SET PT.

$$K_p + \frac{K_I}{s} + \frac{K_r s}{1+T_r s}$$

A

SIGNAL TO
FUEL &
AIR
%

B

$$\frac{K s}{(1+T_1 s)(1+T_2 s)}$$

PRESSURE %

(a)

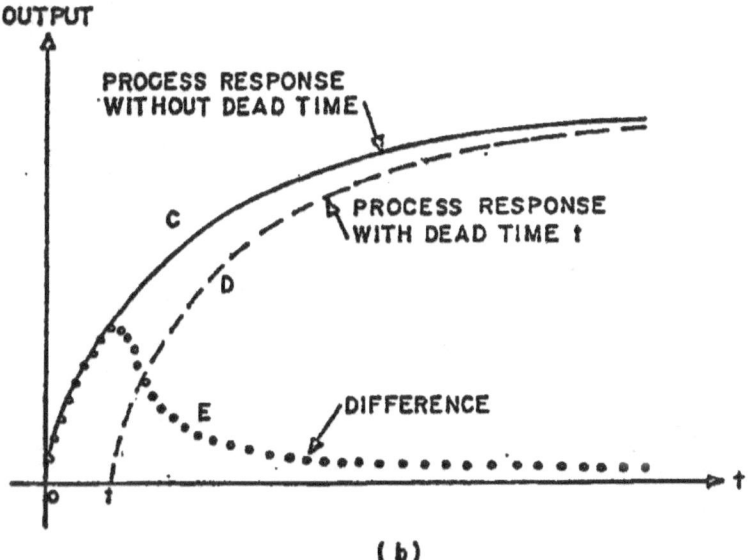

OUTPUT

PROCESS RESPONSE
WITHOUT DEAD TIME

C

PROCESS RESPONSE
WITH DEAD TIME t

D

E

DIFFERENCE

t

(b)

FIGURE 3.13   Special Compensator for Dead Time

(a)   Compensator block diagram
(b)   Response curves describing compensator action

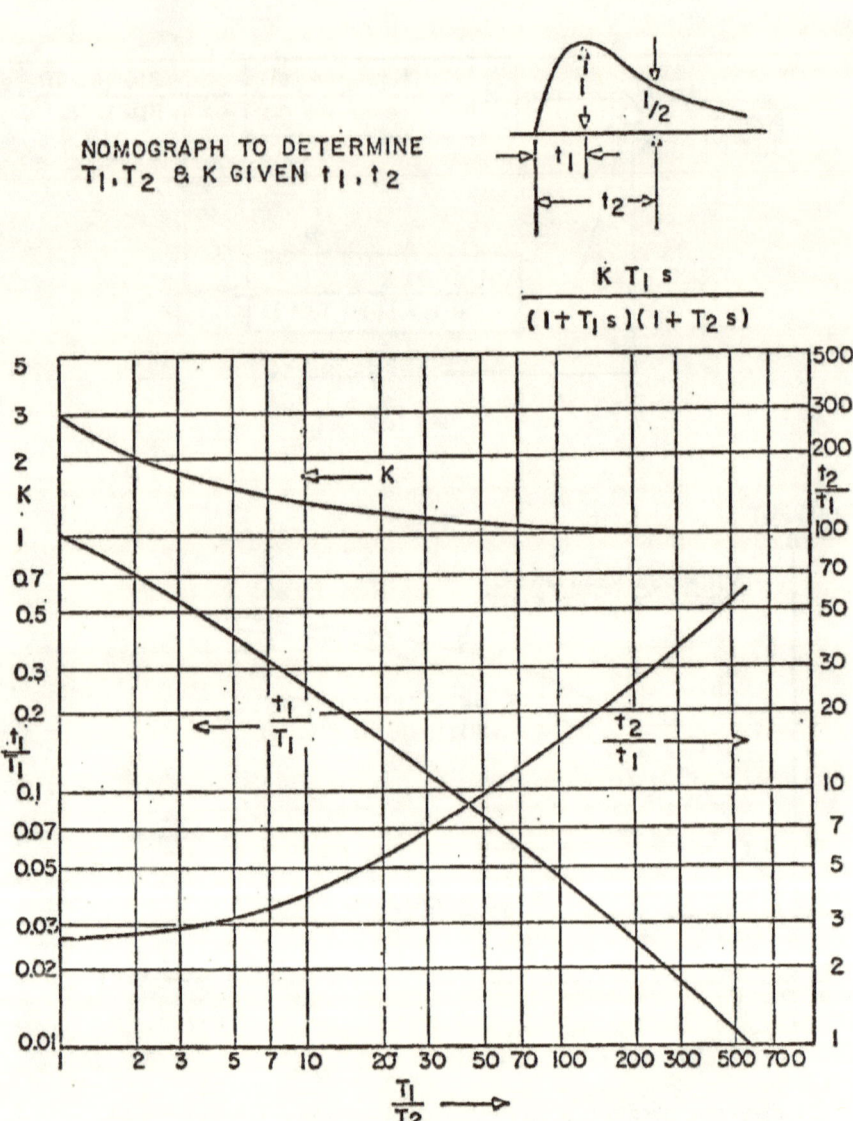

FIGURE 3.14 Nomograph for Calculation of the Parameters
of the Dead Time Compensator of Fig. 3.13(a)

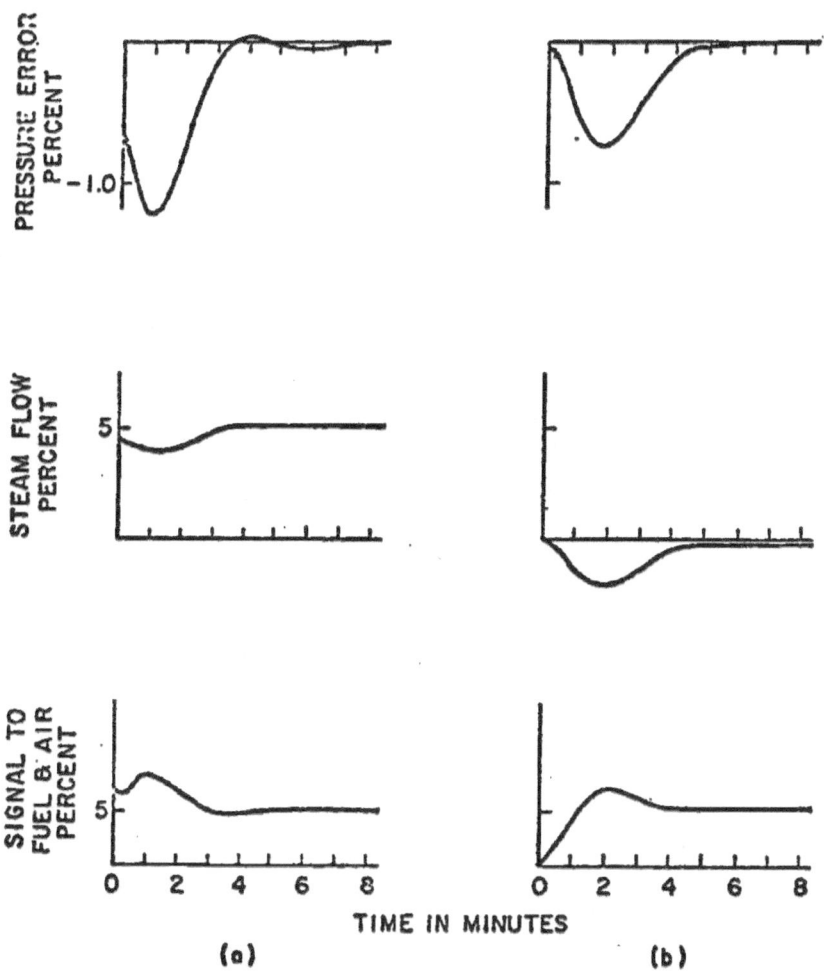

FIGURE 3.15   Performance With Dead Time in the Fuel System
and Control as in Fig. 3.13(a). K_p = 9.6
K_I = 0.1 K_r/T_r = 0.

(a)   Response to a turbine valve step
(b)   Response to a fuel upset

REFERENCES

3.1   de Mello, F.P., and Imad, F.P., "Boiler Pressure Control Configurations", IEEE Conference paper 31 CP-67-12. Winter Power Meeting, January, 1967.

CHAPTER IV

DRUM BOILER FEEDWATER CONTROLS

The primary function of feedwater controls is to maintain drum water level at the desired set-point. In its simplest form one can visualize the process of defining drum level as a simple integration of water flow into the boiler less steam flow out of the drum. The gain of integration is readily calculated from the mass of water that is held for each inch of water level around the normal operating level. Fig. 4.1 shows this simple process model.

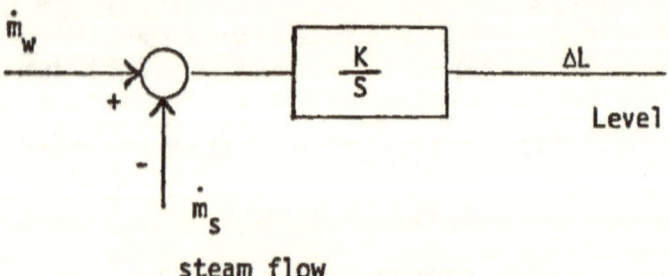

FIGURE 4.1

Since the primary disturbance originates from changes in steam flow, it is logical to structure the controls such as to make feedwater flow follow steam flow, and use the deviation in level as a slow resetting action to bring the required water inventory back to balance.

The use of feedwater, steam flow and level as inputs to the control action gave rise to the term "3-element-control".

From a hierarchical point of view, it is convenient to think in terms of a feedwater subloop control, whereby feedwater flow is made to follow a feedwater demand. This demand in turn is derived from steam flow (feedforward primary signal) and a correcting signal from a controller operating on drum level error (Fig. 4.2)

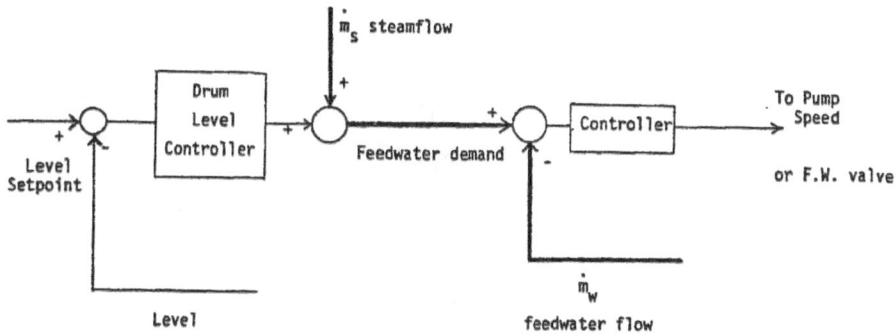

FIGURE 4.2

DRUM LEVEL SWELL EFFECTS

The dynamics of drum level response to feedwater, steam flow and heat to the waterwalls are more complicated than conveyed by the simple representation of Fig. 4.1. Drum swell effects accompany a redistribution of water and steam inventories in the waterwalls with changes in steaming rate. This effect is more accentuated in low pressure boilers (600 psi-1000 psi) because of the greater difference in specific volume between steam and water at these pressures than at the higher pressures (2400-2600 psi). The design of drum internals, i.e., the baffling and the manner in which feedwater is introduced into the drum as well as the degree of subcooling of the feedwater have marked effects on level response.

The peculiar dynamic effects that can arise in control of drum level will be illustrated with a few examples.

Take the case of typical low pressure (600 psi) boilers of 300000 to 600000 lbs/hr capacity. Fig. 4.3 shows the boiler drum, downcomers and the risers or waterwalls where steam is generated.

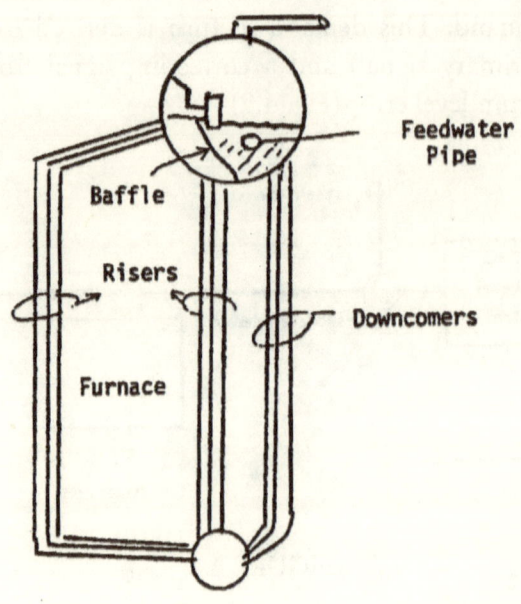

FIGURE 4.3

Feedwater is introduced along a feed pipe running along the length of the drum about 5″ below normal water level. The feedwater could be considerably subcooled (250°F vs. 486°F saturated liquid at 600 psi). With the internal arrangement and baffling shown in Fig. 4.3, the water entering the downcomers is subcooled and is brought to saturation by the heat added in the furnace.

Another arrangement of drum internals is shown in Fig. 4.4 where the feedwater is free to mix with the steaming mixture exiting from the riser tubes and is brought up to saturation temperature mainly by condensing some of the steam generated in the risers.

The response characteristics of level to feedwater flow rate are radically different for these two types of drum internal configuration.

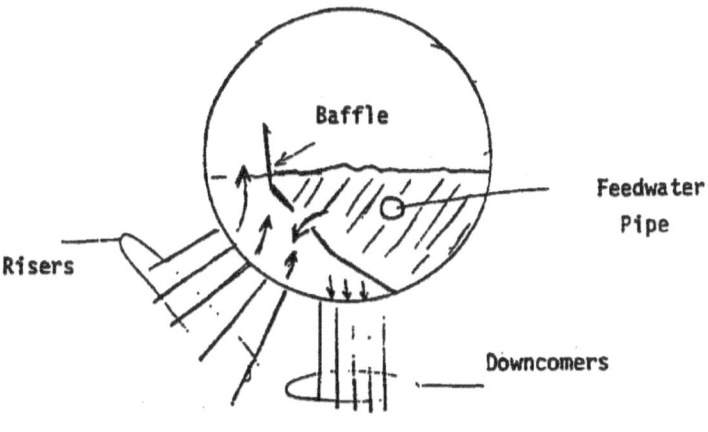

FIGURE 4.4

The process can be characterized as in Fig. 4.5; the control variable is feedwater and the controlled variable is level. Other inputs are heat release and valve opening which set the steam flow out of the boiler.

Response characteristics to changes in feedwater flow demand for the boiler types shown in Fig. 4.3 and Fig. 4.4 are shown qualitatively in Fig. 4.6.

It can be noted that in one case level responds with considerable dead time (case where the feedwater does not mix with the steaming mixture in the drum). Evidently the type of drum level control and settings would have to be radically different for the two cases as can be seen in the following analysis. Drum swell effects for changes in steam flow are illustrated in Fig. 4.7.

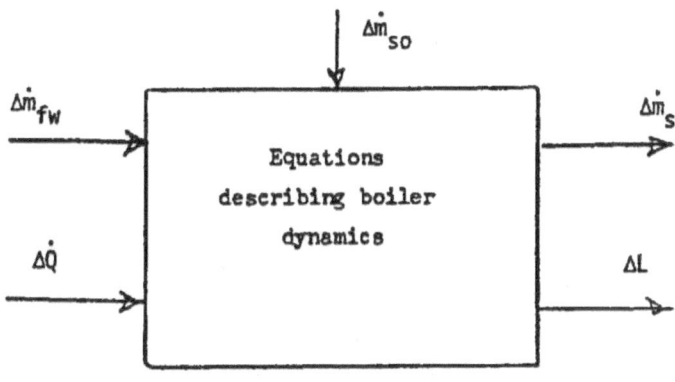

FIGURE 4.5

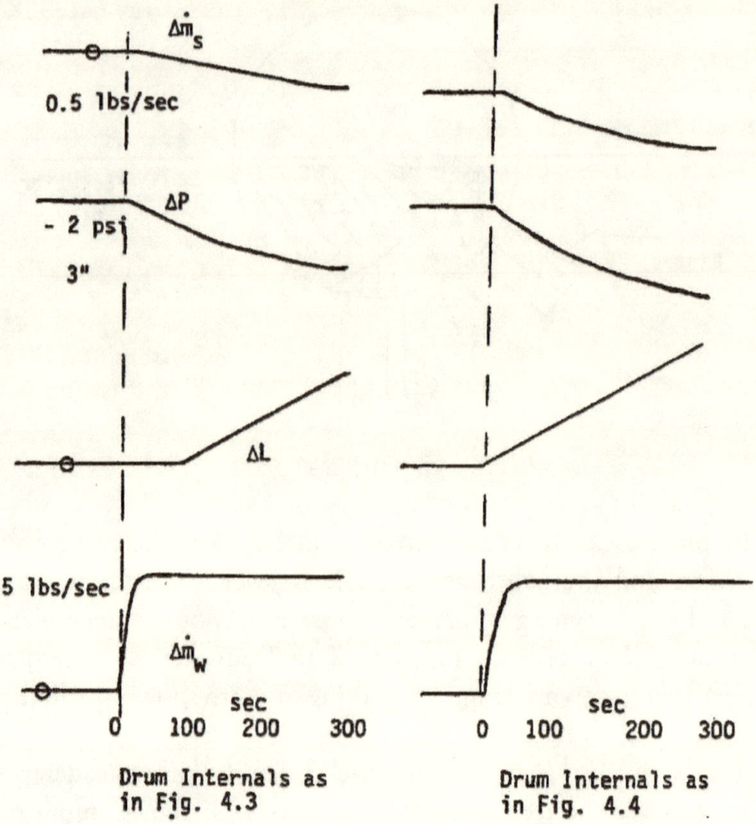

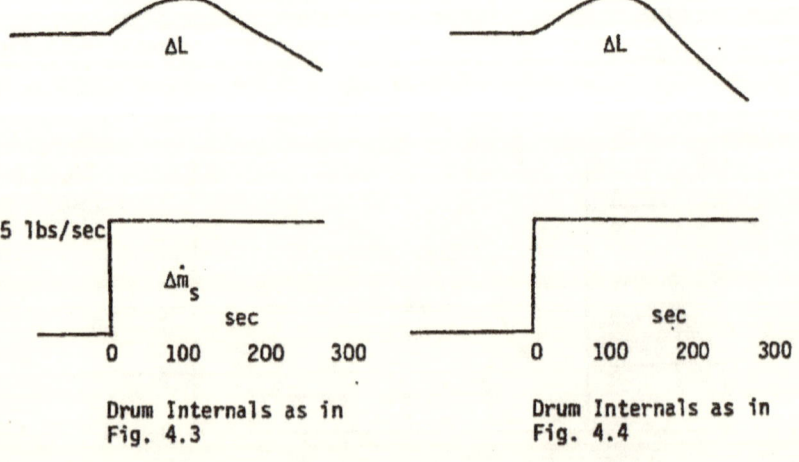

FIGURE 4.6 Response to Changes in Feedwater Demand

FIGURE 4.7 Response to Changes in Steam Flow

DRUM LEVEL CONTROLS WITH COMPENSATION FOR SWELL EFFECTS

The block diagram of the feedwater control system is shown on Fig. 4.8.

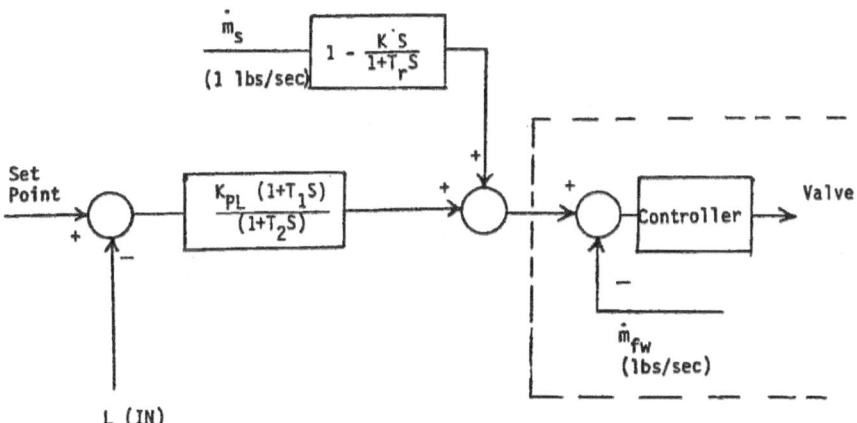

FIGURE 4.8

It can be assumed that the feedwater control subloop within the dashed lines could be tuned to give a response of feedwater to a demand for feedwater characterized by a single 10 sec time constant. Hence, the level control system can be viewed as in Fig. 4.9.

The feedforward anticipatory action from steam flow provides an initial demand in the opposite direction to neutralize part of the swell effect. The tuning of the level control system, however, will be determined by the characteristics of the closed loop shown in Fig. 4.9.

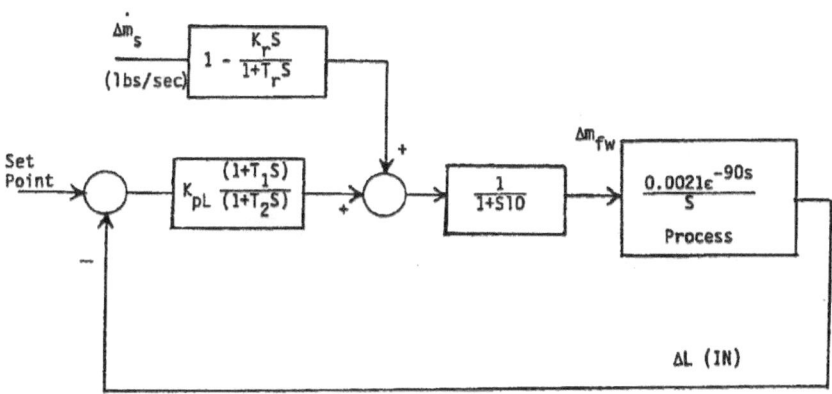

FIGURE 4.9

The stability of the level control system can be analyzed once the nature of the process transfer function $\Delta L/\Delta m_{fw}$ is known. This process function can be described as an integration with dead time for the configuration of Fig. 4.3, i.e.,

$$\Delta L /_{\Delta \dot{m}_{fw}} = \frac{\varepsilon^{-90s}\, 0.0021}{s} \text{ in/ lbs/sec}$$

Fig. 4.10 shows the Bode plot of the open loop function that would yield acceptable performance.

Note that a proportional controller with transient gain reduction has been used. Had the response characteristics not exhibited the dead time as is the case for the drum configuration of Fig. 4.4, a much tighter proportional gain setting could be provided, without need for transient-gain reduction.

Fig. 4.11 shows that with the response of $\Delta L /_{\Delta \dot{m}_{fw}}$ characterized as $\frac{0.0021}{s}$ with no dead time, a simple proportional gain can be applied yielding considerably better closed-loop performance. For the same phase margin of 45°, this gain could be approximately 50, giving an open-loop function $\frac{.105}{S(1+10S)}$. Note that the scale of frequency of W in Fig. 4.11 is 10 times greater than in Fig. 4.10.

The impact of this change in process characteristics is that in the first case the crossover frequency of the drum level control loop is about .0043 rads/sec, while in the second it is 0.1 rads/sec, i.e. a bandwidth 23 times greater in the second case.

Further, a 10% error between feedwater and steam flow (8 lbs/sec) as might be caused by blow down or inaccuracies in metering, would require a compensating error in drum level of 8/7 inches in the first case versus 8/50 inches in the second.

While on this example it is appropriate to note that the transient gain reduction as in Fig. 4.10 yields a system which is subject to poor damping or even instability during major upsets that exercise the feedwater controls through limits. This so-called conditional stability (drum level hunt in this instance) can be readily explained by describing function analysis methods. The effective gain of feedwater subloop and its effective response is reduced (time lag increased) as the oscillation exercises the valves to limits. These effects interpreted on the Bode plot of Fig. 4.10 shows that crossover would occur in the lower frequency region where the phase angle approaches 180°.

Power plant operators often encounter these conditions and have gotten out of the problem by placing the controls on manual, controlling the process until the oscillations are within the linear range of controls at which time automatic control is restored.

Variations on the basic scheme of Fig. 4.2 can be introduced to account for swell effects, and interactions between firing-rate and drum level transients.

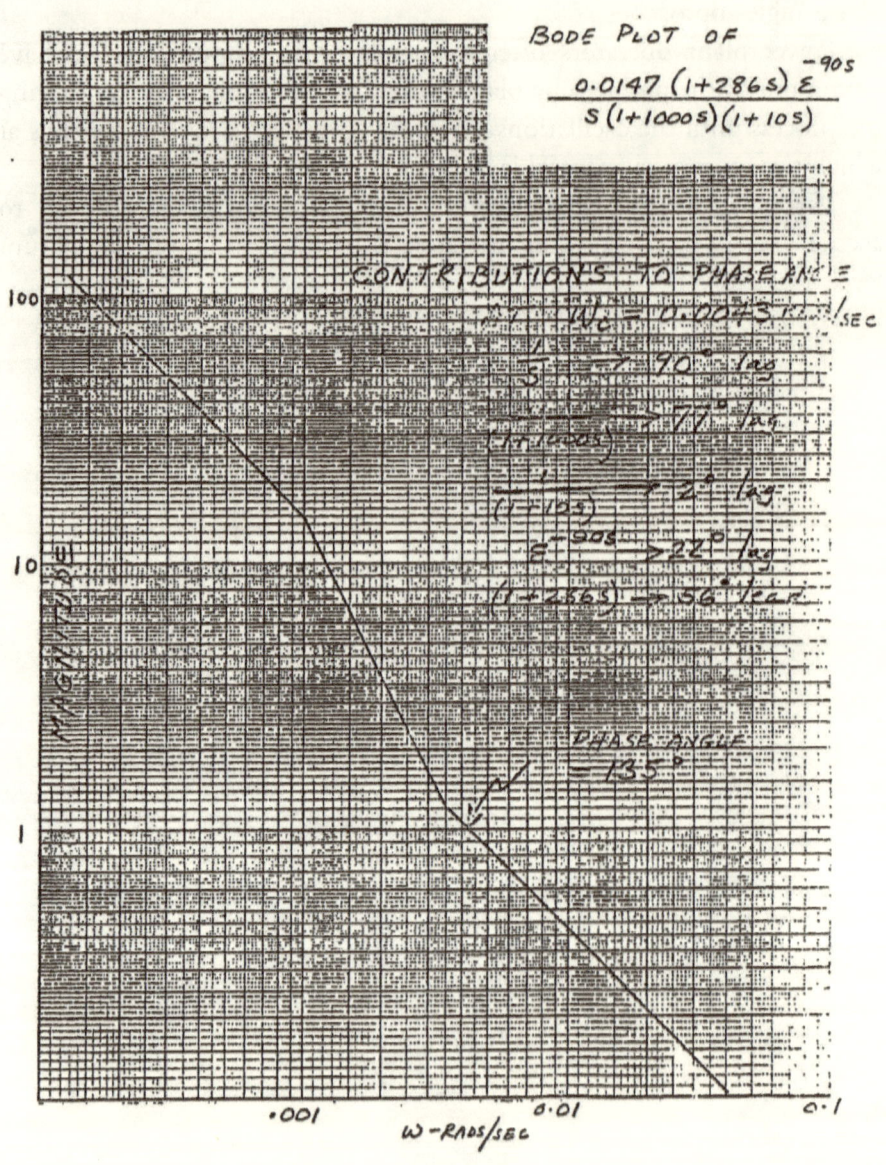

FIGURE 4.10

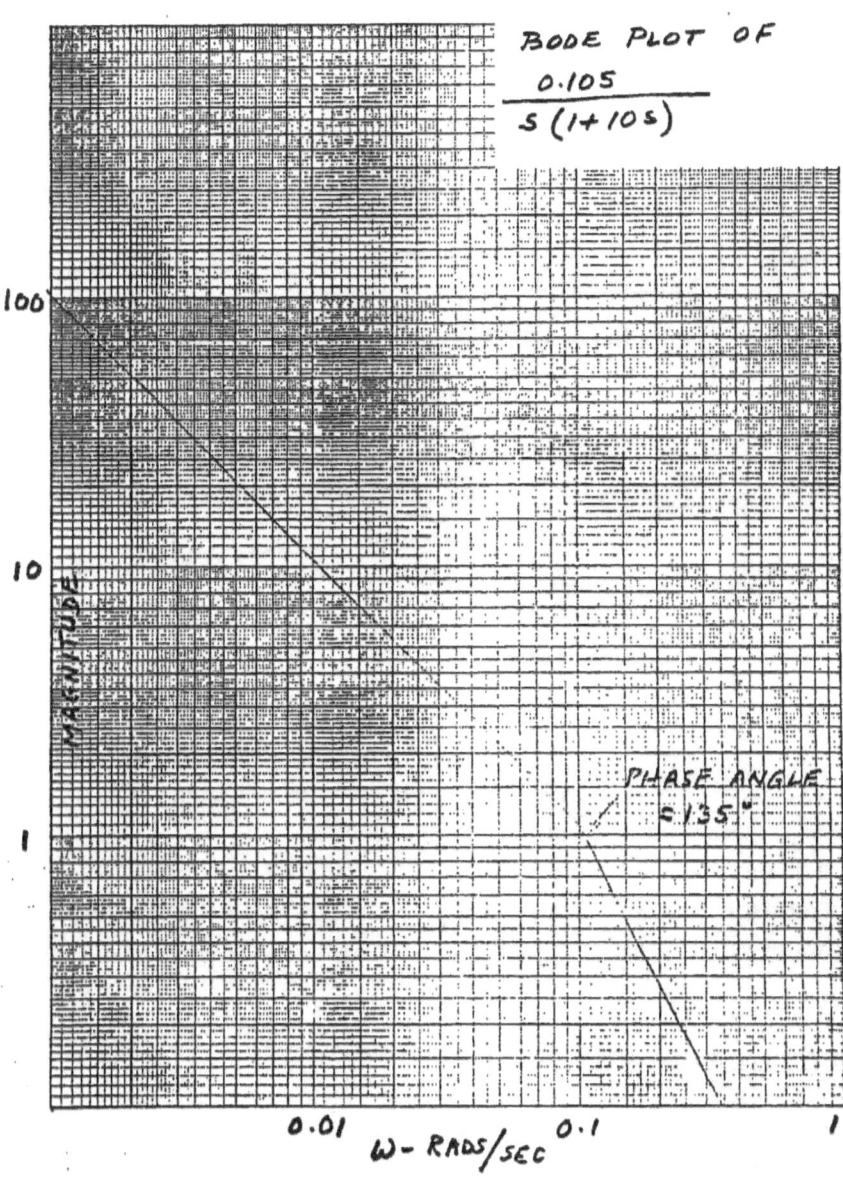

FIGURE 4.11

Fig. 4.12 shows one such scheme. Feedwater demand is again formed from steam flow, although a lag on steam flow is provided to avoid augmenting the swell effect. A rate of change from drum pressure was found to be a good anticipator of swell effects.

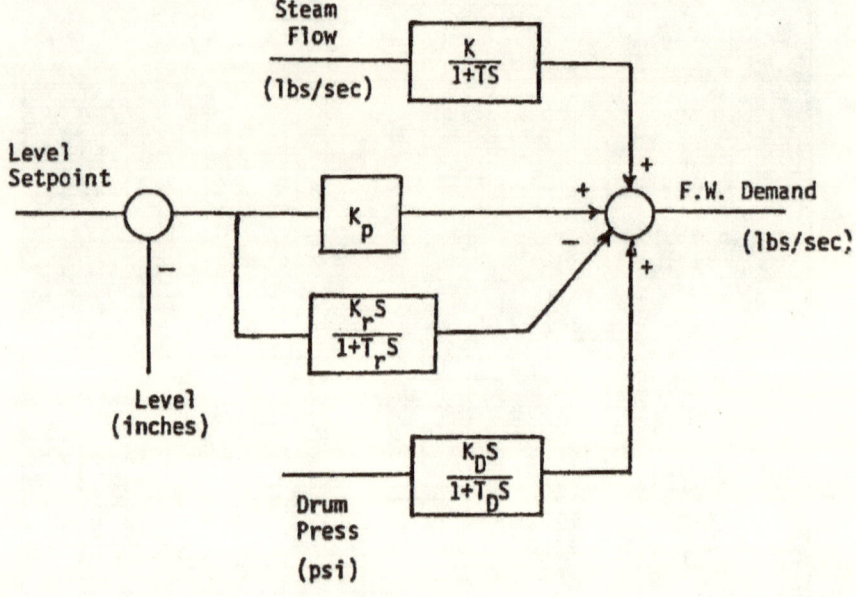

FIGURE 4.12

The feedwater demand is corrected by a proportional control with transient gain reduction operating on level error.

The feedwater demand is then acted upon by a subloop which should be reasonably fast since the response characteristics of flow to changes in feedwater valve position, or changes in feed pump speed, should be fast.

Feedwater Sub-loop Controls

One of the serious difficulties in control relates to process nonlinearities which radically change the gain and/or response characteristics of the process. Control settings which are proper for a given operating point may thus cause poorly damped or unstable operation at other operating points.

While nonlinear characteristics of the main process are unavoidable, one often finds that much of the problem belongs in inadequately designed subloops, with nonlinear valve or damper characteristics. It should be a basic axiom of good control practice that subloops should be designed to be uniformly responsive throughout the operating range and should have a bandwidth greater (preferably by one order of magnitude) than the main process controls.

The difficulties with obtaining a linear response from a feedwater subloop where flow changes are accomplished with changes in pump speed are described in Figs. 4.13 to 4.16.

Fig. 4.13 shows the pump head versus flow characteristics for different pump speeds. On the same figure are superposed the boiler feedwater versus pressure characteristics for 100% and 50% loads.

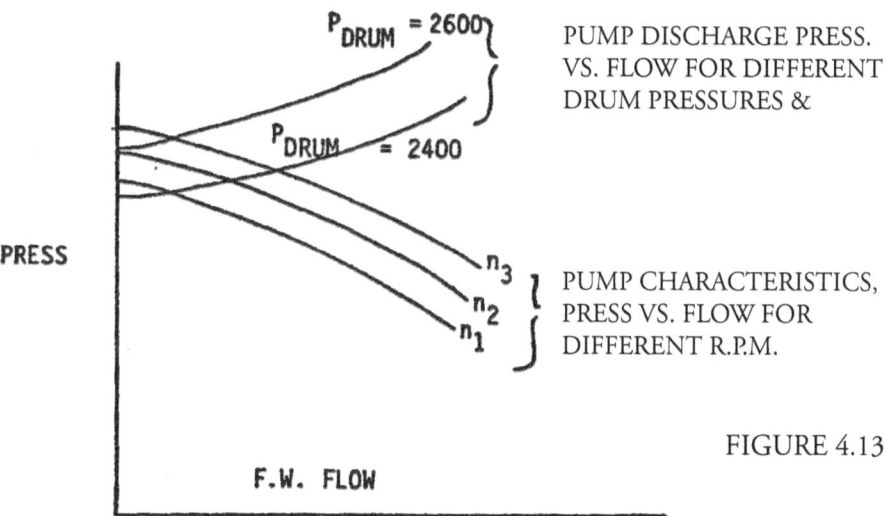

$P_{DRUM} = 2600$ PUMP DISCHARGE PRESS. VS. FLOW FOR DIFFERENT DRUM PRESSURES &

$P_{DRUM} = 2400$

PRESS

n_3 PUMP CHARACTERISTICS, n_2 PRESS VS. FLOW FOR n_1 DIFFERENT R.P.M.

FIGURE 4.13

F.W. FLOW

From the intersection of these pump and boiler pressure characteristics, the pump flow versus pump rpm relationships are shown on Fig. 4.14.

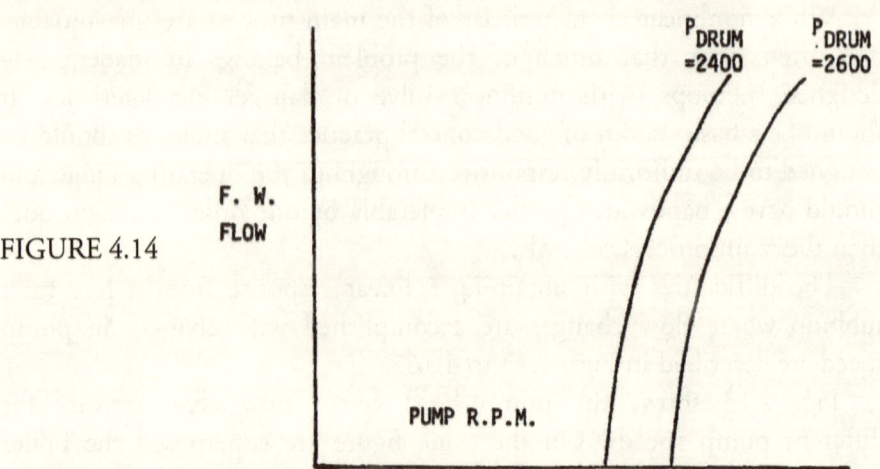

FIGURE 4.14

This figure shows that the pump rpm has to reach a certain threshold value before any flow will occur, a threshold value which is a function of drum pressure. The characteristic also shows that at low flows there is a high gain between rpm and flow. It has been common practice to attempt to compensate this nonlinearity with a cam at the hydraulic coupling scoop tube positioner.

Figure 4.15 shows how the cam could only compensate correctly for one particular drum pressure. This expedient would be satisfactory if constant drum pressure were held, or even if it were held within a fairly narrow band. Unfortunately, during startup, or even to take advantage of variable pressure operation for efficiency optimization, the boiler pressure may have to go through wide ranges and, therefore, a cam which accomplished linearization between signal to scoop tube and flow at one pressure would be inadequate.

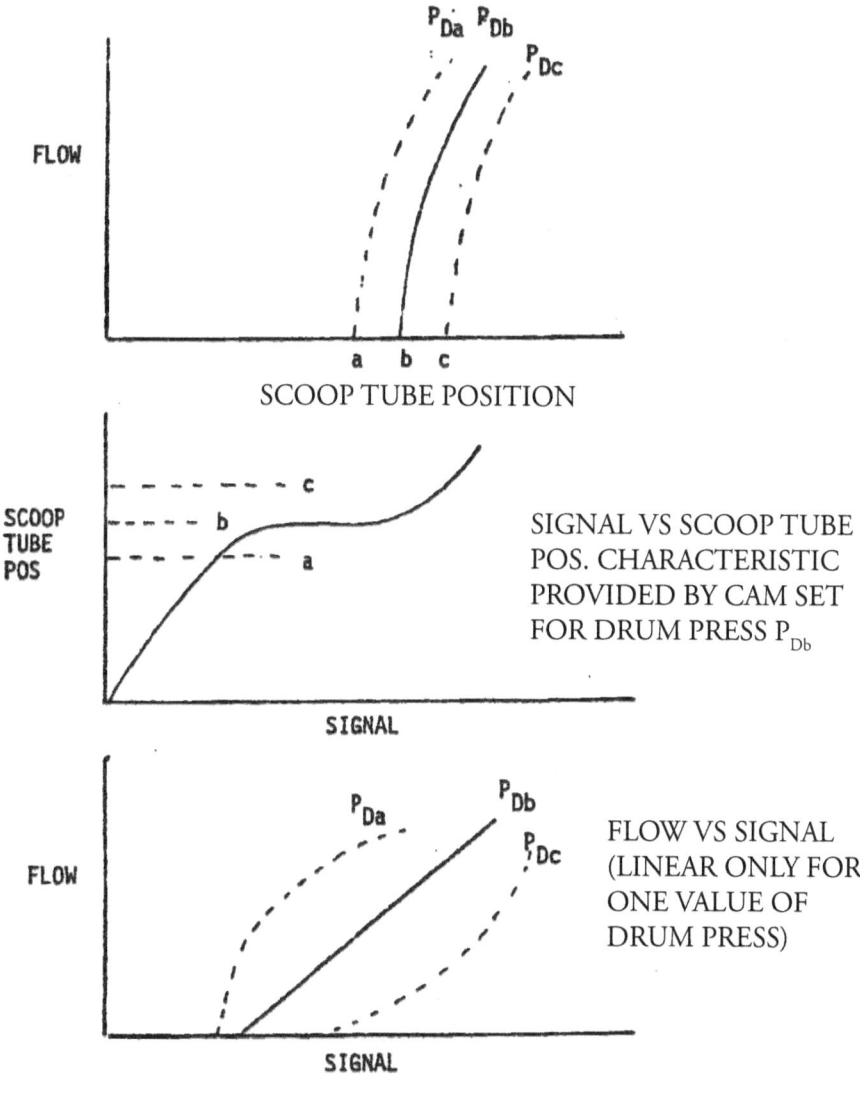

FIGURE 4.15

By providing a function generator and signal proportional to drum pressure, as indicated in Fig. 4.16, the feedwater flow loop is linearized and should operate stably throughout the load range and over significant variations of boiler pressure.

Because of the interaction between feedwater flow and drum pressure, the drum pressure signal is lagged or filtered with a time constant.

The feedwater subloop control as described can thus be adjusted to exhibit a response described by a single 15 sec time constant throughout the operating range.

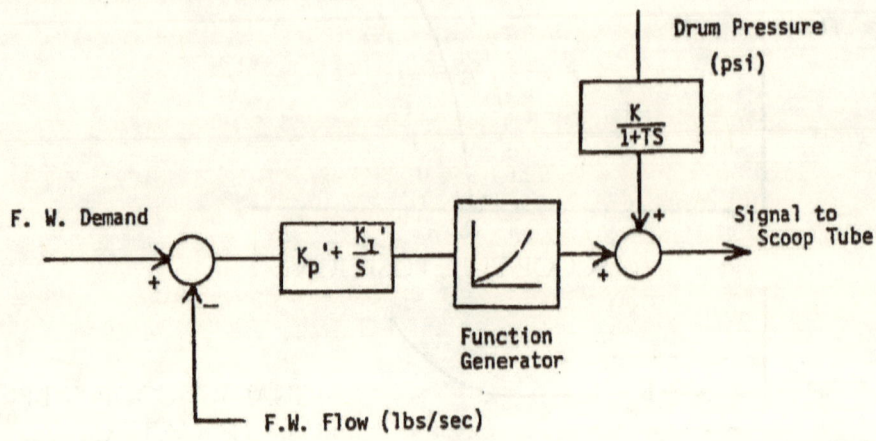

FIGURE 4.16

CHAPTER V

FUEL AND AIR CONTROLS FOR DRUM TYPE BOILERS

Steam generation is controlled by firing rate. For proper and safe combustion conditions, fuel and air should be controlled in coordinated fashion to assure adequate oxygen for combustion.

Too little air can cause a fuel rich mixture and possible danger of explosion. Too much air represents extra heat losses to the stack. Controls should be designed to maintain the desired amount of excess air in the steady state and to provide the proper dynamic action during transients.

5.1   Oil and Gas-Fired Boilers

Control of fuel in oil and gas-fired boilers is relatively easy, as it involves modulation of a fuel valve. The response of fuel flow to valve change is almost instantaneous, hence the fuel flow sub-loop is handled very simply with a proportional and reset controller.

Air flow is usually changed by modulation of fan damper position (FD and/or ID) hence the airflow sub-loop can likewise be handled with simple reset controllers. In the case of balanced draft boilers the control action to the fans must be properly coordinated to hold furnace draft and satisfy airflow demand. Problems associated with draft control are described in Section VI.

Fig. 5.1 shows a commonly used control scheme for coordination of fuel and air controls in oil or gas-fired boilers. Fuel demand is derived from a LO-SELECT station which compares the basic fuel and air demand signal with measured and characterized air flow. In this way one ensures that fuel cannot run ahead of air. Should for any reason air flow lag or not respond to a demand signal, the LO-SELECT station would hold fuel in balance with the available air flow.

The air flow demand signal in similar fashion is the output of a H1-SELECT station which compares demand derived from the basic fuel and air demand with actual fuel flow.

The basic demand signal is modified by a ratio calibration which is either manually set or is directed by a slow auxiliary loop controlling excess air to the desired valve by sensing percent O_2.

The airflow measurement is processed through a function generator to impart the desired nonlinear characteristic between fuel demand and air demand (minimum airflow is usually 30% or higher even when fuel flow is zero).

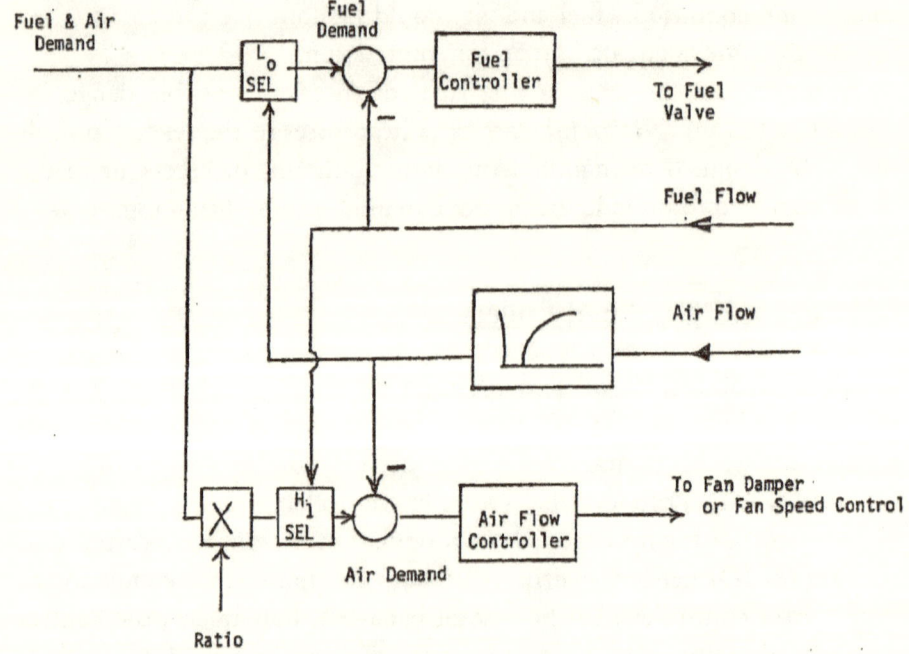

FIGURE 5.1

5.2 Pulverized Coal-Fired Boilers

Pulverized coal-fired boilers provide fuel to the furnace as a mixture of finely ground coal dust and air (Fig. 5.2). Coal is introduced to a coal mill through a coal feeder which drops coal from a hopper to the mill. Rollers in the mill grind the coal to a fine dust which is transported through coal pipes to the burners by primary air. Each mill usually supplies coal to four corner mounted burners at a given elevation. Four or five mills feed the burners at a like number of different elevations.

Firing rate is changed basically by changing the feeder speed, that is, the rate of flow of coal into the mill. Evidently there is a lag between the time feeder speed is changed and actual change of pulverized coal in the furnace. This lag is sometimes quite variable and the dynamics of coal mills are indeed a mystery. Primary air flow is kept constant in many cases. In others, primary air is made to vary in proportion to feeder speed.

The response of fuel in the furnace to changes in feeder speed exhibits considerable lags. These lags can also be variable as a function of the condition of the mill (amount of wear and/or the actual coal inventory of fines in the mill).

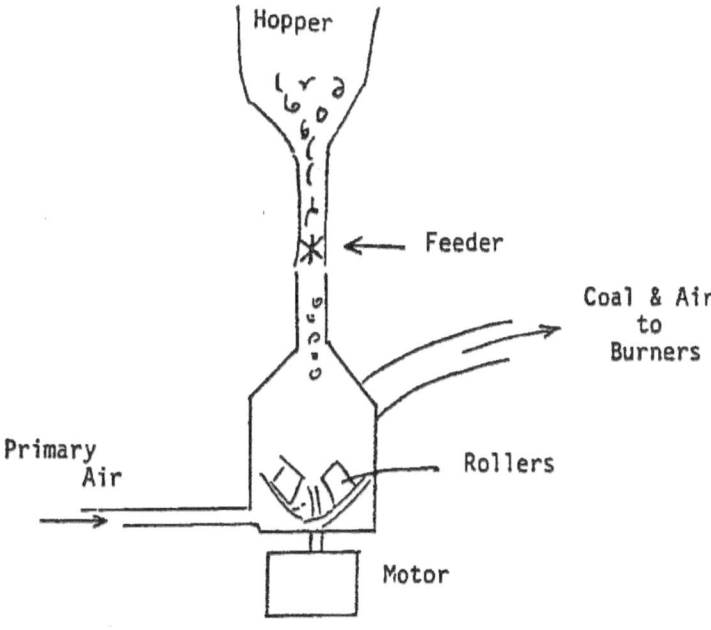

FIGURE 5.2

Conceptually one may think of the pulverizer process in terms of the block diagram of Fig. 5.3 where the fuel flow to the furnace is shown as being proportional to some function of level of fines in the mill.

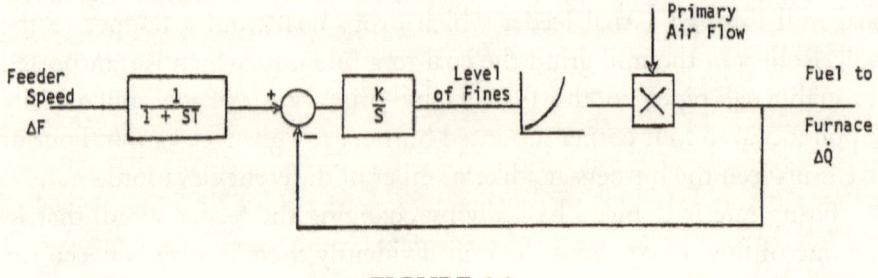

FIGURE 5.3

The net result is that the response of the mill is usually characterized by a dead time in series with a time lag.

$$\frac{\Delta \dot{Q}}{\Delta F} = \frac{\varepsilon^{-ST_1}}{(1+ST_2)}$$

where

$$T_1 = 30 \text{ to } 60 \text{ sec}$$
$$T_2 = 15 \text{ to } 30 \text{ sec}$$

In schemes where some change in primary air flow is effected simultaneous with a change in feeder speed, a small, almost instantaneous change, in fuel flow results, provided the coal mill inventory is not depleted.

The important point is that the response of fuel in the furnace to a change in feeder speed is usually not measurable. An inferential measure of fuel is usually taken as feeder speed. This fact makes the control of fuel and air in coal-fired boilers radically different than for oil and gas-fired boilers.

Fig. 5.4 shows the functional diagram of a fuel control system designed for coal firing. Pressure error works through the pressure controls (proportional and reset with additional stabilization for process lags) to set a fuel demand to a cascaded controller which almost instantly satisfies this demand by sending the proper signal to the various feeders. The purpose of the cascaded fuel controller is to assure the same gain between the pressure controls and the actual fuel in the boiler, irrespective of number of mills in operation.

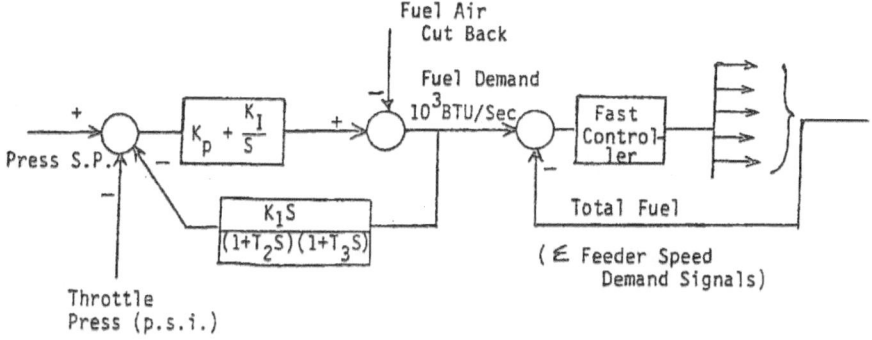

FIGURE 5.4

The feedback function $\dfrac{K_1 s}{(1+T_1 s)(1+T_2 s)}$ around the proportional plus reset controller is a stabilizing function to compensate for process lags and dead time. Its action can be understood with reference to Figs. 5.5(a), (b), and (c). It was also described in Chapter III.

Fig. 5.5(a) shows a control loop with a single time constant process without dead time. A simple proportional plus reset controller can be used on this process to yield extremely tight and stable performance. The proportional gain theoretically can be increased to infinity. Practically, however, even a so-called one time constant process has small additional lags which limit the maximum loop gain or loop response to maintain stability.

Fig. 5.5(b) shows a control loop with a process characterized by a dead time and a single time constant. Here a proportional plus reset controller would give very sluggish performance as the proportional gain and reset settings would have to be lowered significantly for proper stability.

Fig. 5.5(c) shows the same process with dead time as in Fig. 5.5(b), however, in parallel with the process, a stabilization function is added, providing a transient feedback around the controller such that the overall feedback (process plus stabilization) looks like a one time-constant process without dead time. As far as the controller is concerned, it does not know the difference between the case of Fig. 5.5(a) and that of Fig. 5.5(c) and, therefore, it can be tuned just as tight as for the case of Fig.

5.5(a). The resulting control action would be identical but, of course, the closed loop response of the process, although just as stable and responsive, is delayed to the extent of the dead time which cannot be physically overcome.

The stabilization function, as indicated on Fig. 5.5(c) is a rate over two time constants function. Its adjustment can be calculated from the shape of the transient response obtained by taking the difference between the undelayed single time constant process and the actual process. It was provided in anticipation of possible lags in the fuel system.

One method of determining the response of the required stabilization network is to make an open-loop test of the process as seen by the controller, translate the response back to time zero, wiping out the dead time, and then take the difference between the translated response and the actual response as the step response shape that should characterize the stabilization.

The shape of the response for the stabilization often can be approximated by the function $\dfrac{KT_1s}{(1+T_1s)(1+T_2s)}$ (rate over two time constants). Fig. 3.14 in Chapter 3 shows a nomograph for selection

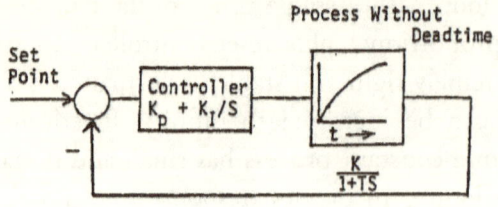

FIGURE 5.5(a)

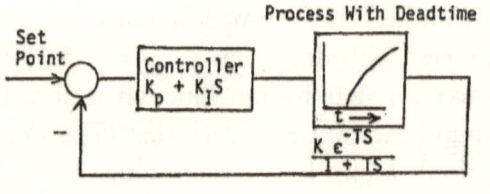
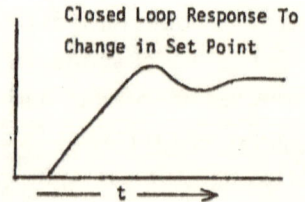

FIGURE 5.5(b)

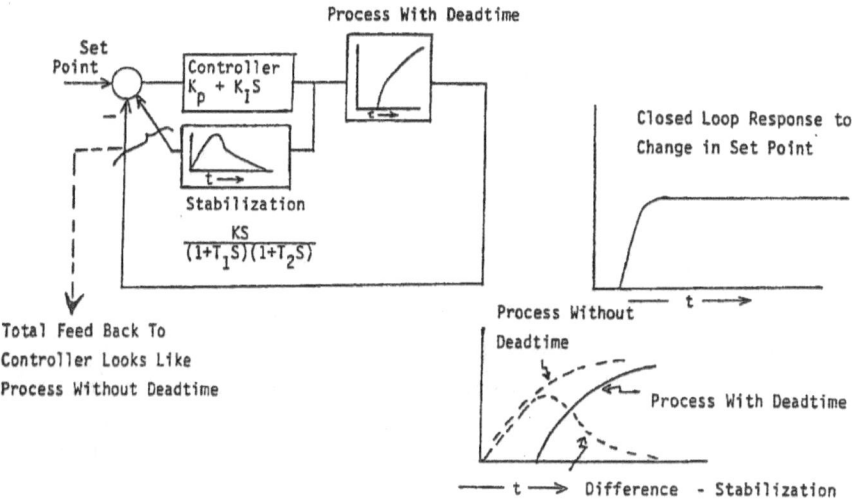

FIGURE 5.5(c)

of T_1, T_2 and k based on the knowledge of the desired shape of the stabilization, characterized by the time to peak t_1 and the time to ½ the peak on the tail, t_2. Knowing t_1 and t_2, the nomograph is entered with the ratio t_2/t_1 on the right-hand side ordinate, and from the t_2/t_1 curve the ratio T_1/T_2 is picked from the abscissa. Next, with this T_1/T_2 value on the abscissa, the value of t_1/T_1 is read off the t_1/T_1 curve on the left-hand side ordinate and so is the value of k off the k curve.

As illustrated in Chapter 3, approximately the same control effect can be obtained with a three mode controller with a high rate adjustment and very low proportional gain.

5.3 Air Flow Controls in Coal-Fired Boilers

The problem of matching air to the fuel in coal-fired plants arises due to the fact that there is no direct measure of the fuel being burned— only inferential measures such as coal-feeder speeds which are only approximately indicative of the BTU being burned (within 10% or so in the steady state). During transients the relationship between coal feeder speed and BTU in the furnace is further complicated by fuel system (pulverizer) lags which can be long and variable.

Because of the relative inaccuracies between feeder speed and actual fuel BTU, it has been the practice in coal-fired plants to set the air flow from measured steam flow which, in the steady state, is a more reliable indication of fuel burned.

The idea of supplementing this relationship between air flow and steam flow in the steady state with transient action from the signal to the fuel system is not recommended because of the possibility of a runaway positive feedback situation. It would be possible to have the signal calling for increases in fuel to momentarily fail to increase the heat release in the furnace (due to fuel system upsets, insufficient number of mills or unpredictably large pulverizer dead time) while it succeeded in increasing air flow. Now, since increased air flow causes a drop in pressure, this would cause the pressure controls to further call for more fuel and air. The air would respond ahead of the fuel decreasing the pressure still further and a runaway positive feedback situation could result. Evidently, where the fuel system characteristics are not completely predictable and repeatable, it is impractical to use the signal to the fuel system to call for transient changes in air flow.

The transient shaping of the air flow can be provided directly from steam flow. This avoids the positive feedback aspects of the previous alternative, but in order to provide good transient matching of air-to-fuel, it relies on repeatable and predictable fuel system characteristics.

A more positive, as opposed to inferential, measure of actual fuel in the furnace can be synthesized from steam flow added to a term proportional to rate of change of drum pressure.

Fig. 5.6 shows a simplified block diagram describing the pressure and flow response to changes in fuel, air and turbine valve on a boiler with control of burner tilts.

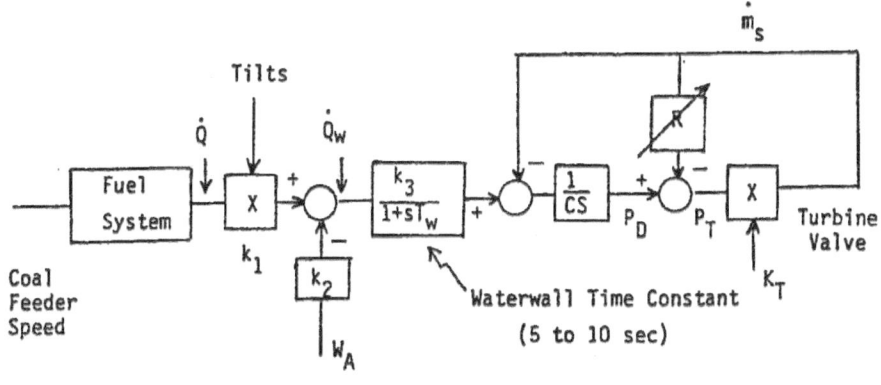

$$\dot{Q} = \text{Total fuel burned}$$
$$\dot{Q}_w = \text{Heat absorbed by waterwalls}$$
$$P_D = \text{Drum pressure}$$
$$P_T = \text{Throttle pressure}$$
$$m_s = \text{Steam flow rate}$$
$$W_A = \text{Air flow rate}$$

FIGURE 5.6

The heat absorbed by the waterwalls is related to the total heat released and air flow,

$$\Delta \dot{Q}_w = k_1 \Delta \dot{Q} - k_2 \Delta W_A \tag{5.1}$$

where k_1 is a function of burner tilt position and slagging conditions, and k_2 may vary somewhat with load point. Other parameters which vary with load point are the friction drop resistance term R and valve position flow coefficient k_T.

Developing linearized small perturbation equations from the block diagram

$$\Delta P_D = \frac{1}{Cs} \left[\frac{k_3}{1+sT} \Delta \dot{Q}_w - k_{To} \Delta P_t - P_{To} \Delta k_T \right] \tag{5.2}$$

$$\Delta P_T = \Delta P_D - R k_{To} \Delta P_T - P_{To} R \Delta k_T \tag{5.3}$$

where subscript "o" denotes steady state operating values and "Δ" denotes changes from these values.

Using 5.1, 5.2, and 5.3 to express ΔP_D in terms of the inputs

$$\Delta P_D = \frac{k_1 k_3 (1 + Rk_{T_0}) \Delta \dot{Q}}{(1 + sT_W)\left[(1 + Rk_{T_0})Cs + kT_0\right]} - \frac{k_2 k_3 (1 + Rk_{T_0})}{(1 + sT_W)\left[(1 + Rk_{T_0})Cs + kT_0\right]} \Delta W_A = \frac{P_{T_0} \Delta k_T}{\left[(1 + Rk_{T_0})Cs + kT_0\right]} \quad (5.4)$$

i.e., change in drum pressure has been expressed in terms of change in fuel burned $\Delta \dot{Q}$, change in air flow ΔW_A and change in turbine throttle Δk_T.

Now, if $\Delta W_A = G(s)\,(\Delta m_s + k_A s \Delta P_D)$, i.e., if air flow is made to follow steam flow plus rate of change of drum pressure through a lead network $G(s)$, then

$$\Delta W_A = G(s)\left[\frac{P_{T_0} \Delta k_t}{1 + Rk_{T_0}} + \frac{K_{T_0} \Delta P_D}{1 + Rk_{T_0}} + k_A s \Delta P_D \right] \quad (5.5)$$

where $\quad \Delta \dot{m}_s = \dfrac{P_{T_0}}{1 + Rk_{T_0}} \Delta K_T + \dfrac{k_{T_0}}{1 + Rk_{T_0}} \Delta P_T$

i.e.,

$$\Delta W_A = G(s)\left[\left(\frac{P_{T_0}}{1 + Rk_{T_0}}\right) \Delta K_T + \Delta P_D \left(\frac{k_{T_0} + (1 - Rk_{T_0})k_A s}{(1 + Rk_{T_0})}\right)\right] \quad (5.6)$$

Substituting in (6)

$$\Delta W_A = G(s)\left[\frac{P_{T_0} \Delta K_T}{1 + Rk_{T_0}} + \frac{k_1 k_3 \left[(1 + Rk_{T_0})k_A s + k_{T_0}\right] \Delta \dot{Q}}{(1 + sT_W)\left[(1 + Rk_{T_0})Cs + k_{T_0}\right]} - \frac{k_2 k_3 \left[(1 + Rk_{T_0})k_A s + k_{T_0}\right]}{(1 + sT_W)\left[(1 + Rk_{T_0})Cs + k_{T_0}\right]} \Delta W_w \right.$$
$$\left. \frac{P_{T_0}\left[k_{T_0} + (1 + Rk_{T_0})k_A s\right]}{\left[(1 + Rk_{T_0})Gs + k_{T_0}\right]\left[1 + Rk_{T_0}\right]} \Delta k_T \right] \quad (5.7)$$

Making $K_A = C$ in above expression, it reduces to

$$\Delta W_A = G(s) \frac{k_1 k_3}{\left[1 + sT_w\right]} \Delta \dot{Q} - \frac{k_2 k_3 \Delta W_A}{\left[1 + sT_w\right]} \quad (5.8)$$

and if $G(s) = 1 + sT_w$, we have

$$\Delta W_A (1 + k_2 k_3) = k_1 k_3 \Delta \dot{Q}$$

$$\frac{\Delta W_A}{\Delta \dot{Q}} = \frac{k_1 k_3}{1 + k_2 k_3} \quad (5.9)$$

Equation 5.9 shows that air flow can be made proportional to heat release $\Delta \dot{Q}$ provided $k_1 k_2$ and k_3 remain constant. The two parameters which are subject to change are k_1 (due to changing tilt position and slagging) and k_1 (due to changing load). The changes in these parameters are not too significant, however. Note that the variable parameters k_T and R fall out of the picture and do not affect the relationships.

Fig. 5.7 shows the block diagram for the air flow controls. The function generator downstream of the measured air flow signal provides the required characterization to yield the proper excess air at the different loads.

The air flow demand signal derived from the steam flow signal is multiplied by a ratio setting which can scale the signal typically by ±5%. The ratio setting can be altered manually or by means of an integral controller responding to measured O2 error trying to hold the average value of 0_2 at the desired level as specified by a demand signal derived from steam flow through a function generator.

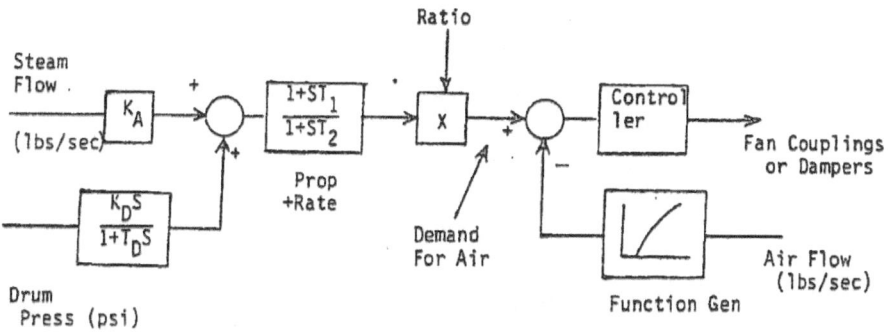

FIGURE 5.7

5.4   Fuel-Air Cutback

The high select, low select scheme of Fig. 5.1 used with oil and gas-fired boilers provides the necessary override action to ensure safe combustion conditions.

As mentioned previously in the case of coal-fired boilers, owing to the lags between fuel demand and fuel in the furnace and the fact that at times the fuel may not correspond to the demand signal (pulverizer stoppage, mills near full capacity, etc.) there is danger that the cutback action may tie airflow into the pressure control loop with undesirable positive feedback effects previously mentioned. In coal-fired boilers it is, therefore, usual to provide a fuel/air cutback controller as described in Fig. 5.8.

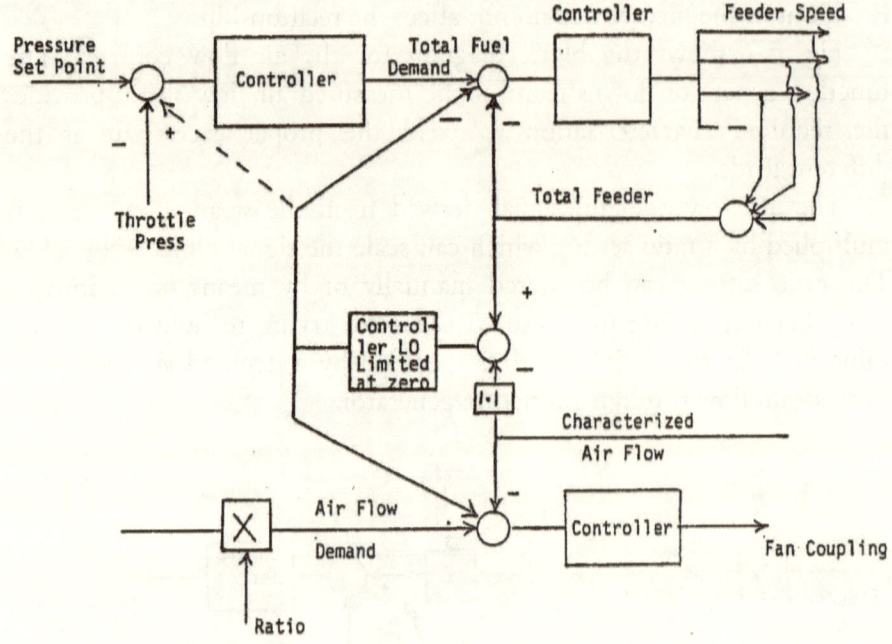

FIGURE 5.8

Characterized air flow is compared with total feeder signal, and the difference is an error signal to the cutback controller. The strengths of the signals from feeder speed relative to air flow are adjusted such that normally a 10% error exists keeping the cutback controller clamped at zero output. When an unbalance exceeding 10% in the fuel rich direction occurs, the cutback controller will simultaneously drive the fuel demand down and air flow demand up, while it feeds an appropriate signal to the pressure controller to prevent windup.

5.5 O_2 Controls

The O_2 controller operates as an input to a scaling multiplier on the demand for air flow and adjusts this multiplier (ratio) within limits (usually ± 5%). O_2 sensors exhibit significant dead times, hence the controller settings must be with slow integration action.

A linear approximation of the O_2 control loop is described in Fig. 5.9 where transmitter signal levels of 16 ma are assumed.

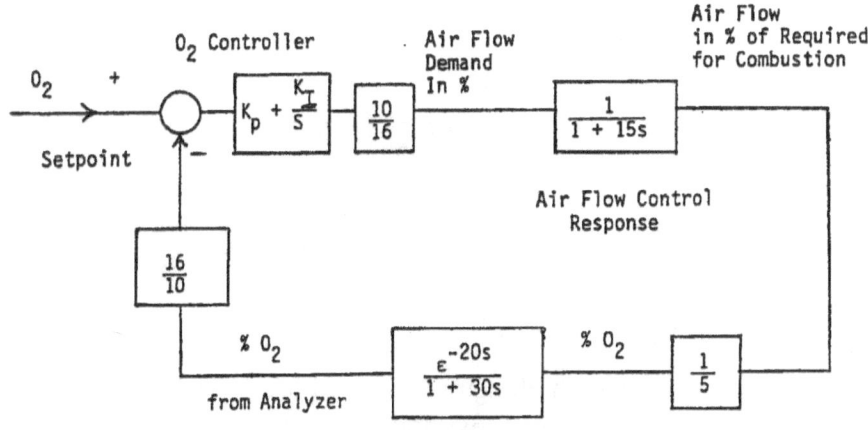

FIGURE 5.9

The full range of the O_2 controller can modify the air flow demand by 10%, hence the gain 10/16. A 5% change in combustion air results in approximately a 1% change in O_2, hence the factor 1/5.

The full range of the O_2 analyzer is 0 to 10%, hence the ratio 16/10.

REFERENCES

5.1   de Mello, F.P., Ewart, D.N., and Stephens, W.M., "Simulation as a Design Tool for Plant Jack McDonough Boiler Controls", ISA, 7th National Power Instrumentation Symposium, 1964, pp. 35-46.

5.2   de Mello, F.P., and Paulson. R.E., "Simulation of Plant Dynamics and Design of Plant Control Systems", Southeastern Electric Exchange Meeting, New Orleans, April, 1966.

CHAPTER VI

FURNACE DRAFT CONTROLS

Most large utility boilers used to be designed with pressurized furnaces. The justification for such designs was the lower installed equipment cost (one set of forced draft fans) and reduced fan maintenance since the FD fans handle clean air. However, problems with maintaining casing-pressure integrity, and concern for personnel safety have made the industry return to use of balanced draft units. Concurrently, clean air standards requiring use of scrubbers, etc., introduce higher draft losses and the need for higher-head induced draft fans.

While the problem of regulating the fans to supply the required airflow and, at the same time, maintain the desired furnace draft during normal operation is elementary, special control requirements are associated with limiting transient pressure excursions during major upsets such as main fuel trips. A discussion of the factors involved will be based on studies of a 600 MW oil-fired unit reported in reference (6.1). Although this unit has unusual requirements, the example serves to illustrate the nature of control and override logic involved in furnace draft controls.

As shown in Figure 6.1, combustion air is supplied to the boiler by two constant speed motor driven forced draft fans, with inlet vanes control. There are two sets of Induced Draft fan trains; each train consisting of a primary I.D. fan discharging to a secondary I.D. fan inlet. The secondary I.D. fans discharge through a scrubber system to the stack. The primary I.D. fans are constant speed, motor driven, with control on inlet vanes. The secondary I.D. fans are speed controlled through hydraulic couplings. Because of the head loss requirements created by high draft loss boiler design, particulate scrubbers and future sulphur removal scrubber, these Induced Draft fans are considerably oversized for a unit of this capacity. Each of the fans are driven by 7000 HP motors and each set can develop a total head of 76 inches W.C.

Combustion air flow measurements are made by piezometer rings located in the inlet of the two forced draft fans. Furnace draft taps are just below the furnace roof, with each of three transmitters connected to its own dedicated tap.

This oil-fired high-draft loss boiler design has created severe problems for the electric operating companies. Of initial concern was the large induced draft fans, with their high inertia and a long coast-down time, continuing to pull a significant draft after the forced draft fans had stopped rotating following the tripping of all fans.

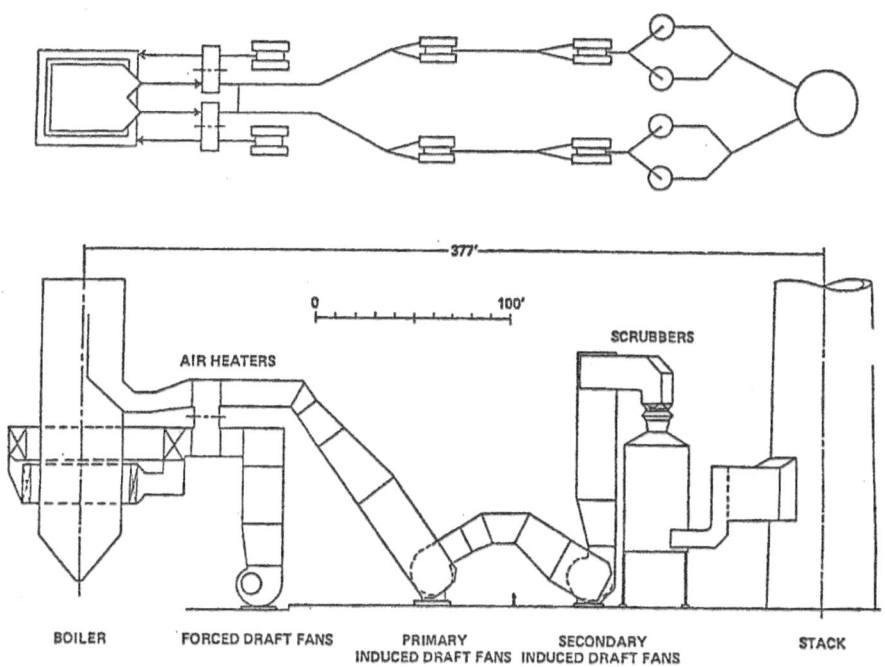

FIGURE 6.1

On further consideration, the boiler implosion potential from sudden loss of fire appeared even more critical; especially because of the use of series sets of two (2) induced draft fans discharging to four (4) flue gas scrubbers.

The most critical element in minimizing an implosion potential was found to be that of time. Studies disclosed that maximum negative pressure occurred approximately 1.7 seconds after loss of flame and loss of flame occurs at an estimated one second after loss of fuel. This allows approximately 2.7 seconds to take corrective control action to counter the anticipated high negative pressure unless the time could be extended.

PROCESS CHARACTERISTICS

The dynamic behavior of the boiler air-gas path was simulated by a digital computer. The simulation study had two objectives:

First,   to indicate the pressure excursion profiles that could be expected following a master fuel trip disturbance.

Second,   to indicate the relative merits of alternative F. D. and I. D. automatic control arrangements.

The simulation model accounts for the following effects:

- gas flow head loss according to the square law, $P = CpV^2$.
- variation of gas temperature through the boiler due to combustion heat release and to convective and radiative heat transfer.
- variation of gas specific volume according to the gas law, $P = pRT$.
- variations in the mass of gas residing in furnace sections.
- variations in fan characteristics with gas density, fan speed, and damper position.

Before discussing model results, it is useful to review the fundamental requirements of furnace air flow control.

AIR FLOW CONTROL REQUIREMENTS

For the furnace and fan configuration shown schematically in Figure 6.1, air and gas flow through the furnace and ducting is maintained by the combined action of forced draft and induced draft fans. Opposition to this flow is contributed by air heaters, tube sections, furnace dampers, duct losses, and scrubbers, and is distributed along the entire length of the flow path.

It is useful to visualize the pressure distribution along the boiler air gas path as shown in Figure 6.2a. This shows positive pressure upstream of the furnace, negative pressure downstream of the furnace and then positive pressure again to force the flue gases through the scrubber to the stack.

The scrubber is an adjustable orifice device designed to hold a constant head loss in steady operation, but the head losses through the remainder of the flow path vary as the square of air/gas flow. Thus, if Figure 6.2a applies for full load, the pressure profile at part load should be as shown in Figure 6.2b. Note that, to maintain balanced pressure in the furnace, the net head developed by both F.D. and I.D. fans must be reduced.

Now consider the case where the F.D. and I.D. fan controls are imperfectly manipulated to reduce air flow to a part load value. The pressure profile will be as shown in Figure 6.2c; the slope of the pressure profile and the total head loss will be essentially unchanged from Figure 6.2b, but the whole profile will be biased up or down.

Examination of Figure 6.2 shows that:

- The net head of F.D. and I.D. fans must be adjusted in unison to regulate air flow while maintaining furnace pressure at a given value.
- The net head of the F.D. and I.D. fans must be adjusted in opposition to regulate furnace draft while holding a given air flow.

In accordance with these characteristics, reduction of unit load under normal operating conditions involves simultaneous gradual reduction of the net head of both F.D. and I.D. fans. This is achieved by closing inlet vanes on the F.D. fans, closing inlet dampers on the primary I.D. fans, and reducing the speed of the secondary I.D. fans.

It is to be noted now that the above control rules apply for normal operation where the system is undergoing relatively gradual changes. The control requirements following sudden loss of fire are significantly different. The nature of these different requirements is best illustrated by reference to simulations of fuel trip disturbances as presented in the next section.

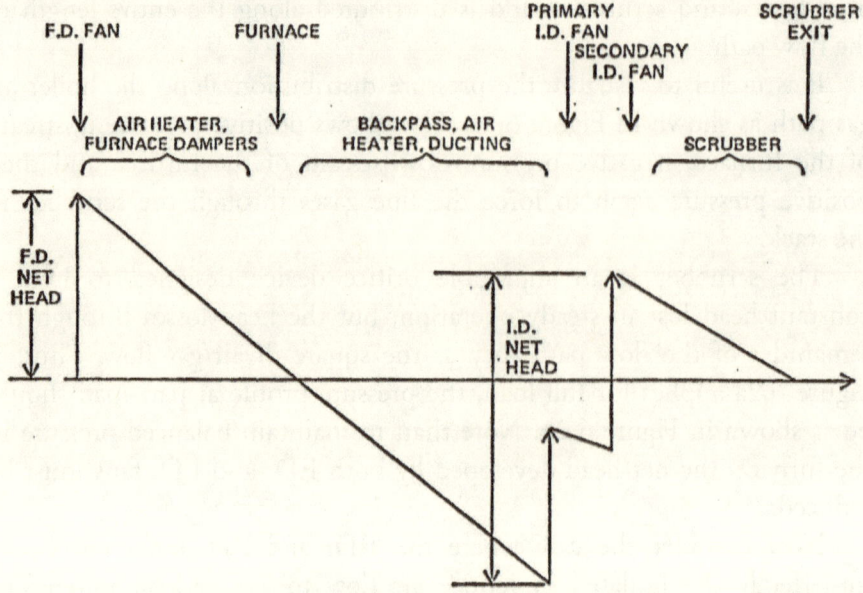

FIGURE 6.2a Full Load, Air Flow

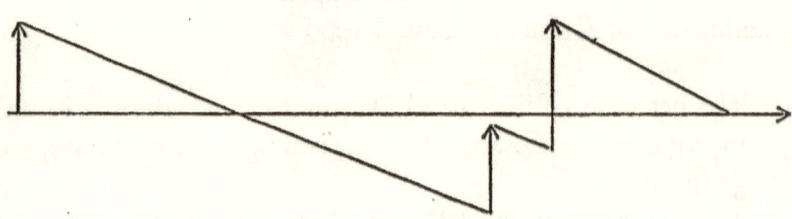

FIGURE 6.2b Reduced Load and Air Flow,
Properly Coordinated Fan Adjustment

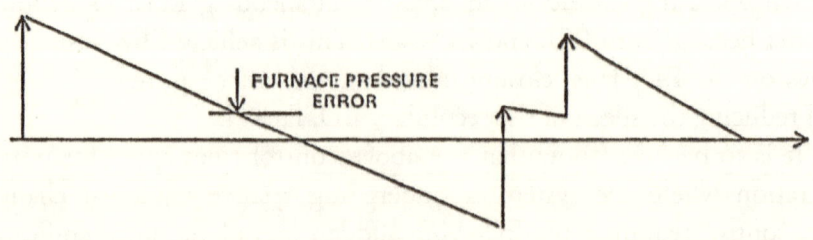

FIGURE 6.2c Reduced Air Flow, Improperly Coordinated Fan Adjustment

FIGURE 6.2

SUDDEN DISTURBANCE BEHAVIOR

Figure 6.3 shows a simulation of the boiler's response to a sudden complete shut-off of fuel flow in the condition where all air flow and furnace draft controls are locked. There is an immediate and rapid decline in the temperature of the gas within the furnace. The pressure in the furnace is governed by the gas law: $Pv = mRT$ where volume, v, and the gas constant, R, are constant. The rate of change of the furnace mass inventory is dependent on the air and gas flows which are governed by fan and head loss characteristics. The mass inventory adjustment rate does not immediately match the rate of decay of temperature and there is, following from the gas law, a rapid fall in pressure inside the furnace.

Figure 6.3 shows an increase in furnace air inflow as furnace pressure and temperature fall. There is a corresponding decrease in gas outflow. Both of these flow changes aggravate the disturbance since they create decreased net F.D. head and increased net I.D. head, hence depressing the already low furnace pressure. The furnace inventory adjustment "catches up" with the temperature change process after about 2 seconds and the furnace pressure then commences its return to nominal value.

This behavior shows that initial movement of the I.D. fan controls in the closing direction will ultimately be needed to regulate air flow at the desired post-disturbance value, and will tend to counteract the transient dip in furnace pressure. More importantly it shows that, although they will ultimately be required to close in correspondence with the I.D. fan controls to maintain the new air flow, the F.D. fan controls should be moved in the opposite (opening) direction during the initial transient following a sudden firing rate reduction.

This requirement for different directions of motion of the F.D. fan controls during sudden firing rate transients, as opposed to gradual load changes, is a key factor in selecting between the two commonly used control arrangements described below.

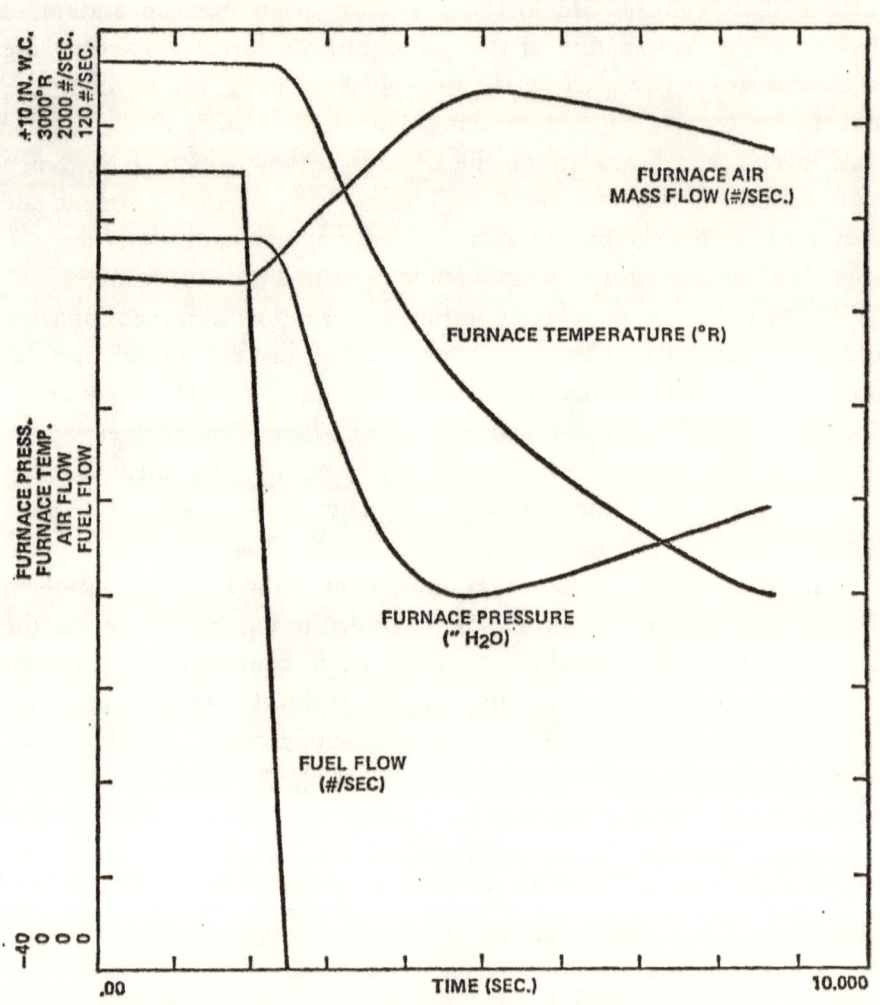

Rapid fuel shut-off, fan dampers locked.

FIGURE 6.3

AIR FLOW CONTROL ALTERNATIVES

Two feedback control arrangements that have commonly been used to meet the control objectives are shown in Figure 6.4. Both achieve the desired coordination of fan control actions by assigning one set of fans to regulate air flow and the other to regulate furnace draft. The difference is that they have reversed responsibilities for the F.D. and I.D. fan control loops. Simulation runs covering several different boilers have shown that both arrangements can give satisfactory regulation of air flow and furnace draft when following load ramps as rapid as 100% per minute.

The behavior of the two control arrangements following sudden disturbances of the load reference is quite different, however. Scheme A, which makes the F.D. fan control sensitive to load reference, will tend to close the F.D. fan inlet vanes immediately following a sudden downward disturbance of load reference and fuel flow, hence aggravating the negative furnace pressure excursion. In contrast, Scheme B will respond to the sudden reduction of load reference by closing the I.D. fan dampers, but the associated negative furnace pressure caused by the fuel flow reduction will cause it to initially open the F.D. fan inlet vanes. It will close the F.D. vanes only after the negative furnace pressure excursion has passed.

Thus, Scheme B is regarded as being preferable to Scheme A because it inherently gives the correct initial response to potentially damaging fuel trip disturbances while Scheme A would have to depend on override logic to prevent inappropriate closure of the F.D. fan vanes.

The other factor affecting the selection of Scheme B is that the F.D. fan control, usually has the more direct effect on furnace pressure since there is less buffering volume and head loss in the ductwork between the F.D. fan and the furnace than between the furnace and the I.D. fans. Now, furnace pressure control is the highest priority requirement; it should be assigned to a fast-acting inner control loop, with the air flow control being handled by an outer control loop. Since stable control design requires the inner loop to be faster acting than the outer loop, it is desirable to assign the more direct control means, the F.D. fan vanes in this case, to furnace pressure control.

The recommended control arrangement incorporates the essential features of Scheme B for the regulation of furnace draft and air flow, together with high gain overrides to drive both F.D. and I.D. fan controls in the compensating directions indicated by Figure 6.2 whenever the deviation of furnace draft from set-point exceeds preset emergency limits.

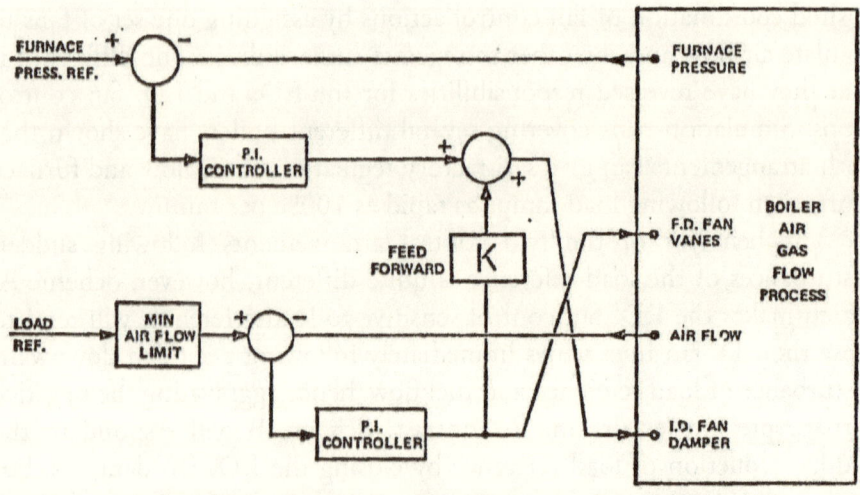

SCHEME A, I.D. FAN CONTROLS FURNACE PRESSURE

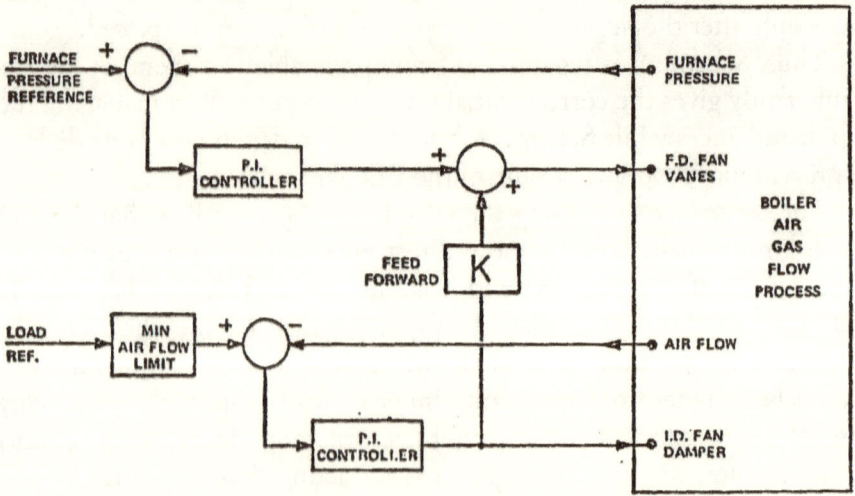

SCHEME B, F.D. FAN CONTROLS FURNACE PRESSURE

FIGURE 6.4

ORIGINAL CONTROL SYSTEM

The original control system configuration as it was installed, before any specific requirements for coping with furnace implosion were imposed on it, is shown on Figures 6.5 and 6.6.

The original air flow and furnace draft controls on this unit were implemented with pneumatic control hardware throughout. The forced draft inlet dampers are each powered by one 10″ × 16″ pneumatic piston drive. The two primary I.D. fan inlet vanes required, for each fan, two 10″ × 16″ piston drives, set-up in a master/slave configuration. Operation of each of the secondary I.D. fans hydraulic coupling control levers was accomplished with one 8″ × 16″ piston drive to each lever.

The original system assigned the duty of maintaining air flow to its required set point to all three sets of fans in parallel (the forced draft H/A stations could be considered air flow master stations). The furnace draft controller had plus/minus trim adjustment on the primary I.D. fan inlet dampers available to it to maintain balance between fans. No specific provisions were made in this system to cope with the implosion possibilities inherent with a main fuel trip. With no F.D. fan running, an automatically set maximum I.D. inlet vane position equivalent to about 30% air flow is established.

Initial checks on the original pneumatic draft control system disclosed that there was a time lapse of eighteen (18) seconds from the time the furnace pressure transmitter felt a change until the primary induced draft fans inlet vanes, used for furnace draft control, began to move. The vane drives required thirty (30) seconds to go full travel.

Booster relays were installed in the inlet vanes drives and the time for full travel, without load, was reduced to approximately 5-6 seconds. Installation of electronic transmitters and electric transmission to the pneumatic control system and from the control system to the drives reduced the overall loop time to 3.5 seconds. To gain more time, sequential elevation tripping of fuel was utilized as recommended in Reference 6.2.

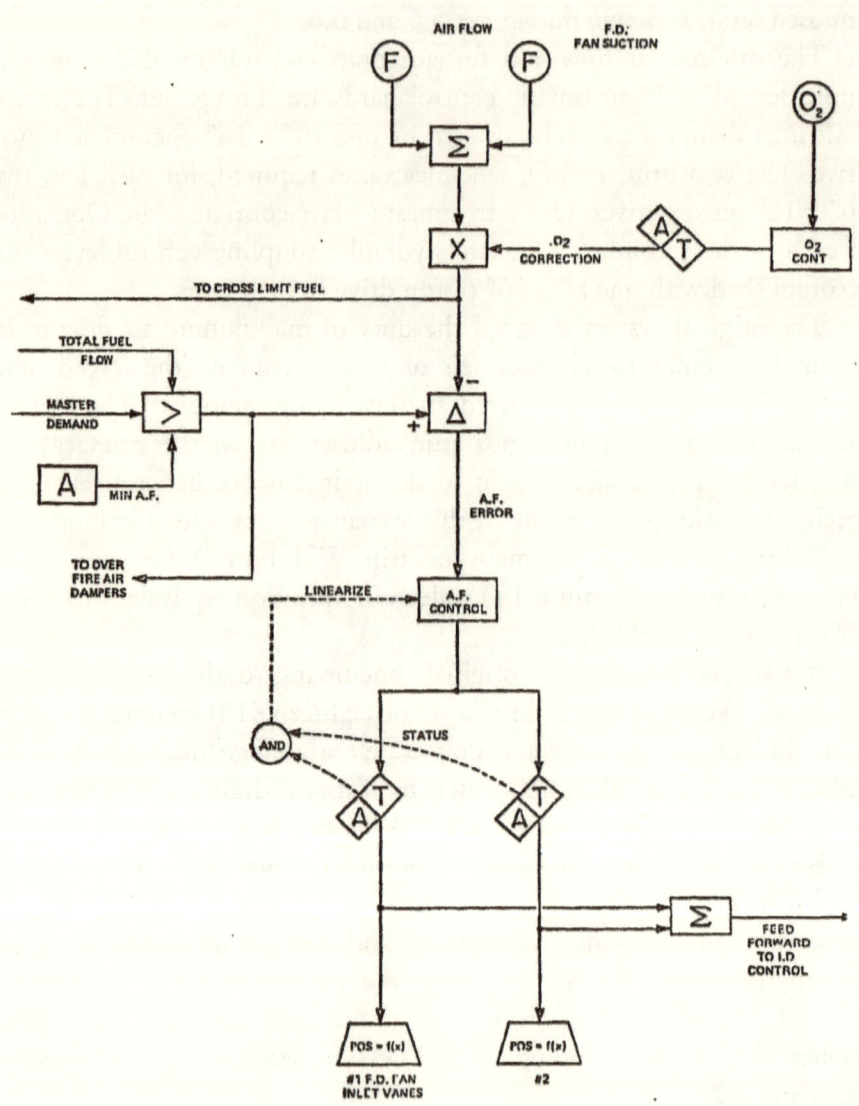

FIGURE 6.5

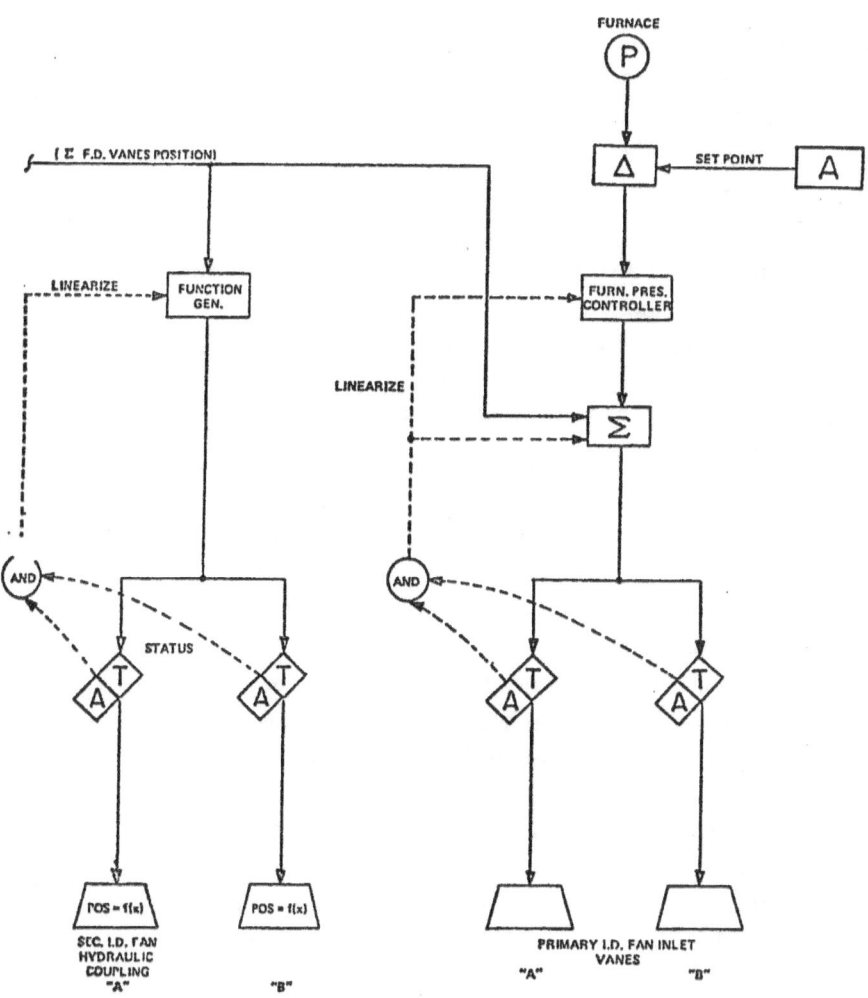

FIGURE 6.6

REVISED CONTROL SYSTEM

Based on computer model studies, it was found advisable to do most of the control using primary induced draft fans and to use the secondary I.D. fans to supply the additional head to push the exhaust gases through the scrubber to the stack. The studies led to the control revisions described below:

a) Primary I.D. Fans Control (Figure 6.7)
    The air flow demand and the air flow measurement, with O_2 correction, are developed just as in the original system to index the air flow controller. However, in this revision, the air flow demand is also fed-forward as a demand for total primary I.D. fans inlet damper position. This reduces the duty of the integral component of the air flow controller to a calibrating trim adjustment, since the load matching duty in positioning the vanes is carried by the feed-forward demand. Utilizing average vane position demand (from below the H/A stations) as a negative feedback against this demand, a much faster and more accurately timed system linearization for one-two fan operation is obtained.

    This same averaged primary I.D. vanes signal acts as the feed-forward demand to the F.D. fans control. Separately, each primary I.D. fan position demand indexes its corresponding secondary I.D. fan speed control. Thus, air flow control is applied, in parallel to all three pairs of fans.

    Back-up safety features, and special action taken specifically to help contain furnace pressure within bounds on a M.F.T., come into the system below the primary I.D. fans H/A stations, so they will be effective even if the fan loops are on manual.

    If furnace pressure increases to a value higher than the normal control range (set at +3″), an increasing signal from the furnace pressure dead-band controller (discussed later in the F.D. fan control sub-loop) will take over and increase the opening of the vanes at the high signal auctioneers. On an out-of-range low furnace pressure (set at −4″) a runback takeover at the low signal auctioneers will close the vanes.

Entering the summers below the auctioneers, is a transient negative signal that is triggered by the same logic that initiates a M.F.T. This negative signal, promptly reduces the primary I.D. fan vanes position by 35%, and then wipes itself out over a 30 second period.

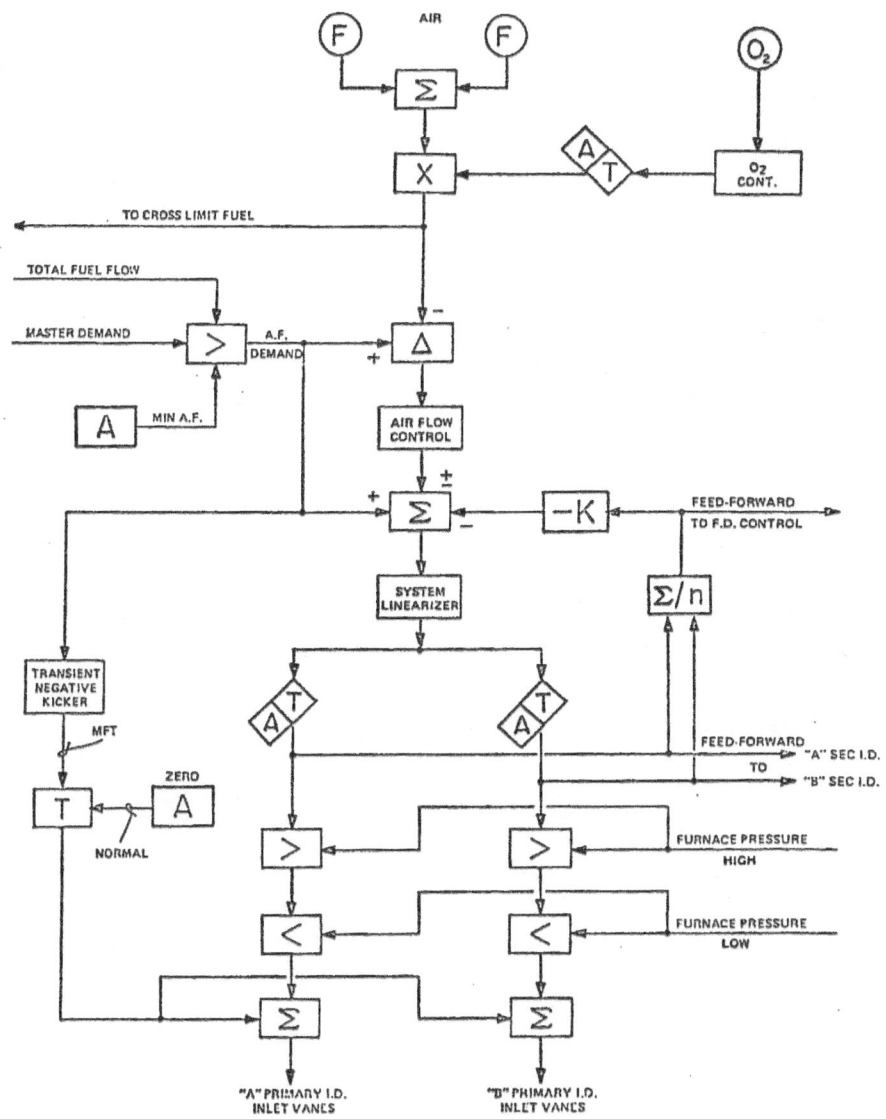

FIGURE 6.7

b) Secondary I. Fans Control (Figure 6.8)
    Since the primary and secondary I.D. fans form physically, and operationally, series pairs, the control system is set up on that basis. The "A" primary I.D. fan vanes positioning signal becomes the feedforward, through a function generator, to the "A" secondary fan hydraulic coupling control. The pair of "B" fans are treated in the same way. Empirically determined from operational data, a load referenced set point for the duct pressure between the two fans is compared with the actual duct pressure to provide a balancing trim to keep the two fans working together properly. This duct pressure corrective action also provides the desirable feature of a self-timing derivative type action to the high inertia, slow responding secondary fan speed loop. "Out-of-range" furnace pressure overrides are applied to the secondary I.D. fans also.

c) Forced Draft Fan Control (Figure 6.9)
    The averaged primary I.D. fan vane positioning signal acts as the feedforward to the F.D. fan vane position control. Calibration of this feedforward based on the steady state relationship between I.D. and F.D. vanes is provided.

    Because of the importance of the furnace pressure measurement, it was decided to use three transmitters feeding into a median selecting circuit. In this way, the loss of any one transmitter can be accepted without loss of automatic control. This verified furnace pressure signal from the selecting circuit is used to index the furnace pressure controller. Because of the load related feedforward from the primary I.D. fan vane positions the furnace draft controller has only a trim duty to perform under normal conditions.

    In the dead-band controller, furnace pressure is compared against both high and low pressure set points. Should a sufficient upset occur to drive furnace pressure to one or the other of these limits, the appropriate override action will be taken at the low and high signal selectors below the F.D. fans H/A station. These same signals, inverted, are used in the primary and secondary I.D. fan circuits, as previously discussed.

    This "out-of-range" furnace draft controller is of prime importance as a safety backup to the normal control. For this reason it comes into the system below the Hand/Automatic stations. Introducing this backup control loop below the H/A stations, however, forces an evaluation of failure modes that could occur. The two extremes would be failures that create 100% positive or 100% negative outputs from the "backup" controller. Either case would be disastrous, since the I.D. and F.D. controls would be driven inversely to extreme positions.

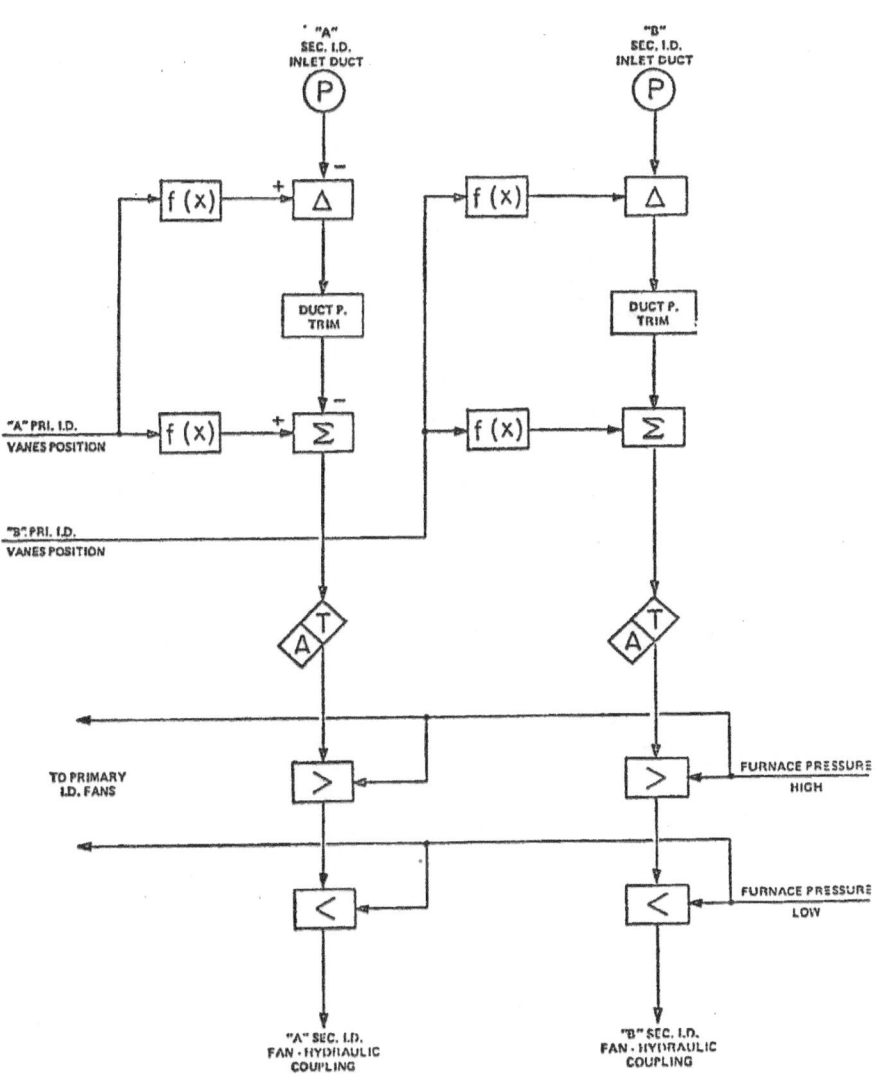

FIGURE 6.8

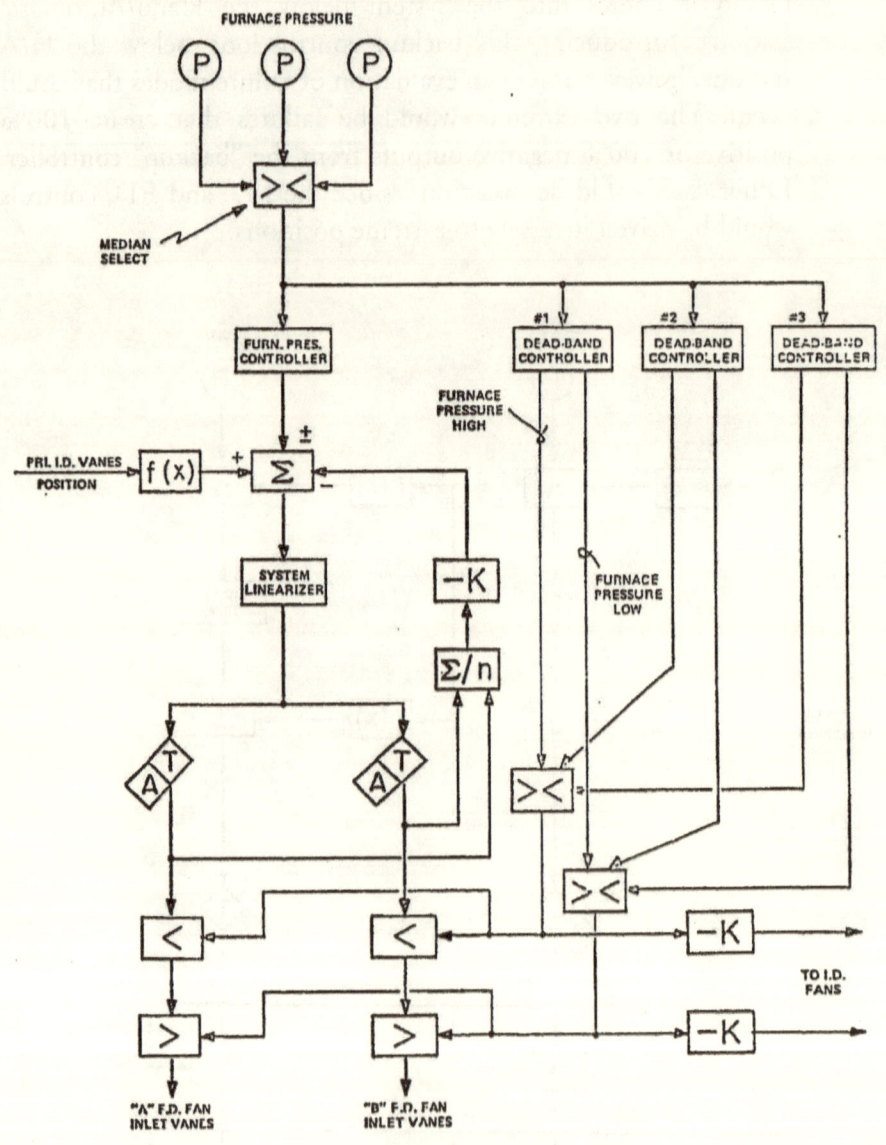

FIGURE 6.9

To protect against such a catastrophic failure situation, the dead-banded furnace draft controllers are triplicated and the three controller outputs are auctioneered in a median signal selecting circuit.

In this manner a failure in any one control loop, or furnace pressure transmitter, can be accepted without system upset. Monitors alarm to the operator any such failure, and alert him to the fact that one of the channels of control is faulty.

d) Operational Characteristics

In terms of the overall control philosophy, the major change made in this revision is to assign furnace pressure supervisory control to the F.D. vanes, instead of to the primary I.D. fan vanes as originally designed. Air flow control is still assigned in parallel, to all three sets of fans. The operational characteristics of this revision under normal operating conditions are apparent from the system's description just discussed. We can postulate a M.F.T. and follow, qualitatively, the control actions of the system.

Two immediate actions are triggered from, and coincident with, the M.F.T. contact:

1. The master demand (and thus the air flow demand) is locked up at its existing value.
2. The transient negative signal is inserted in the primary I.D. fan vanes positioning signal, closing these vanes, at their fastest rate, about 35% to offset the anticipated furnace negative pressure.

In spite of this kicker action to the primary I.D. fans, the furnace pressure will start going strongly negative as the furnace temperature collapses. This starts the forced draft dampers opening by normal furnace pressure control action.

The combination of lower furnace pressure and further open F.D. fan vanes increases the air flow as measured on the suction side of the F.D. fans.

The increase in air flow starts the air flow controller driving all fans downward.

The feedforward action to the F.D. fan loops in the reducing direction starts countering the opening action of the furnace draft controller, and continues as the reduction in the furnace draft requires progressively less F.D. fan.

When measured air flow has been reduced to match the "locked-in" demand, all I.D. fans will be operating at a lower level and the forced draft fan dampers will be back at essentially the same position they started from at the time M.F.T. was triggered.

CONTROL RESPONSE SIMULATIONS

A simulation study was made with controller models and logical statements added to the basic furnace model to represent the proposed control system in detail. The controls were tuned by making an initial series of small disturbance runs and adjusting the controller proportional and integral gains for good damping and response in normal operation. Simulations of major boiler trip disturbances were then made by suddenly stepping the load reference to zero and reducing fuel flow according to the desired shut-off sequence.

One representative set of runs showed the effect of controls on furnace pressure transients. All runs shown here used the initial operating condition where the flue gas scrubbers are fully opened to give minimum head loss; both fan trains are operating, the secondary I.D. fan motors are in operation, and the primary I.D. fan motors are de-energized.

A simulation of a full load trip from this initial operating condition where all fuel flow is shut off in two seconds with all controls locked showed a maximum furnace draft excursion of 20 inches of water. One typical series of simulation runs was made to show the extent to which control action could reduce this excursion.

Figures 6.10a and b show the response to this disturbance when the F.D. (draft) control is active in its normal mode and the I.D. control is in manual at fixed damper position. The negative furnace pressure excursion of 18 inches w.g. is slightly less than that observed with no control action and the initial negative pressure excursion is followed by a significant positive excursion.

A limitation of the basic control scheme as used to yield the results in Fig. 6.10 is that it relies on the regulating controller whose gain settings are optimized for normal operation, and does not provide the strongest possible control action in emergency situations. Gain values that would give maximum rate control action in boiler trip transients would result in poorly damped response to normal small disturbances.

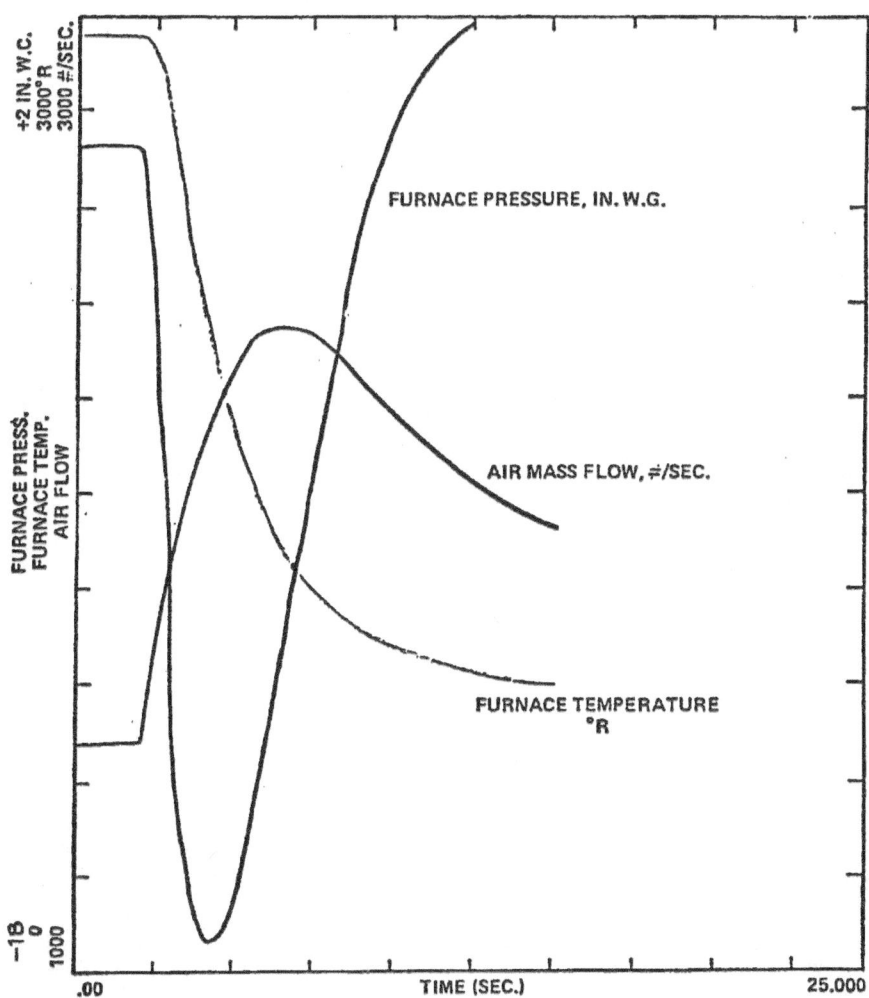

Rapid fuel shut-off F.D. vanes-auto I.D. vanes-locked

FIGURE 6.10a

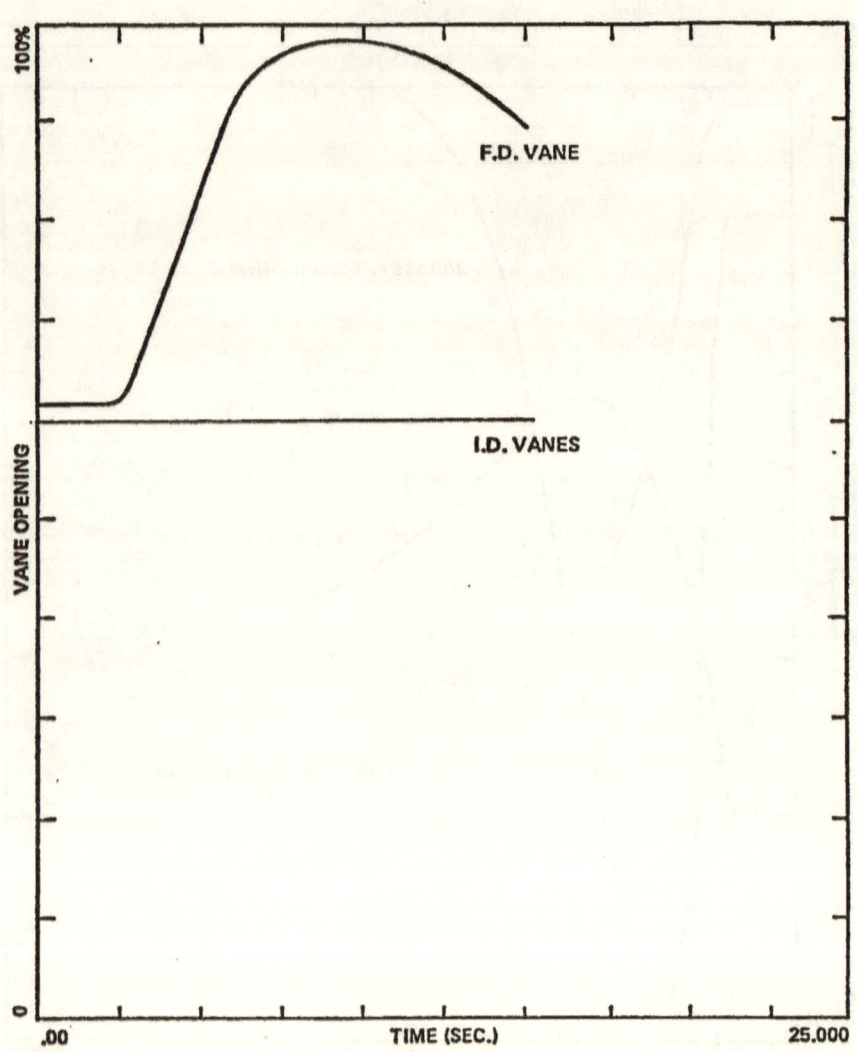

Rapid fuel shut-off F.D. vanes-auto I.D. vanes-locked

FIGURE 6.10b

This limitation of the normal regulating control is overcome by the use of high gain overrides which become active when the furnace pressure deviates from set point by more than 4 inches w.g. The effect of these overriding controls is evident from Figure 11 which shows the response to the same transient as in Figure 10 but with the high gain overrides set to drive the F.D. fan vanes strongly in the opening direction and I.D. fan dampers strongly in the closing direction when the furnace pressure falls below -4 inches w.g. The overriding controls give a significant reduction of the initial transient, and are able to return the furnace pressure rapidly to the -4 inch w.g. value. As the rate of change of furnace temperature decreases the normal regulating loop is able to regain control and return the furnace draft smoothly to its set point value.

Figure 6.11a and b showed that the use of fast acting override control action provides a strong reduction of the furnace draft excursion, but that the first very rapid fall in furnace temperature following complete fuel shutoff in two seconds cannot be compensated by control action because of the stroking speed limitation of the damper and vane actuators. If the primary I.D. fans had been in operation, giving a greater maximum available I.D. head, the initial negative furnace pressure excursion shown in Figure 11 would have been significantly greater.

The most effective method of limiting the first downward swing of furnace pressure has been found to be the use of programmed reduction of fuel flow to reduce the initial rate of fall of furnace temperature.

Figures 6.12a and b shows the response to a Master Fuel Trip when both F.D. and I.D. controls are in automatic mode and the emergency overrides are activated. In this case the "hydromotor" fuel valves on the four levels of No. 6 oil guns were closed in succession at 0, 2, 4, and 6 seconds after receipt of the MFT signal. The staged fuel shutoff is shown to reduce the rate of fall of furnace temperature to a value where the overriding controls can hold the initial furnace pressure excursion very close to the -4 inch w.g. level.

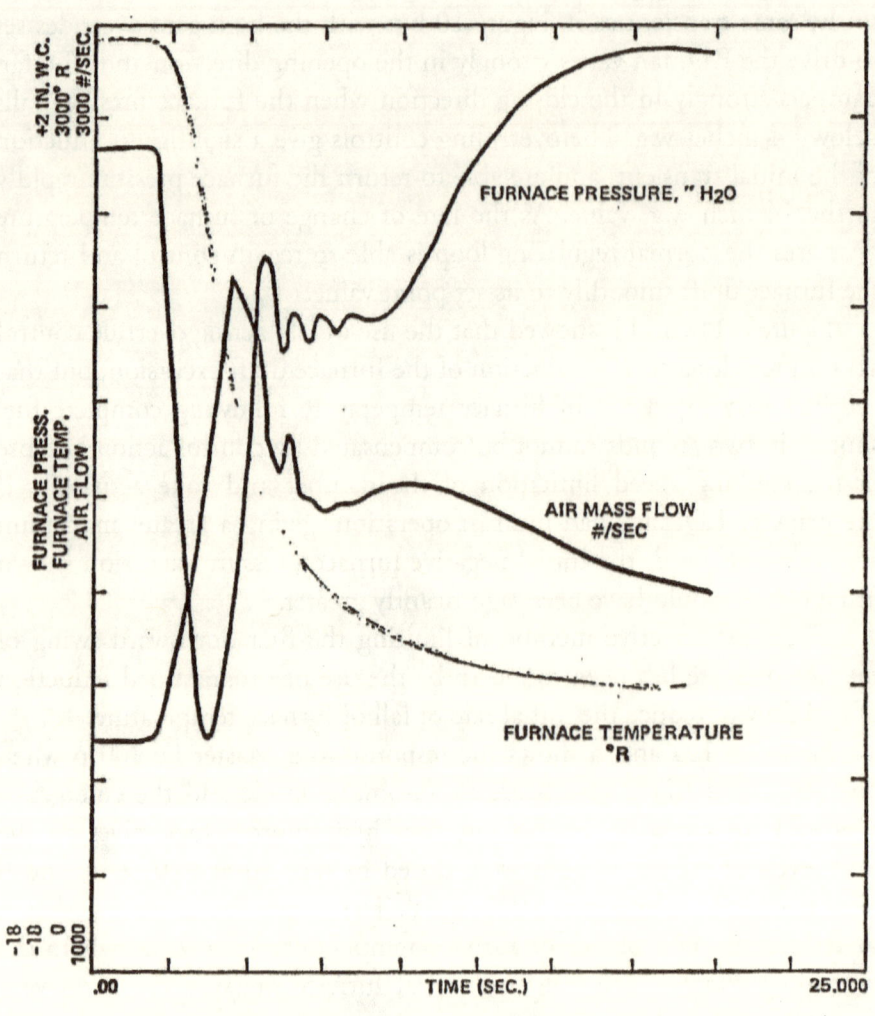

Rapid fuel shut-off over-ride controls active

FIGURE 6.11a

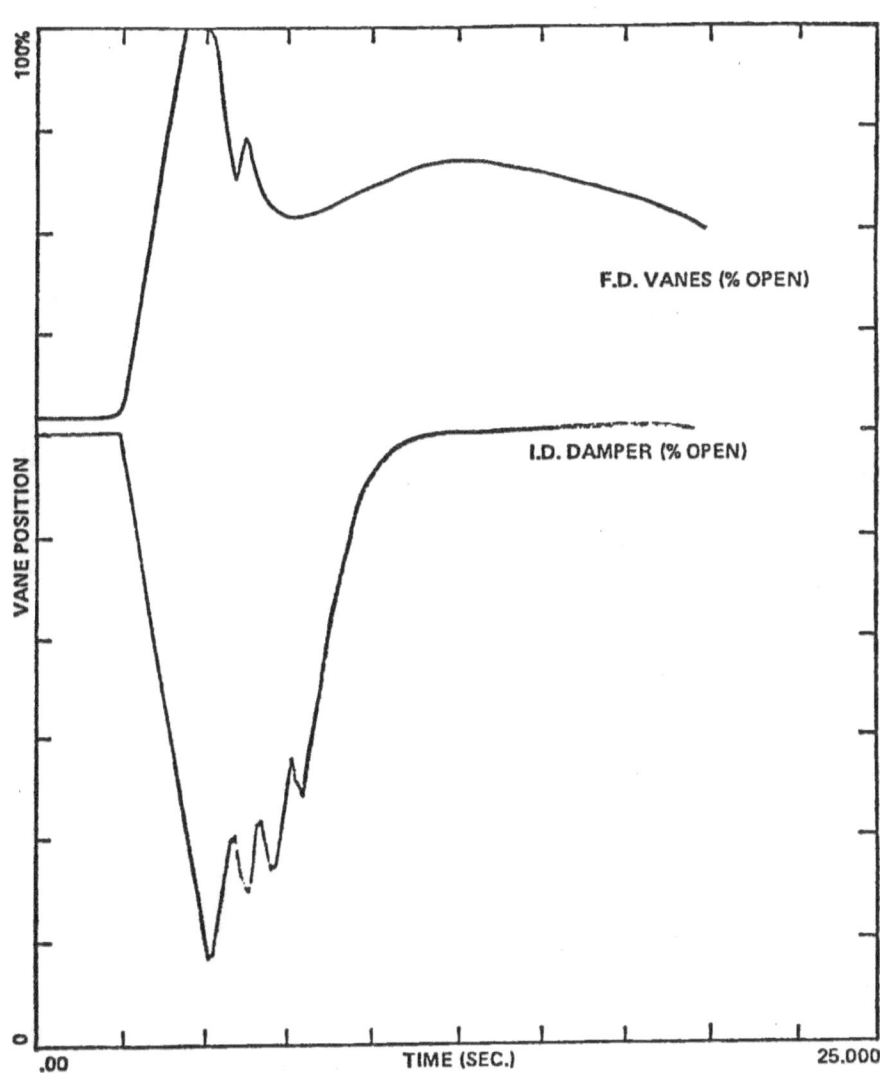

Rapid fuel shut-off over-ride controls active

FIGURE 6.11b

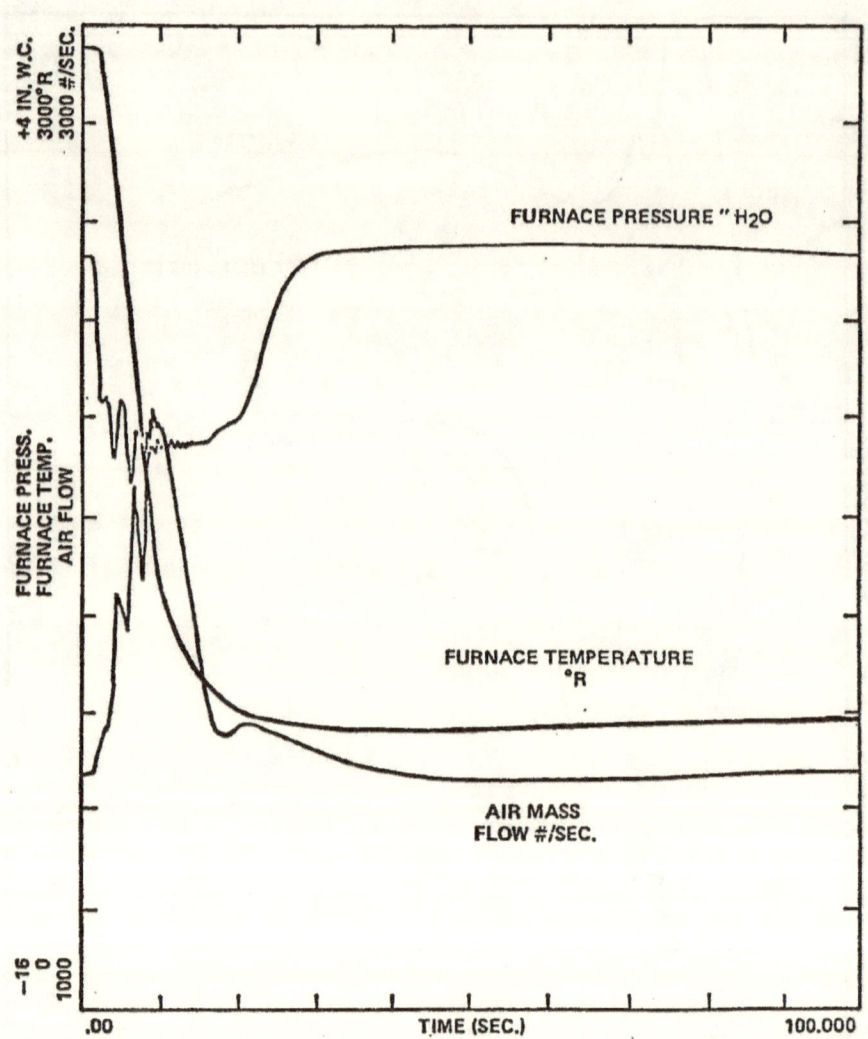

Staged fuel shut-off full automatic fan controls

FIGURE 6.12a

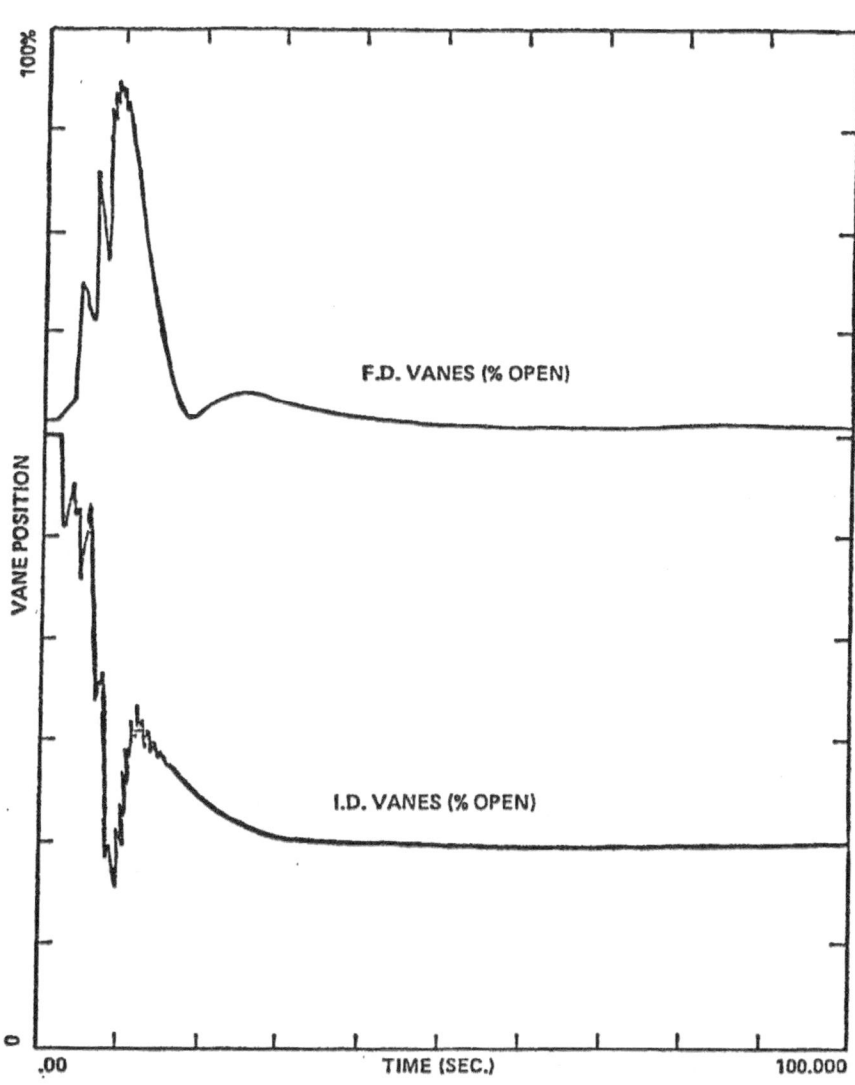

Staged fuel shut-off full automatic fan controls

FIGURE 6.12b

IMPLOSION/EXPLOSION

It might be argued that the control system, advocated here could increase the possibility of boiler explosions. The rationale for this argument seems to be grounded in N.F.P.A. standards for prevention of furnace explosions; specifically in the statements contained in paragraph 563 of standard 85D quoting directly:

> 563, Recommended Procedure For Purging After an Emergency Trip:

> "If the fans are operating after the trip, continue in service. Do not immediately increase the <u>air flow</u> by deliberate manual or automatic control action. If the air flow is above 25% of full load air flow, it may gradually be decreased to this value for a <u>post-firing purge</u> of at least five minutes. If the air flow is below 25% of full load air flow at the time of trip, it shall be continued at the existing rate for five minutes and then gradually increased to 25% of full load air flow and held at this value for a <u>post-firing purge</u> of five minutes."

This paragraph has been widely interpreted and implemented as a requirement to lock the forced draft dampers in the position existing at the time of trip, and allow the induced draft dampers (or speed) to go toward shut on furnace pressure control. Some interpretations have called for locking the I.D. fan actuation also. To us, this seems completely at odds with the problem to be solved. The overall intent of this paragraph, as stated in its heading and several times in the body of the text, is to establish a procedure for <u>post firing purge</u>. Earlier, in this same NFPA publication, "purge" is defined in Chapter 3 by:

> "<u>A flow of air</u> through the furnace, boiler gas passages and associated flues and ducts which will effectively remove any gaseous combustibles and replace with air."

And on page 33 by:

> "Purge air flow shall be equal to or greater than 25 percent of full load <u>volumetric</u> air flow for a period of 5 minutes."

Studying these direct quotations raises the question: Just what "air flow" does paragraph 563 recommend not be increased? In the control system, when we refer to "air flow", we mean the air delivered by the forced draft fans to the burner windbox. After a fuel trip, "purge air flow" is not equal to this "forced draft air flow" until furnace draft is very nearly restored to its set value. Referring to Reference 2 we find reported on page 10 the situation after a fuel trip, with all dampers locked. The plots of furnace pressure and I.D. suction pressure following the trip show that from about 2 seconds to 12 seconds after the trip, furnace pressure was reduced to very nearly equal I.D. fan suction pressure. In other words, for about 10 critical seconds there was <u>no purge air flow</u>. The "forced draft air flow" entering the furnace during this time was all used to restore volumetric inventory lost when the fuel was tripped off. An additional piece of information illustrated by this same set of response curves is that, even with all fan dampers locked, the "forced draft air flow" increases as the furnace pressure plunges to its minimum value. This is recognized in paragraph 563 by the phrase "deliberate automatic or manual control action." While the suggested control system temporarily increases "forced draft air flow" on a controlled basis following a fuel trip, a control philosophy that locks up the forced draft damper differs from this only in degree, since both approaches result in a temporary increase in "forced draft air flow" while lost inventory is being restored. Only when the volumetric unbalance is restored can "forced draft air flow" be considered as "purge air flow". The only control action that would result in no change in F.D. air flow would be a closure of the FD dampers. This appears to be a sure invitation for a boiler implosion. Another reason sometimes stated for not increasing air flow is that an increase in air flow will pick up combustible material that has settled out in low-velocity areas, and that this can create an explosive mixture. Our proposed solution increases, transiently, only "forced draft air flow" in an effort to restore air flow through the furnace boiler gas passes, flues and ducts (the defined "purge air flow"). There is no sudden increase in flow rate through those parts of the boiler where combustible material could collect; the increase occurs only in the clean air side of the boiler. In fact, if we do not take the steps we do to keep the furnace pressure from plunging to minus 15 to 17 inches w.c., we can visualize more combustibles being stirred up by the inrush of air through observation doors and any other points of leakage. These paths are not the normal flow path, and there would be more tendency for accumulation pickup by these random air flow paths.

In summary, the suggested control configuration and override logic will reduce the probability of implosion following M.F.T., without increasing the possibility of a furnace explosion. The actions taken positively account for the temporary imbalance created by the rapid inventory upset in the furnace, giving optimum short-term protection. Control action quickly returns the forced draft dampers to the position they were in prior to the trip, thus restoring and assuring the continuous post-trip surge air flow as required.

CONCLUSIONS

A careful review of furnace control and performance by both simulation and test has shown that large furnace draft excursions can be contained by:

a) Timed fuel shutoff over a period of 6 to 8 seconds following initiation of boiler trip.
b) Fast-acting actuators for all fan control vanes.
c) Assigning the furnace draft control duty to the fans (FD or ID) having the prompter impact on furnace draft; in most boilers this will be the forced draft fans.
d) Direct acting override controls responding to out-of-range furnace pressure, applied to both F.D. and I.D. fans, combined with a transient closing bias of I.D. damper position initiated by master fuel trip.

The control system, properly designed and tuned to handle master fuel trip situations, also provided improved control under normal operation.

REFERENCES

6.1 Vollmer, H.D., Undrill, J.M., Crim, H.G., Jr., "Draft Control System Design, Simulation and Test for a High Implosion Potential Boiler", American Power Conference, Chicago, IL., April, 1977.

6.2 "Recommendations for Furnace Enclosure Damage Prevention"—Combustion Engineering Co. position paper.

6.3 Euchner, P.C., Undrill, J.M., "Furnace Implosions, An Analysis and Proposed Protective Systems", American Power Conference, Chicago, IL., April, 1975.

CHAPTER VII

STEAM TEMPERATURE CONTROLS

Steam temperature is controlled by a number of methods. In the case of superheat steam, spray water is usually employed. This method is used only as an override action in the case of reheat steam temperature control, as use of reheat spray water represents a cycle loss.

SUPERHEAT TEMPERATURE CONTROLS

Fig. 7.1 shows a schematic of the superheater showing its inlet, downstream of a spray water injection point, where thermocouple T_{AS} is located, a midpoint temperature sensing thermocouple T_{IM} and an outlet temperature thermocouple T_{SO}. In most superheaters a midpoint sensing thermocouple is not provided. Shown on Fig. 7.1 also is the response characteristics of the temperatures at the three locations to a step change in spray flow.

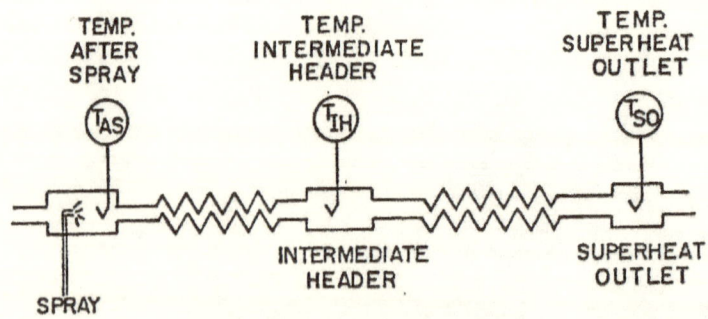

FIGURE 7.1   SECONDARY SUPERHEATER

128

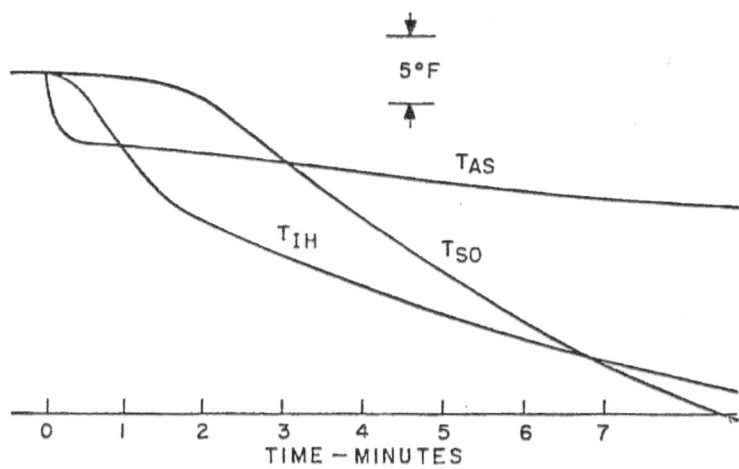

PREDICTED RESPONSE OF SUPERHEATER TEMPERATURES
FOR A STEP CHANGE IN SPRAY FLOW (8LBS/SEC)

The closer the spray point to the superheat outlet, the smaller is the
lag between spray flow change and superheat outlet temperature change.
However, as can be noted in Fig. 7.2, the closer the spray point to the
outlet the higher is the temperature of tubing upstream of the spray point,
and hence the higher is the cost of the heating section tube material.

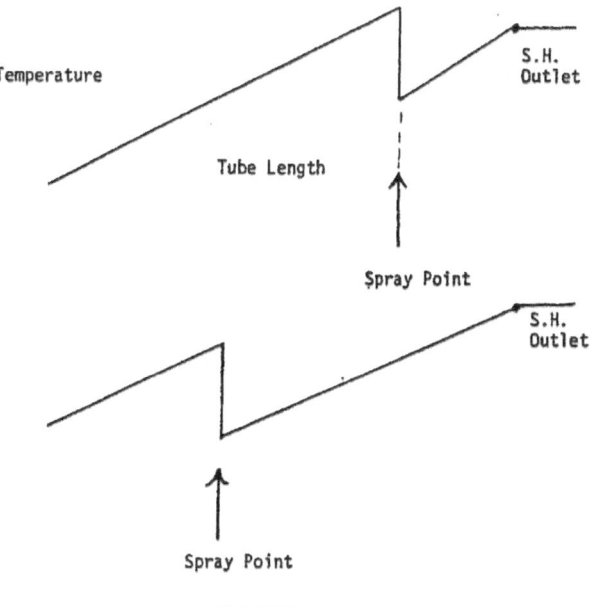

FIGURE 7.2

The heating section between drum and spray point is called the primary superheater, and the sections from that point to boiler outlet, secondary superheater.

The example of spray location at very close to saturation temperature shown in Fig. 7.2 is on a 2500 psi boiler where the inlet of the secondary superheater is at 710° and outlet 1000° F. The mass of metal between the spray point and outlet is in the order of 700000 lbs for this 500 MW boiler.

Fig. 7.3 shows the control system configuration selected for superheat temperature control.

Basically, control of outlet temperature is done by modulating the demand for inlet temperature. This demand for inlet temperature is set by proportional and rate feedback from midpoint temperature T_{IH} and from control action based on deviation of T_{SO} from set point. The use of the feedback from midpoint temperature is for purposes of minimizing temperature deviations due to heat flux disturbances along the length of the superheater. The effects of such disturbances originating in the first section of the superheater would be sensed sooner at the midpoint than at the outlet.

The procedure of tuning up this system is as follows. First, the temperature after spray loop is tuned by having the main steam temperature controller open (on manual) and having the feedback from the midpoint temperature open.

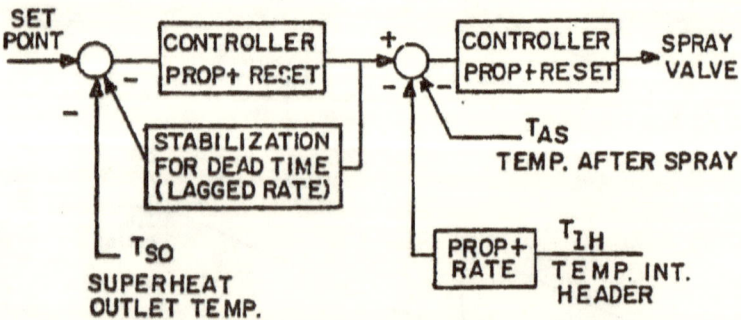

SUPERHEAT TEMPERATURE CONTROL CONFIGURATION
WITH ANTICIPATORY FEEDBACK FROM
INTERMEDIATE HEADER TEMP.

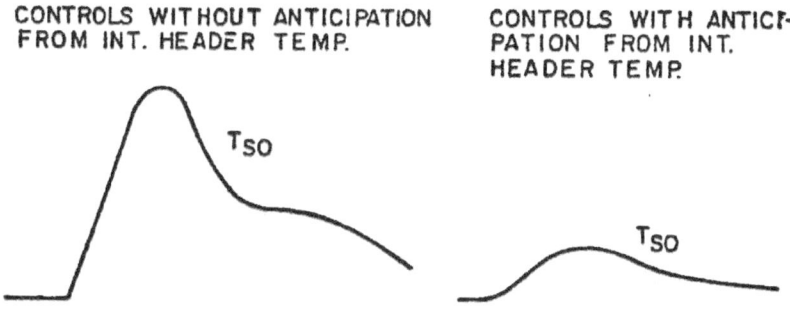

CONTROLS WITHOUT ANTICIPATION FROM INT. HEADER TEMP.

CONTROLS WITH ANTICI- PATION FROM INT. HEADER TEMP.

T_{SO}

T_{SO}

SUPERHEAT OUTLET TEMPERATURE DEVIATIONS
FOLLOWING HEAT FLUX DISTURBANCES

FIGURE 7.3

This loop can be tight since the only lag involved is the thermocouple time constant in series with the small valve delay. The feedback from temperature after spray is essential to linearize the control action. If control were merely to the spray valve, a given change in valve position would yield a very different effect as function of steam flow rate and head across the spray valve.

In order to set proper characteristics in the main steam temperature controller (gains and stabilization settings), it is necessary to determine the process characteristics as seen by this controller, i.e., the response of T_{SO} to a change in demand signal from this controller.

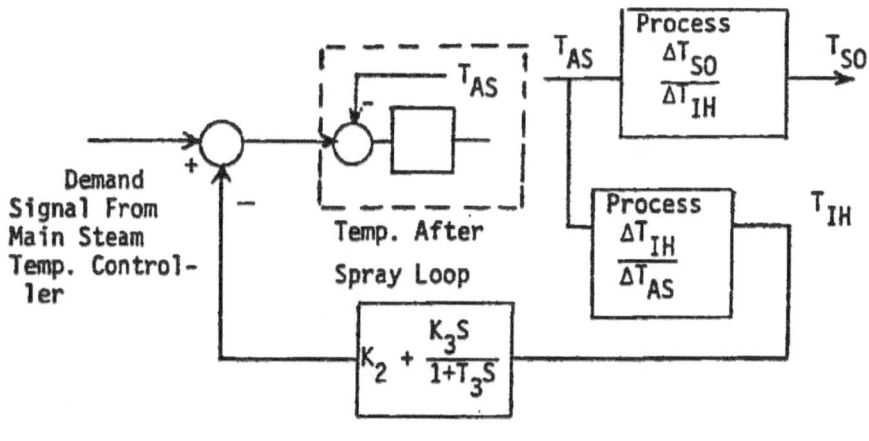

FIGURE 7.4

As can be seen on Fig. 7.4, this response characteristic includes the effects of feedback from the intermediate temperature ΔT_{IH}. Hence, the first step is to adjust this feedback to an optimum setting with the main temperature controller open.

This is done by adjusting parameters K_2, K_3 and T_3 (Fig. 7.4) of the feedback with the object of minimizing the deviations on ΔT_{IH} and ΔT_{SO} as caused by a process disturbance introduced downstream of the spray point.

The effect of this disturbance can be simulated by introducing a change in demand signal out of main steam temperature controller (on manual) and increasing the proportional and rate gains from midpoint temperature until the effect of this demand signal on deviations in T_{IH} and T_{SO} are reduced considerably. For instance, as can be seen on the open loop response curves of Fig. 7.1, a given change in temperature after spray produces eventually almost twice this change in outlet temperature. Introducing feedback from midpoint temperature, a change in demand out of the main steam temperature controller can be made to reduce considerably the change in outlet temperature which is equivalent to minimizing effects of upsets downstream of the spray point on outlet temperature.

Finally, having both feedback loops from T_{AS} and T_{IH} adjusted, the main steam temperature controls can be adjusted as explained previously in the section covering dead time compensation for the pressure controls. The process overall response, as seen by the main steam temperature controls, would include the action of both feedbacks and can be obtained again by recording the effect on T_{SO} of a change in demand from the controller.

As is discussed in Chapter XII covering the modeling from first principles, the response characteristics of temperature are a function of tube length, mass and flow rate. The time delays increase as flow rate decreases. Since it is impractical to implement adaptive controller parameter changes in analog control systems, one must tune the system to account for the largest lags which implies adjustment for part load conditions.

REHEAT TEMPERATURE CONTROLS

There are several methods of controlling reheat temperature by changes in the distribution of heat absorbed in the furnace (waterwalls) relative to the heat absorbed in the reheat and superheat sections.

One of these methods consists in changing the tilt angle of the burners in boilers equipped with this feature. Another way is to change recirculation through a recirculation fan that introduces gas, from downstream of the economizer into the furnace and increases the mass flow rate through the boiler. In this manner the furnace temperature is decreased with consequent decrease in heat to the waterwalls while the heat transferred by convection through the back passes is increased.

A third method is to change distribution of heat in parallel sections of the back pass through dampers. By placing sections of the primary superheater in one path and sections of the reheater in the other, the heat distribution to these can be altered by damper control.

A representative control configuration is shown in Fig. 7.5. Reheat temperature error through proportional, reset and rate controls sets the gas recirculation damper position through a function generator. Rate anticipation from steam flow is provided and can be used if found necessary. If a repeatable relation exists between steam flow and recirculation damper, then a feedforward signal from steam flow can be used. In this example, burner tilts are programmed as function of steam flow through an appropriate lag.

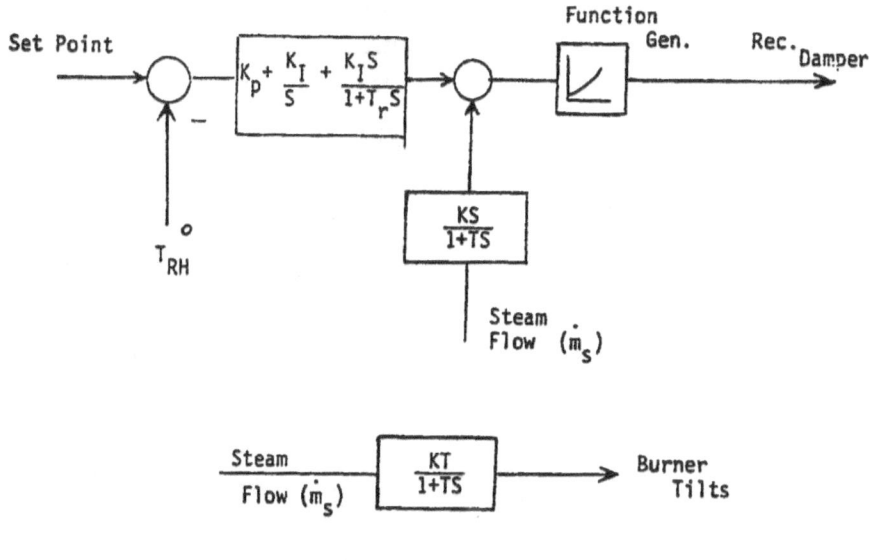

FIGURE 7.5

REFERENCES

7.1 de Mello, F.P., and Paulson, R.E., "Simulation of Plant Dynamics and Design of Plant Control Systems", South Eastern Electric Exchange Meeting, New Orleans, April, 1966.

CHAPTER VIII

MISCELLANEOUS LOOP CONTROLS

There are a number of miscellaneous controls which can be treated independently. Generally they are handled by simple two mode controllers. Examples of such miscellaneous controls are condensate flow control and primary air temperature control.

CONDENSATE FLOW CONTROL

Control of condensate flow is basically to hold the de-aerator level. Since the draw down from the de-aerator is due to feedwater flow, it is logical to have condensate flow follow feedwater flow, with trimming control from de-aerator level.

Fig. 8.1 shows a typical control system for condensate flow control, including the scaling for transmitters which operate on a 16 ma signal level.

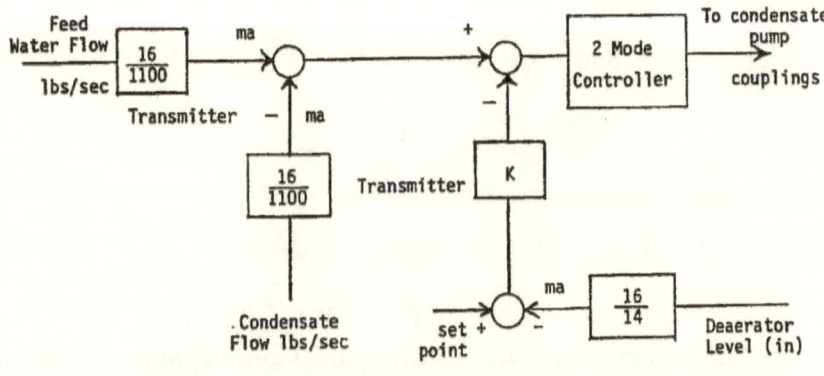

FIGURE 8.1

Fig. 8.2 is a block diagram including the process response (response of condensate flow to signal to pump couplings and response of de-aerator level to error between feedwater flow and condensate flow). For tutorial purposes, units are identified including the scaling between process variables and controller signal levels.

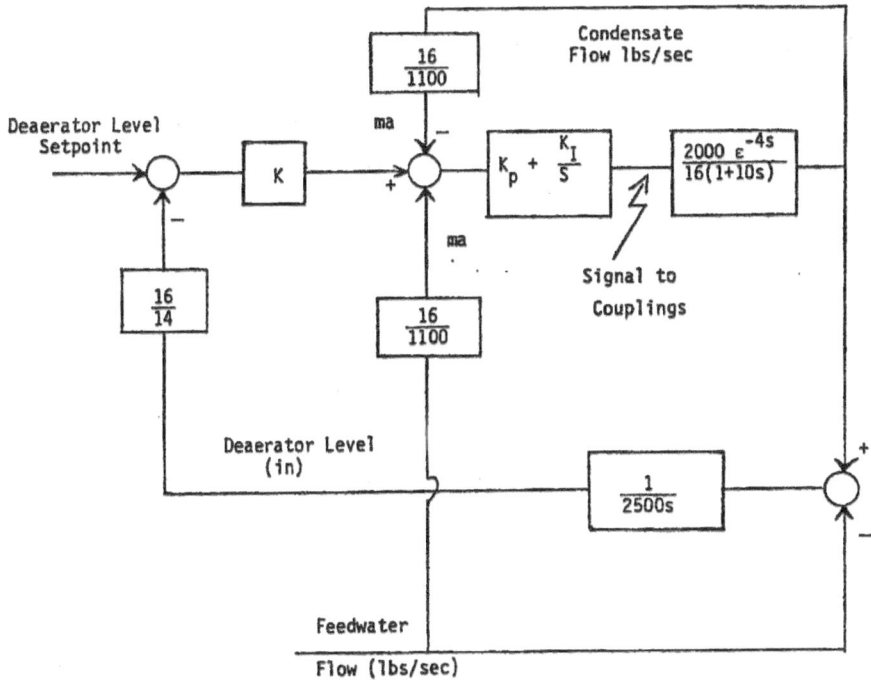

FIGURE 8.2

For instance, from the pump curves the gain of flow to signal to couplings is established as 2000 lbs/sec per ma. The flow response is characterized by a dead time of 4 sec in addition to a time constant of 10 sec.

In adjusting controller parameters, we recognize a sub-loop (flow control) and an outer loop (level control).

First, the flow control sub-loop is adjusted to yield well damped but fast response.

The process function is:

$$\frac{2000}{1100} \frac{\varepsilon^{-4s}}{(1+10s)}$$

Reasonable tuning values can be derived either by time response methods as explained in Appendix F or by frequency response techniques. A proportional gain K_p = 0.5 and an integral gain K_I of 0.05/sec is recommended.

This corresponds to a proportional band of 200% and repeats/min equal to $\dfrac{0.05 \times 60}{0.5} = 6$.

With such settings the closed loop response of condensate flow to a change in demand should be characterized by a 10 to 15 second time constant.

The remaining parameter is the proportional gain K of the deaerator level control. This gain is set considering the stability of the outer loop which can be described as in Fig. 8.3.

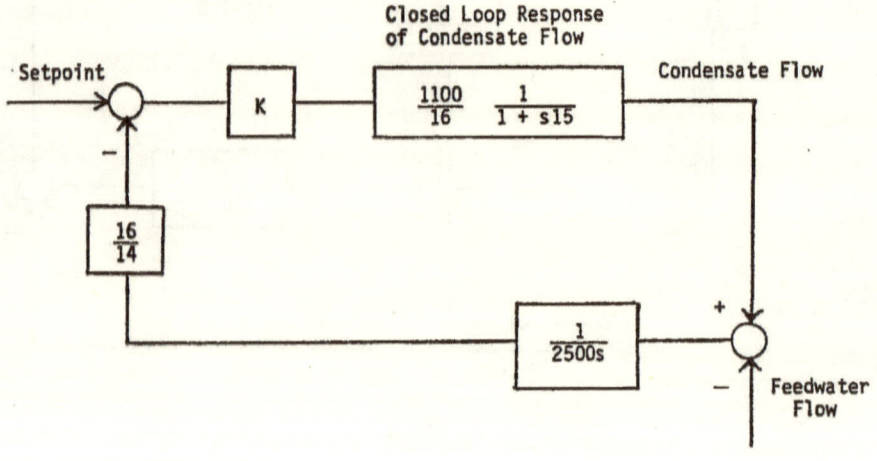

FIGURE 8.3

For a well-damped response one should aim to keep crossover at frequencies below 1/15.

Aiming at a crossover at W_c = 1/30 rads/sec, K should be approximately $\dfrac{14}{1100} \times \dfrac{2500}{30} = 1.06$.

The phase angle at crossover is 90° for the function $\dfrac{1}{2500s}$ and 26.6° for the function $\left(\dfrac{1}{1+15s}\right)$ i.e., a total of 116.6°.

PRIMARY AIR FLOW CONTROLS

Cold and hot air dampers are controlled to (1) provide the required amount of primary air flow to each individual mill, (2) maintain the desired temperature of the mixture of cold and hot air.

Fig. 8.4 shows a common configuration of primary air flow controls. The flow loop acts to open or close both hot air and cold air dampers. The temperature loop acts to open one damper while simultaneously it closes the other to modify the ratio of hot air to cold air as necessary.

The flow loop response is fast, basically made up of the damper response. The temperature loop on the other hand exhibits the thermocouple response time constant, in the order of 30 sec, hence the temperature loop must be adjusted by an order of magnitude slower than the air flow loop.

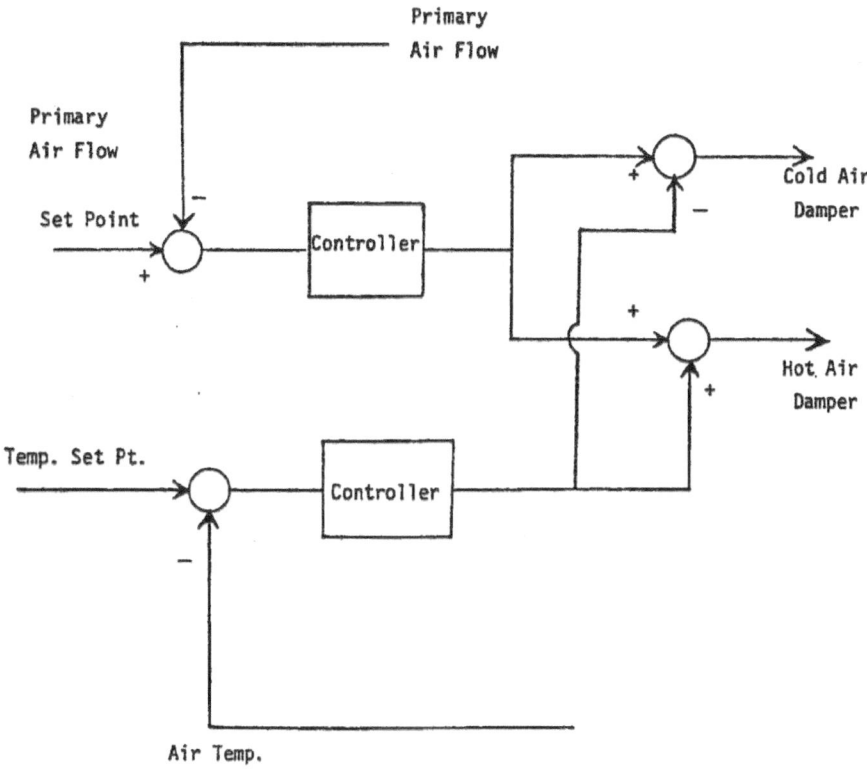

FIGURE 8.4

For ideal non-interaction between the loops the system of Fig. 8.5 would be logical. It is not used because of the expense and complication of introducing multipliers in analog systems, and because, with the two loops fairly well decoupled in bandwidth, interaction is not a problem.

Obviously such sophistications would not be any problem in direct digital control systems (Chapter XI).

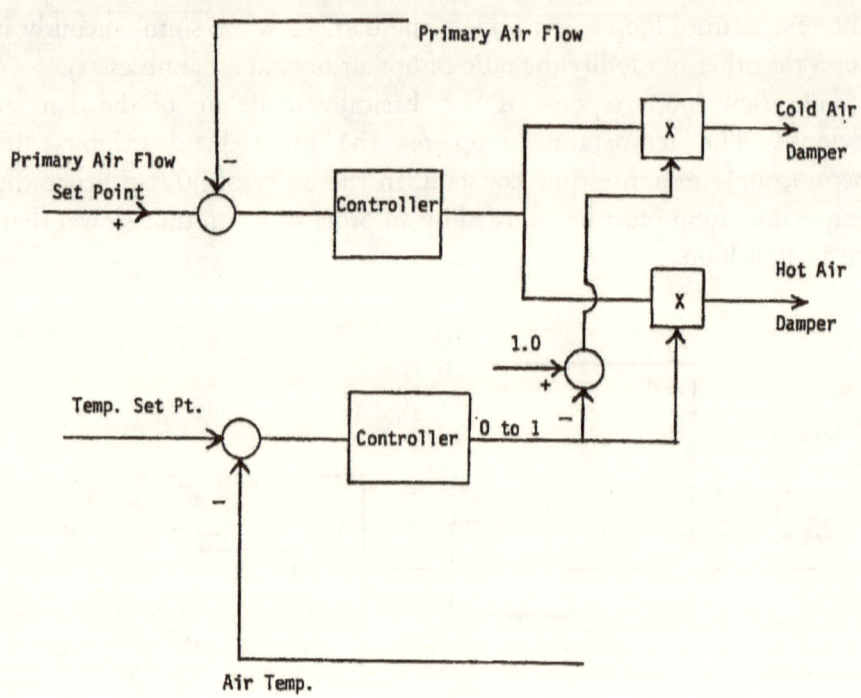

FIGURE 8.5

CHAPTER IX

CONTROLS FOR
ONCE-THROUGH BOILERS

Before considering control requirements for once-through boilers, it is instructive to examine some of the physical aspects and differences associated with various types of steam generators.

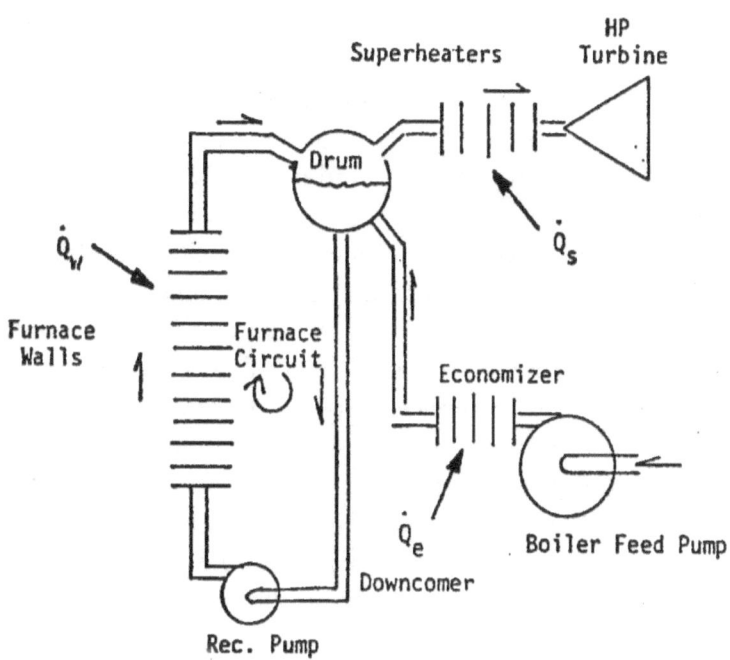

FIGURE 9.1

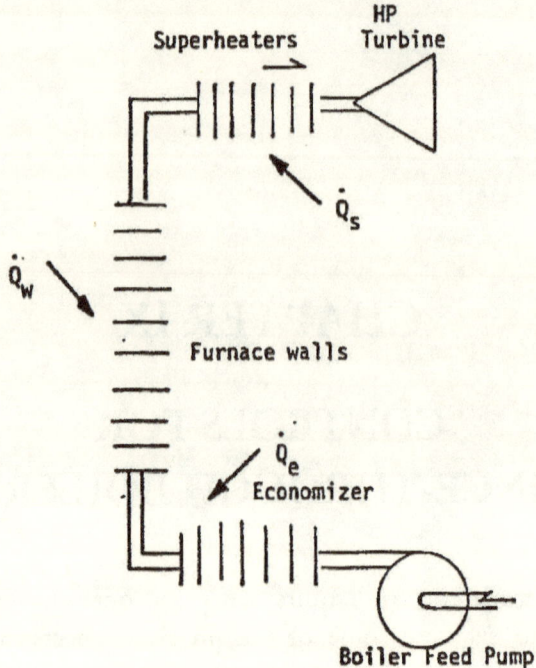

FIGURE 9.2

Figs. 9.1 and 9.2 are schematics of the two basic types of steam generators in use today. Fig. 9.1 indicates the fluid circuitry associated with the so-called conventional or drum-type unit. In this system the pressure level is always below the critical point of steam, 3206.2 psi (i.e., steam has latent heat).

Feedwater is preheated in the economizing section and flows into the drum, downcomer, water wall or furnace circuit. Since steam generation occurs in the water wall sections, and the recirculation flow rate around the furnace circuit is five to ten times that of the generated steam flow, depending on the load the unit is carrying, the water wall exit flow is a low quality mixture of steam and liquid, and the furnace circuit remains essentially in a saturated condition at the existing drum pressure. Mechanical separators in the drum insure that negligible moisture is carried over into the superheating sections. The steam is superheated to rated steam temperature conditions in the superheating sections and passes on into the high pressure turbine.

At this point some comments on the response characteristics and dominant effects which occur transiently in this type of unit are appropriate. Consider the response for a change in the turbine throttle valve position or equivalently, load. The increase in steam taken from the boiler results in a decreasing boiler pressure. This decreasing pressure is arrested by increasing the steam generation in the furnace walls through increased fuel input. The large change in the specific volume of the fluid in flashing from liquid to steam is the physical mechanism by which pressure is restored. Note that changing feedwater flow results in essentially just a change in drum level. Thus, in conventional units, combustion is used for pressure control, and feedwater basically follows the steam flow to insure that an adequate supply of liquid is in the boiler at all times.

Another important aspect of this type unit, particularly from an analytical modeling standpoint, is the fact that the fluid properties and conditions in each section of the boiler under normal operating conditions are well defined; that is, liquid in the economizer, saturation conditions in the furnace circuit and superheated steam in the superheaters. This permits the equations describing the fluid properties for a particular condition to be used throughout a given section (i.e., furnace circuit, superheater, etc.), without the complex problem of trying to describe the properties of the fluid if it were to undergo drastic changes in properties in a particular subsection.

Consider Fig. 9.2 which shows the fluid circuitry of the once-through boiler. These units may operate under subcritical and supercritical conditions (i.e., above 3606.2 psi). Note that the flow of fluid through the boiler is continuous, and that as heat is added along the fluid path, the fluid properties change in a continuous manner.

The pressure level in this type unit is basically specified by the flow being forced through the entire boiler circuit by the feed pump. Changes in fuel input will have essentially a transient effect due to the expansion or condensation of the fluid within the boiler until the new operating conditions are achieved. Since these units operate in the higher pressure ranges, the changes in the fluid specific volume are relatively smaller than conventional units, thus diminishing the effect of heat addition on pressure. In this type unit then, feedwater has a dominant effect on pressure.

Another important difference in characteristics and, hence, control requirements of once-through versus drum-type boilers relates to the fact that the steady state flow rate of fluid in the furnace tubes for once-through boilers is proportional to load in the normal operating range of 30 to 100%, whereas it is almost independent of load or of feedwater flow in the case of drum-type boilers. In drum boilers, changes in feedwater flow or steam flow do not materially affect the fluid flow rate in water wall tubes and, hence, there is no concern of major imbalance between heat flux to the tubes and heat absorption to the inner fluid. For once-through boilers, however, it is important to keep a close match between the once-through flow rate, i.e., feedwater flow and fuel flow, since only a few seconds of a major imbalance can result in rapid rise of tube temperatures and burn out. Fuel and feedwater flow must, therefore, be moved in unison and corrections in the ratio of fuel to feedwater should be made within well-defined limits in a slow reset recalibration mode to correct for temperature deviations.

BOILER RESPONSE CHARACTERISTICS

The important first step in the formulation of a control philosophy and selection of a control configuration is to obtain a good description of the plant's responses to various input disturbances. These responses are often referred to as "plant open loop characteristics".

Figs. 9.3 through 9.6 extracted from reference 9.1 show the nature of boiler variables' responses to changes in turbine valve position, pump speed, spray valve change and change in heat flux in the gas path.

These response characteristics are for a subcritical once-through boiler. A supercritical boiler will have essentially the same type of response characteristics, with even stronger coupling between feedwater flow rate changes and pressure changes.

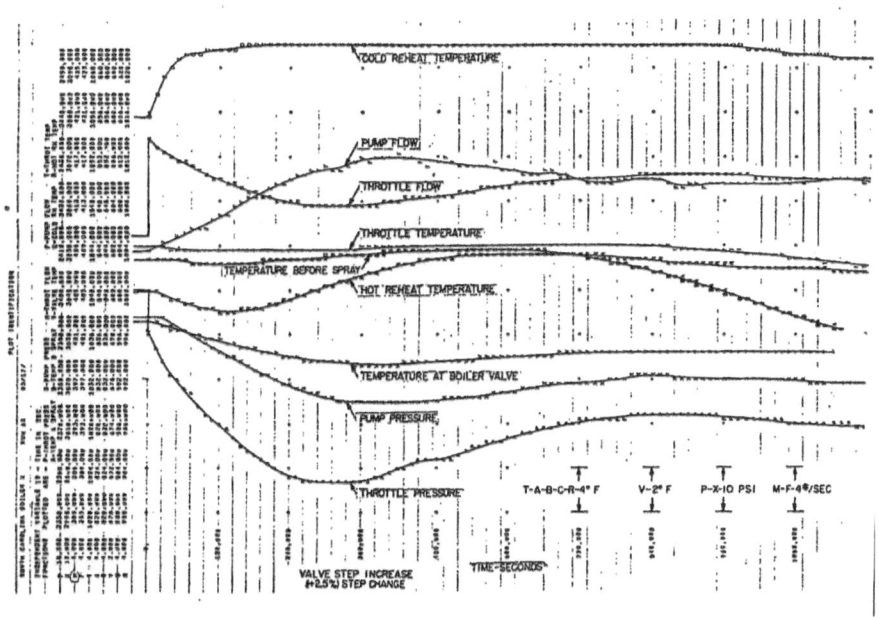

FIGURE 9.3

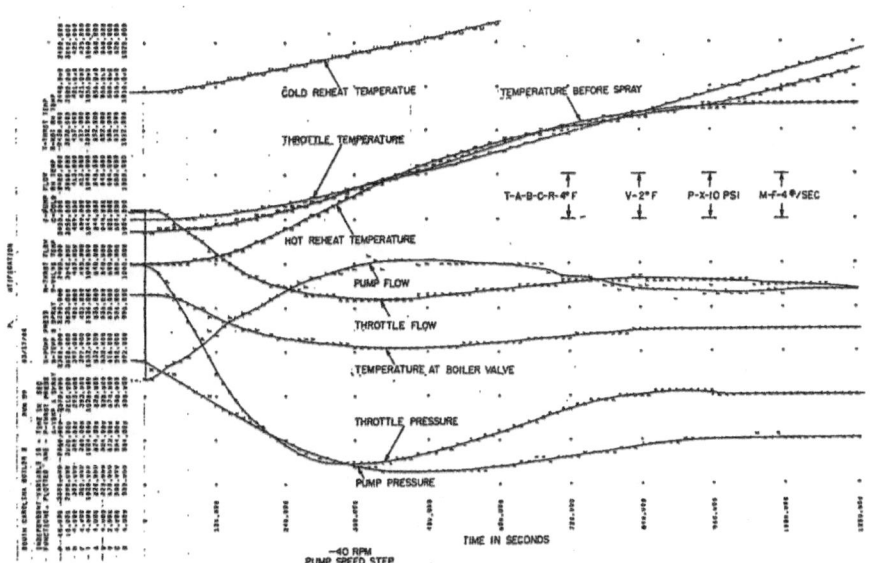

FIGURE 9.4

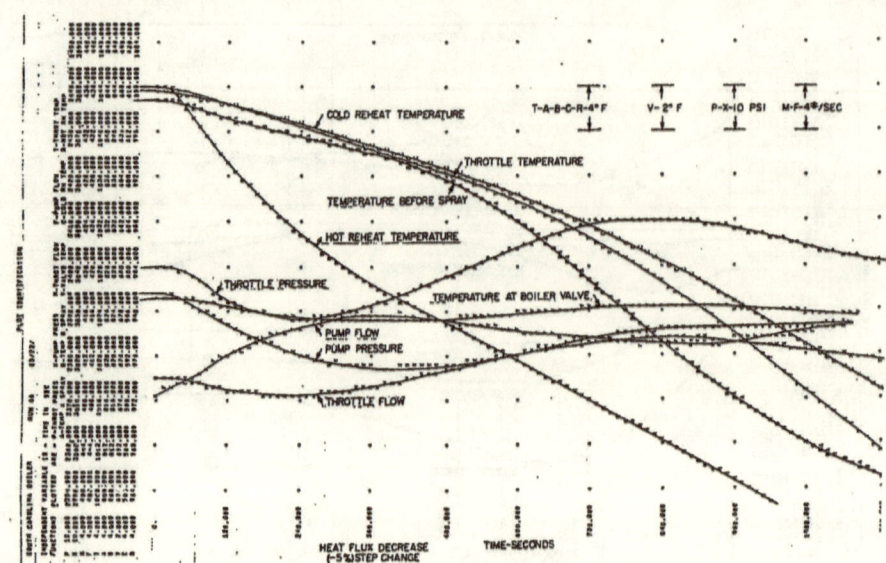

FIGURE 9.5

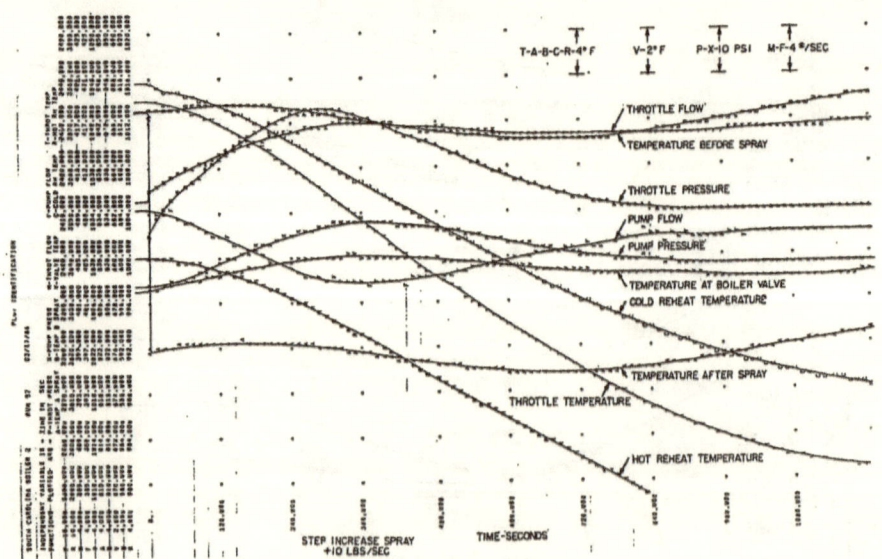

FIGURE 9.6

BOILER-TURBINE CONTROLS

The main objectives of boiler-turbine controls are:

1. To attend to the turbine demands for energy as expediently as possible within the storage and auxiliary system capabilities of the plant.
2. To maintain process conditions, pressure temperatures and flow rates (feedwater, fuel and air) at optimum and safe values under load demand disturbances as well as other upsets such as fuel disturbances, losses of auxiliaries, etc.

The demands on control performance are obviously related to the severity of load change response solicited from a given steam generating unit. Several philosophies of control have evolved in attempts to accommodate the need or capability for rapid load changes versus the need to control critical plant variables (temperatures and pressures). A review of these philosophies is appropriate as an introduction to principles of control of once-through boiler and in particular the coordination of boiler-turbine controls.

CONVENTIONAL PRACTICE

Most conventional drum-type steam units are operated in the boiler following mode wherein changes in generation are initiated by turbine control valves, and the boiler controls respond with necessary control action upon sensing the changes in steam flow and deviations in pressure. In this mode the turbine has access to the stored energy in the boiler and load changes within reasonable magnitudes occur with fairly rapid response, as shown on Fig. 9.7. This characteristic is beneficial from a governing standpoint improving the quality of frequency control, and it is characteristic of conventional steam systems to maintain a fairly narrow band of frequency due to this fast governor action.[9.3] It is also beneficial in arresting frequency sags following large upsets. The boiler controls must be responsive and stable to withstand this mode of operation.

TURBINE FOLLOWING MODE

The turbine-follow mode, on the other hand, involves use of the turbine control valves to regulate boiler pressure. This can be done with practically no time delay so that under this mode boiler pressure suffers virtually no transient deviations, and no use is made of stored energy in the boiler. Steam flow through the turbine and, therefore, turbine power follows closely the amount of steam generation, i.e., the input to the boiler. Control of boiler variables is easy under this mode of operation, although the response of turbine power is considerably slower. The units' power output does, however, reflect both the intentional and unintentional variations in fuel input which in the case of coal-fired units can be a problem. The turbine following mode is lacking in the main functional requirement of a power plant which is to meet electrical load demands as expediently as possible. Governing action would be quickly washed out by the action of the pressure controls.

Fig. 9.8 shows response characteristics of a turbine following plant where the change in load demand actuates a corresponding change in fuel and feedwater input, and the turbine valve maintains constant boiler pressure. The output power is delayed by the lags in fuel system and storage time constants of the boiler. The dimension of the control task is reduced appreciably as boiler input becomes an independently specified variable rather than a process dependent quantity involving feedback.

COORDINATED BOILER-TURBINE CONTROLS

Recognizing the advantages and disadvantages of both previously discussed modes of operation and control and the need for varying degrees of compromise between the desire for fast response to load changes and the desire for boiler safety and good quality of control of steam conditions, a logical control mode is one that offers an adjustable blend of the two previously described schemes. This need for coordination of the boiler turbine is more pressing in the case of once-through boilers where the process interactions between feed pump, firing, and turbine valve are closer coupled than was the case with drum-type boilers with their large saturated liquid storage and insulating effect of having saturation pegged at the drum.

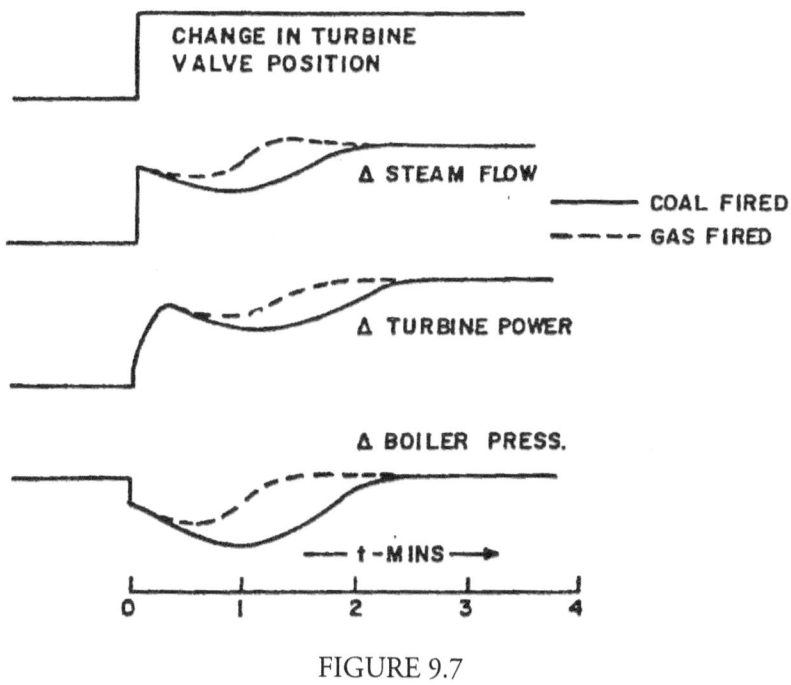

FIGURE 9.7

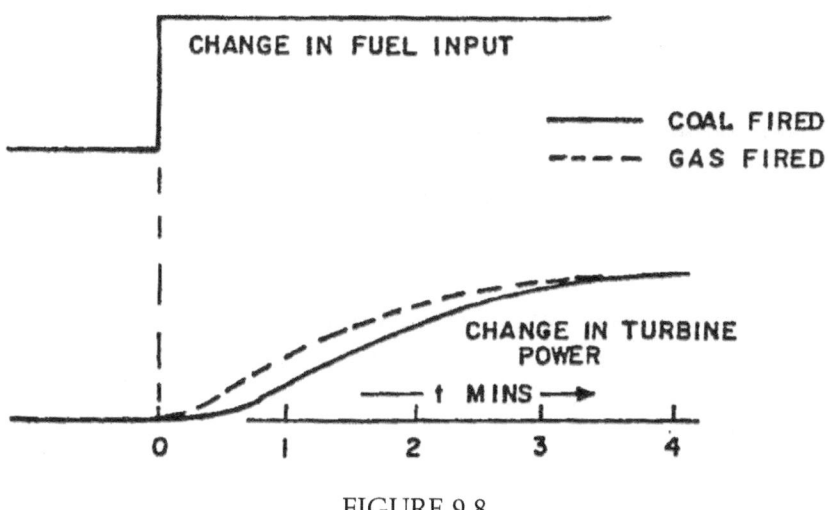

FIGURE 9.8

A type of coordinated controls that has been described in the literature is shown functionally in the schematic of Fig. 9.9. To achieve this coordination, it is necessary to make the speed changer motor respond to intelligence other than and in addition to load control signals. This is done by integrating the load control pulses to develop a signal indicative of demand for MW which can be altered by other inputs. This signal is modified by a frequency deviation bias matching the unit's governor droop characteristic to develop the desired MW. Comparison with the unit's actual output develops the MW error. The desired MW signal as well as the MW error signal are sent to the boiler controls. Turbine speed changer position is directed to reduce a combination of MW error and pressure error to zero while the boiler controls likewise are directed to reduce the pressure error biased by the MW error to zero. The sense of the cross-coupled MW error and pressure error biases is in the direction such that a positive MW error (MW lower than demand) would cause the turbine valve to open and the boiler controls to call for more feedwater fuel and air, whereas a positive pressure error (pressure lower than set point) would call for closing of the turbine valve while simultaneously increasing the feedwater, fuel and air.

Analysis shows that the control systems come to rest when both the MW error and pressure errors are restored to zero and transiently the cross coupling is in the direction to attenuate interacting effects of the MW feedback and pressure feedback loops. Depending on the cross coupling strength between pressure and MW loops, the load response can be adjusted to any degree between that shown on Fig. 9.7 and that of Fig. 9.8.

Another variation of coordinated controls where the turbine valve is primarily the pressure regulator and boiler inputs are actuated basically from load demand is to introduce a transient offset in the pressure set point from load demand. With such an arrangement, on a step increase in load demand, the pressure set point moves down with a timed washout causing the turbine valve to open immediately.

One important aspect of these methods is the role of the turbine valve as the basic boiler pressure regulator. Relying on turbine valve for this role has drawbacks as follows:

1. Constant motion of speed changer motor due to the normal process noise, fuel disturbances, etc., which can be significant in the case of coal-fired plants. Pressure deviations would be eliminated at the expense of MW deviations.

2. Reliance on turbine valves for pressure control removes the incentive to do a good control job on the boiler input variables. If responsibility for good and stable pressure control is not assigned to the boiler input variables, this can become a problem under the condition of valves wide open, at which point the turbine valves are beyond control range.

3. In those cases where a transient offset in pressure set point activated from load demand is not provided, the use of turbine valves for pressure control does materially reduce the load response capability of the unit.

IMPROVED METHOD OF BOILER-TURBINE CONTROL COORDINATION

Keeping in mind the disadvantages of continuous pressure control by means of turbine valves and the over-all objectives of quick load response within reasonable limits of boiler storage capability, we have developed the scheme of Fig. 9.10 which meets these objectives.

The features of the configuration of Fig. 9.10 may be understood in the light of the following considerations applying to the basic scheme of Fig. 9.9.

The MW feedback loop without adequate cross coupling from pressure error presents stability and potential runaway problems if for some reason the boiler fails to keep in step with the demand. That is, without biasing action from pressure error, the MW loop will cause valve over-travel in attempting to satisfy MW demand, irrespective of boiler pressure. An analogous situation occurs with the use of first stage pressure feedback, whereas this condition would be avoided if a suitable indication of valve position (if linear with flow) were used.

Examine now the significance of combining pressure error to the MW demand. To prevent interaction of the pressure and MW loops, the gain of the pressure bias cross coupling term would be adjusted such that the turbine valve would not attempt to correct the component of

MW error due to the pressure not being at set point, but would allow the pressure loop to accomplish this component correction. This means that the cross coupling gain would be $\dfrac{MP}{P} = K$ or the equivalent turbine valve opening. Since MW in the steady state are proportional to valve position times pressure, the combination of MW added to the negative pressure error gained by the non-interacting value K_v (Fig. 9.10) which is proportional to load level, can be interpreted as turbine valve opening, i.e.,

$$MW = K_v P = K_v (P_o + P_e)$$

or

$$(MW - K_v P_e) = K_v P_o$$

where K_v is proportional to valve opening, P_o = pressure set point and P_e = pressure error.

To the extent that the pressure error cross coupling term is gained by K_v, the combination of MW and this pressure error term can be replaced by the equivalent turbine valve opening. Any added strength of this pressure error beyond K_v inhibits turbine response and provides some degree of pressure regulation by means of the turbine valves.

In view of the above considerations and in keeping with basic philosophy of taking advantage of stored energy in the boiler within safe limits, the scheme of Fig. 9.10 accomplishes the required degree of non-interaction between the MW loop without inhibiting turbine valve motion as long as pressure error is within a safe band. The response of this system to load demand pulses as well as to governing signals through the turbine is compatible with that of conventional units—exhibiting fast response for limited changes. This type of response is also compatible with dynamics of load-frequency controls.

As shown on Fig. 9.10, the load demand, which can be derived either from the load control signal or set manually, establishes the desired turbine demand index (TDI) which is equivalent to turbine valve position.

This index is derived from first stage pressure by multiplying it by the ratio of pressure set point to actual pressure. It is, therefore, the first stage pressure that would result for the given valve position at the pressure corresponding to pressure set point. Aside from convenience of measurement, this index avoids the nonlinear relationship between valve position and flow, and thus forms an ideal feedback signal instead of actual valve position, which was mentioned previously as being desirable. Essentially, TDI is a direct measured indication of the turbine flow demand compensated for throttle pressure deviations from set point.

The difference between demand and actual TDI acts on the speed changer controller to move the turbine valves. The demand signal for TDI is slowly calibrated by integrated MW error from demand. This has no significant dynamic effect and is a convenient feature especially under base load operation where a MW demand level can be set and the proper TDI will be called for regardless of condenser, heater and other conditions which may affect the steady-state relationship between TDI and MW.

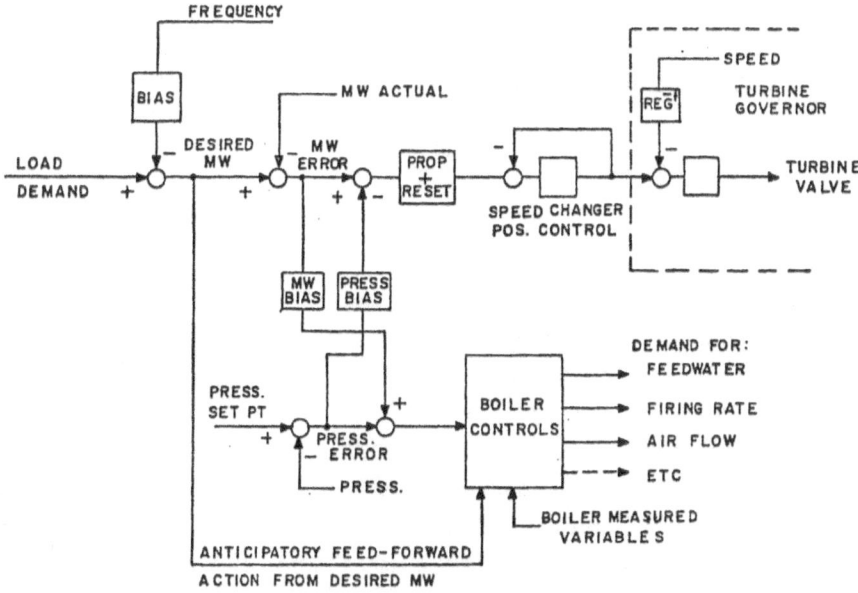

COUPLING OF TURBINE LOAD CONTROLS
WITH BOILER CONTROLS

FIGURE 9.9

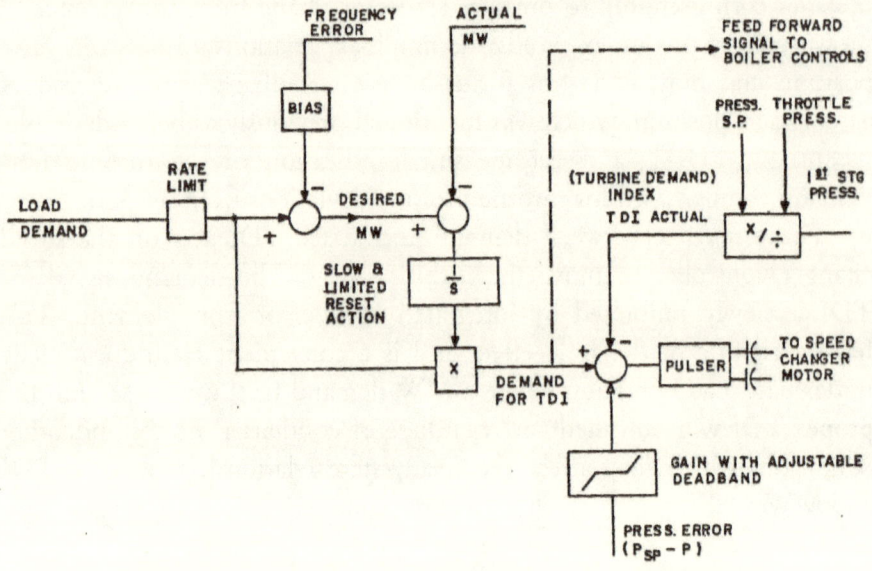

CONTROL SYSTEM COORDINATING TURBINE
LOAD DEMANDS WITH BOILER PLANT CAPABILITIES

FIGURE 9.10

Pressure error, if within an adjustable dead-band, does not inhibit turbine valve response. However, if an excessive pressure error occurs, indicative of changes in load in excess of boiler capabilities, this error will act to regulate pressure with a fairly narrow proportional band. This control action is governed by the deviation of the actual pressure from the pressure set point ± the pressure dead-band such that the action is smooth, coming in and going out of operation.

The value of the dead-band can be adjustable. For the case where the fuel system is responsive and reliable (oil and gas), a large dead-band (±100 psi) can be safely used. On the other hand, should the fuel system exhibit frequent upsets, such as pulverizer plugging, etc., it may be necessary to reduce this dead-band to a lower value in the interest of plant safety.

COORDINATION OF FUEL AND FEEDWATER CONTROLS

Unlike conventional steam generators, the once-through unit does not have a large reservoir of recirculating liquid in the waterwalls or boiling regions of high heat absorption rate, thus making them more subject to burnout conditions. It is therefore important that a reasonable balance between feedwater flow and heat input into this region be maintained. This requirement is the source of difficulty in the practice of basically firing the once-through unit to maintain throttle temperature with a combination of feedforward and feedback action while maintaining pressure with the feedwater control loop. Transient mismatches using this basic philosophy can be severe, tending to burn out conditions when feedwater starvation with excess fuel occurs and tending toward positive feedback pressure effects when an excess of feedwater to fuel causes depressurization resulting in further increases in feedwater flow. It thus becomes necessary that the feedwater and fuel controls derive their primary demand signals from the same origin where only the dynamics of relatively well-known auxiliaries will cause mismatches for which compensation can be provided. The provision and continual use of runback controllers to provide this protection is unreliable, complicated, results in poor and unpredictable control action and is in general an undesirable control policy.

CONTROL AND COOORDINATION OF BOILER INPUTS

Whether or not recourse is taken to continuous use of turbine valves to assist the boiler control (and this is controllable by the setting of the dead-band on the cross-coupled pressure error), the fundamental requirement for successful once-through boiler controls is the proper transient coordination of the inputs to the boiler, namely, feedwater flow, firing rate, spray flow and gas distribution dampers.

The need for close match of fuel and feedwater in once-through boilers dictates that the dominant command to these two subloops be from the same source, and other corrective action to alter the ratio of fuel to feedwater should be of a slow limited nature.

If the only objective in control were to regulate pressure and temperature without concern for keeping fuel and feedwater transiently in step, a configuration such as in Fig. 9.11 would be adequate.

Basically, the pressure controls, operating on pressure error, set the demand for feedwater flow. There is no particular need for anticipatory action for this loop since the pressure error gives an indication of the integrated flow deficiency or surplus in addition to the almost instantaneous deviation proportional to load change. This instantaneous pressure deviation following a change in valve opening is proportional to the load level. It is an inherently adaptive process characteristic since the change in fluid storage in the boiler for a given change in load is approximately proportional to the load level at which the change occurs. Pressure error is thus an inherent self-adapting process intelligence which can be used effectively in feedback configurations.

In the configuration of Fig. 9.11, the primary action to the fuel and air controls is derived from the desired turbine demand index through a proportional term and a rate term to account for changes in thermal stored energy in the boiler. Feedforward action for the fuel and air is necessary since indications from temperature error would be extremely tardy.

Steam temperature error operates with proportional and rate action on demand for temperature after spray. A cascaded loop controlling temperature after spray moves the spray valves to yield very effective transient control of main-steam temperature. A slow reset controller operating on error between spray valve position and a desired valve position (50%) restores the spray valve to set point by modifying fuel and air demand. This demand is also modified by slow reset action on main steam temperature to provide final trimming for proper balance between firing rate and feedwater flow.

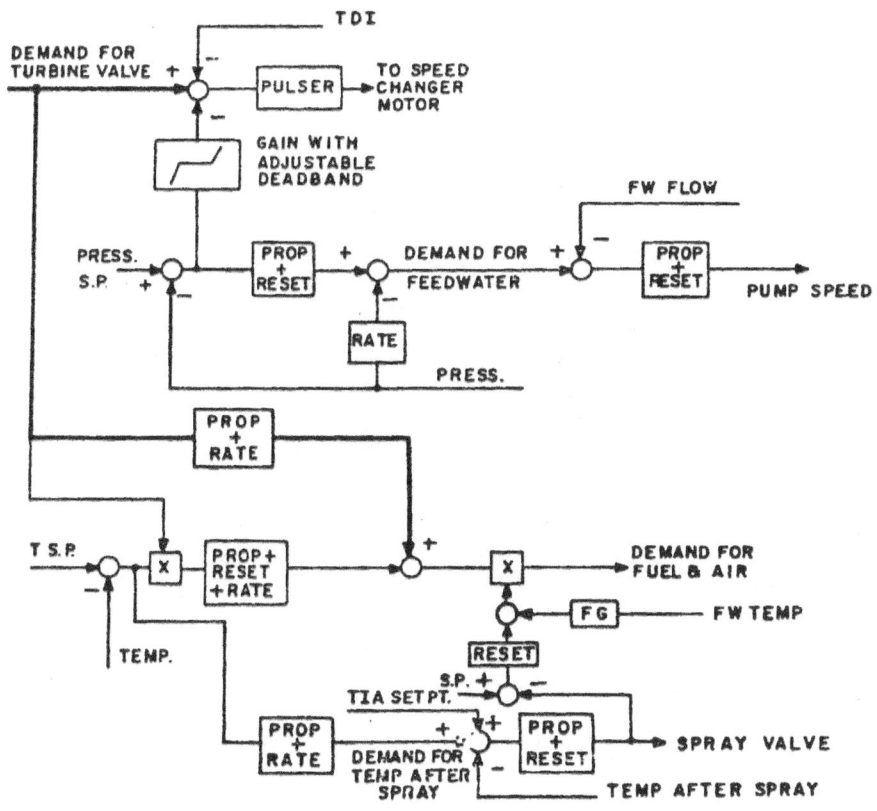

EXAMPLE OF CONTROL CONFIGURATION WHICH PERMITS
TRANSIENT MISMATCHES OF FUEL AND FEED WATER

FIGURE 9.11

Although the configuration of Fig. 9.11 can be adjusted for good control of pressure and temperature, it is lacking in the important requirement of close matching between firing rate and feedwater flow to avoid damage to the furnace circuits. As can be seen from control performance shown on Fig. 9.12, there is considerable transient mismatch between feedwater flow and firing rate. The difficulties associated with this scheme relate to making firing rate and feedwater flow follow each other when each is under the influence of different primary signals, (pressure controller output for feedwater and TDI demand for fuel and air). Reliance on runback controllers to keep the two variables in step with each other did not yield satisfactory results.

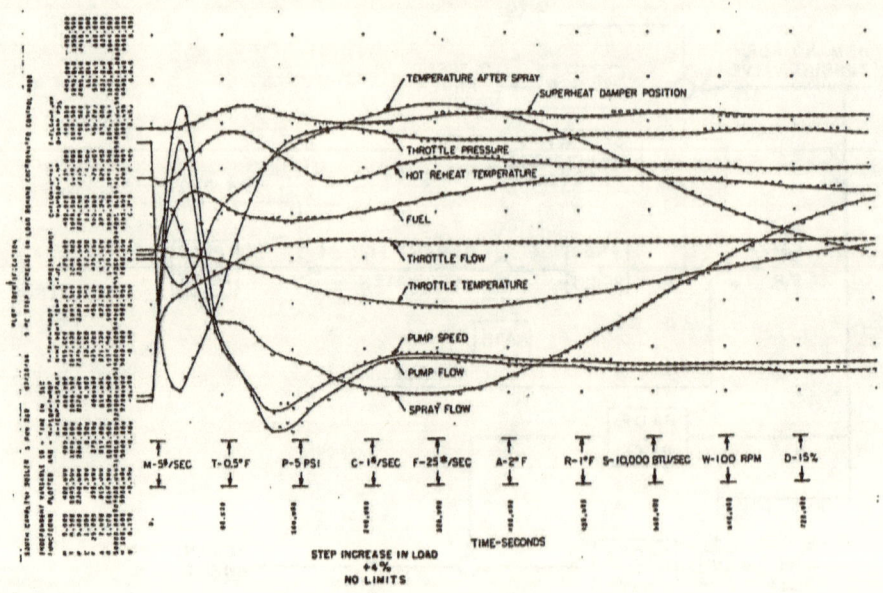

FIGURE 9.12

FUEL AND TEMPERATURE CONTROL COORDINATION

The most effective and expedient means of temperature control by far is through the well accepted and proven means of de-superheating spray. It is therefore reasonable that this means of temperature control be relegated the responsibility of controlling temperature deviations under transient conditions. It should be noted, however, that spray control is only a means of transiently redistributing the energy levels in the boiler and will have no effect on the steady-state throttle conditions for a given total feedwater flow (i.e., through flow plus spray flow) and fuel input. Thus, final trimming to attain a desired throttle condition and load (i.e., flow) must be done through the fuel controls. Since the spray control is so effective, this trimming control action may be in the form of relatively minor control action on fuel derived from temperature error.

Although spray is effective in temperature control, caution must be exercised to insure that the control capacity or limit of the spray system is not exceeded by excessive transient duty. In order to insure that the spray valve remains in control range, additional intelligence to the fuel system must be provided to indicate the extent that an additional or deficient amount of spray is necessary to maintain required throttle conditions. The deviation of the spray flow from its normal mid-control range value may be interpreted as an equivalent change in fuel demand required to eliminate excessive spray flow deviations and to bring the spray back to its prescribed mid-control range value. Since the change in pressure head across the spray valve and the amount of spray required at a given flow are essentially compensating effects, the deviation of the spray valve from its prescribed midcontrol range position forms a desirable and direct indication of the required modification of fuel demand to eliminate spray flow deviations from its prescribed value, which is a fixed percentage of through flow.

The need to use information on spray valve deviation from set point (mid-range) to modify fuel and air demand can be demonstrated in simulation runs where no use of this information is made to modify fuel. In these runs temperature control will be excellent for several minutes until the spray valve runs out of control range and then large deviations follow. This is understandable since in the limit, if perfect control could be achieved by means of spray valve action, the temperature error would be essentially zero and there would be essentially no information to modify fuel through the temperature controller.

It should also be noted in the spray control in Fig. 9.11 that in steady state the temperature after spray is required to return to a prescribed fixed value. This imposes a constraint on the temperature control system which cannot be satisfied if the spray valve is to be reset to a given position. The result will be a slow and sustained oscillation of the throttle temperature as the reset action from spray valve position interacts with the reset action of the temperature controller. This is corrected by feeding back rate of change of temperature after spray with a slow washout rather than temperature after spray.

This same basic philosophy of main-steam temperature control can be applied with the use of temperature difference before and after spray as intelligence in lieu of spray valve position.

PREFERRED CONFIGURATION

The problems of the configuration of Fig. 9.11 are avoided in the arrangement of Fig. 9.13 where a common primary demand signal for feedwater and firing rate is derived from the pressure controls. In Fig. 9-13 we have included in dotted lines a feedforward signal from TDI demand or from TDI, contributing to the demand for feedwater fuel and air.

Some advantages of using this feedforward signal result from the reduction in the amount of reset action needed from the pressure controls making it somewhat easier to obtain stability for these controls.

The reduction of integral action and feedback gains in the pressure controls for the case where proportional feedforward from TDI is used makes these controls less responsive to other upsets. However, unlike the case of pressure control on coal-fired drum type units where unmeasured upsets frequently occurring in the fuel system can only be corrected through action from pressure error (Ref. 3.1), the reliability of the feedwater flow sub-loop makes similar pressure upsets less significant in the once-through case where pressure control is primarily through feedwater flow.

A feedwater signal from TDI to reheat damper position, also marked in dotted lines on Fig. 9.13, was included in studies of control performance, but was found to be harmful in increasing transient deviations of reheat temperature. This is because of the complex nature of the reheat temperature transient which cannot be minimized by a simple proportional feedforward action on the damper. The damper actually has to go through a down and up and then down motion to minimize the temperature error, and the end point of this trajectory (intelligence from proportional feedforward action) is of little importance in minimizing the peak deviation in temperature.

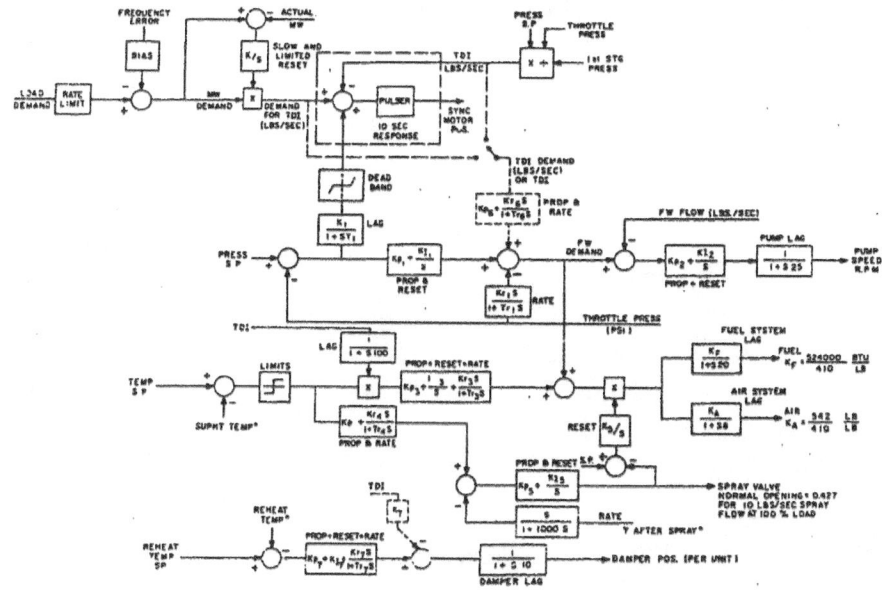

FIGURE 9.13

Fig. 9.14 shows control performance predictions for a ramp increase of 5% min. for 5 min. starting from 36% load point for a once-through subcritical unit.

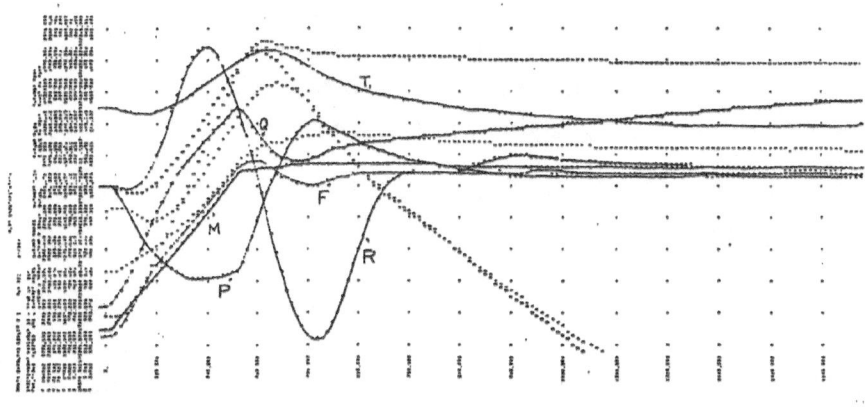

FIGURE 9.14

ANTICIPATORY AND SAFETY FEATURES

For the sake of clarity, we have not included in the diagram of Fig. 9.13 a number of secondary features which are normally part of the control system. These features are generally detailed in drawings outlining the hardware layout.

Anticipatory compensation for feedwater enthalpy changes are provided by automatic calibration of the fuel demand signal. This calibration through a multiplier provides the right compensation independent of the flow rate and can be calculated quite accurately based on the feedwater enthalpy deviation relative to the total enthalpy pickup of the main steam in the boiler.

Other features include load demand runback action from controllers monitoring the error between feedwater and feedwater demand as well as fuel cutback action based on the monitoring of the fuel to feedwater ratio, and fuel to air ratios.

The use of the O_2 measurement has been shown in a conventional arrangement with a controller acting on O_2 deviations from set point to modify the air demand through a multiplier (ratio). Although this is a perfectly acceptable scheme, especially for drum-type boilers where the correct amount of fuel is being continually maintained by pressure control action, we believe that in the case of coal-fired once-through boilers, deviations in O_2 are more likely to be due to fluctuations in the fuel which can persist for a longer time until temperature control correction can act. Hence, it would appear preferable to use O_2 controller action to modify, within limits, the fuel demand rather than the air demand.

STARTUP CONTROLS

The need to have recirculation through the furnace tubes at low flow rates and the process of pressurization to keep fluid below the boiling point and various other requirements of temperature matching during turbine roll place special control requirements for once-through boilers involving manipulation of boiler throttling, boiler extraction, superheater bypass and other valves.

An example of the control requirements will be given based on a supercritical once-through design by Combustion Engineering.

Fig. 9.15 shows a simplified flow diagram of a Combustion Engineering supercritical boiler, including the startup water separator and the valves associated with it. Also, note the condensate demineralizer in the main flow path. This is a full flow demineralizer system, which is necessary for continuous water cleanup. The startup of a supercritical boiler involves the operation of the water separator (sometimes called flash tank) and the startup valves.

The CE combined circulation system utilizes a low through-flow capacity startup system. Water wall protection is attained by the use of the boiler circulating pumps which provide ample flow through the water walls by recirculating during the startup phase. The feedwater flow is generally set at a nominal 5%.

For either hot or cold startups, the boiler is filled and vented up to the BT-BTB (boiler throttling and throttling bypass) valve complex. At this time water wall pressure is controlled by boiler extraction valve BE. The water wall pressure set point is gradually increased as a function of water wall temperature from an initial 1000 psig to 3500 psig. The firing rate is adjusted to limit the furnace exit gas temperature to 1000°F with the additional limitation that the water wall outlet temperature rate of change does not exceed 400° per hour.

As the water wall outlet temperature increases, steam will start to flash in the startup water separator. This steam is initially discharged through the SP (spillover) valve into the condenser. At the discretion of the operator, the steam is emitted to the superheater through the steam admission SA valve. This steam cools the superheater sections and warms the main steam lines. The steam and water mixture exists through valve S.D. (steam drain). By proper manipulation of SP, SD, and SA valves, a close match of steam temperature to turbine metal temperature can be obtained and the turbine can be rolled when proper pressure and temperature conditions have been achieved. Desired steam pressure for rolling the turbine may be obtained by adjusting the startup separator pressure set point which modulates the SP valve as a simple back pressure control. The water drain valve WD functions as a level control valve for the water separator.

When the turbine has attained full speed and is uniformly heated, the feedwater flow may be increased to 10% to provide for adequate steam flow for synchronizing and initial loading of the turbine generator. After synchronizing, SD valve is closed and the firing rate is increased to yield approximately 785° water wall outlet temperature at which point the separator "goes dry", i.e., all fluid passing through the BE valve flashes to steam and is re-admitted to the superheater through the SA valve. When this condition is achieved, the control of water wall pressure is transferred from the BE valve to the boiler throttle bypass valve BTB. The BE valve

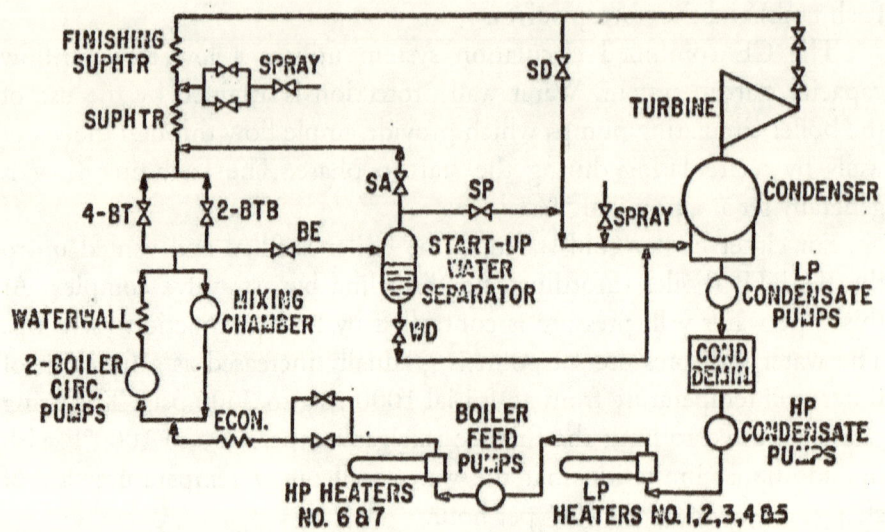

BOILER TURBINE CYCLE CE SUPERCRITICAL

FIGURE 9.15

is then closed and the unit thereby achieves once-through operation. As load on the unit is increased (increased flow) the BTB valves are unable to control the water wall pressure and the BT (boiler throttle) valves open causing the BTB valve to close. There are several BT and BTB valves in parallel. When the BT valves are all open, the BTB vales will again be opened. The BTB and BT valves control the furnace water wall pressure above critical pressure at all times. During this initial loading operation and up to a load index of approximately 30%, the turbine control valves are placed in a fixed position corresponding to this 30% load point. Therefore, during this period, generation is a function of the throttle steam pressure rather than the turbine valve position. Steam pressure reaches approximately 3500 psig at 30% load. After this point, turbine control valves participate in the control of generation.

Startup is considered complete when the throttle steam pressure is at rated condition, turbine control valves are regulating steam flow to the turbine, and all BTB and BT valves are fully open.

Fig. 9.16 illustrates the pressure and temperature changes of a Combustion Engineering supercritical boiler in a warm startup after an 11 hour shutdown. This was a 500 MW supercritical oil fired boiler.

From a description of the startup sequence it can be appreciated that it involves a number of flow and pressure controls operating on various valves. Considerable sequencing logic must be provided.

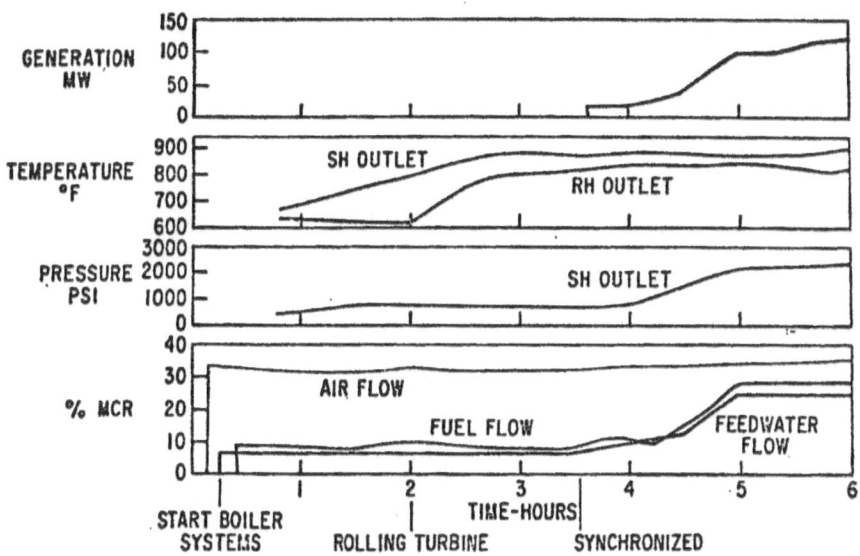

WARM STARTUP AFTER 11-HOUR SHUTDOWN 500-Mw
SUPERCRITICAL UNIT-OIL FIRED

FIGURE 9.16

REFERENCES:

9.1 Ahner, D.J., de Mello, F.P., Dyer, C.E., and Summer, C.V., "Analysis and Design of Controls for a Once-Through Boiler Through Digital Simulation", ISA 9th Power Instrumentation Symposium Proceedings, 1966, pp. 11-30.

9.2	de Mello, F.P., and Paulson, R.E., "Simulation of Plant Dynamics and Design of Plant Control Systems", presented at Southeastern Electric Exchange Meeting, New Orleans, LA, April 14-15, 1966.

9.3	de Mello, F.P., Concordia, C., Kirchmayer, L.K., and Schulz, R.P., "Effect of Prime Mover Response and Governing Characteristics on System Dynamic Performance", American Power Conference, 1966.

9.4	de Mello, F.P., and Ewart, D.N., "MW Response of Fossil-Fueled Steam Units", IEEE Working Group on Power Plant Response to Load Changes, Joint Power Generation Conference, September, 1972.

9.5	CE Coordination Control System for a CE-Sulzer Combined Circulation Steam Generator.

9.6	Price, L.V., Hunter, R.H., and Morse, R.H., "Control System Considerations for the 575 MW Lignite-Fired Big Brown Steam Generators", ISA Power Instrumentation Symposium, May, 1972.

9.7	Durrant, O.W., Loeser, J.K., "Boiler-Turbine Control System for Application to Universal Pressure Boilers", IEEE Conference Paper CP 63-1410 presented at National Power Conference, Cincinnati, Ohio, September 22-26, 1963.

9.8	Argersinger, J.I., Laubli, F., Boegli, E.F., and Scutt, E.D., "Development of an Advanced Control System for Supercritical Pressure Units", IEEE Conference Paper CP63-1409, presented at National Power Conference, Cincinnati, Ohio, September 22-26, 1963.

9.9	Morgan, W.S., and Grimes, A.S., "Load-Frequency Control of Supercritical Units on the A.E.P. System", IEEE Conference Paper 31 CP 66-59, presented at 1966 Winter Power Meeting, New York, January 31-February 4, 1966.

9.10	Adams, J., Clark, D.R., Louis, J.R., Spanbauer, J.P., "Mathematical Modeling of Once-Through Boiler Dynamics", IEEE Transactions, PAS-84, February 1965, p. 146.

CHAPTER X

DIRECT DIGITAL BOILER CONTROL

Technological progress in digital computers has opened up a logical area of application of these devices in direct digital plant controls including boiler control. There are strong technical factors, such as costs per function and their trends, as well as sheer logic and computational capability which favor the increasing use of digital computers in direct control.

Once the inputs to the control and outputs to the actuators are defined, the computation task of defining control action is practically independent of computer hardware. Almost any modern minicomputer system or process computer system has practically unlimited computation capability to accommodate any reasonable control strategy. Before examining methods of applying computers in control, it is well to point to the salient features of digital computers in the context of control, features which make their use increasingly atractive for boiler control.

COMPUTATIONAL AND LOGIC CAPABILITY

Compared with the functions accomplished by analog equipments, it can be said that the digital machine of modern vintage has practically an infinite computational and logic capability. This capability permits implementation of adaptive and nonlinear control and execution of complex logic and computation where required.

LOW INCREMENTAL COST PER FUNCTION

This is particularly true in today's generation computers with their extremely fast cycle time and low memory cost.

COMMUNICATION AND IMPLEMENTATION THROUGH SOFTWARE

This feature is attractive from several standpoints. It makes it easy to change control laws and configurations, a need felt in plant controls. It eliminates the need for transforming mathematically defined control laws into analog circuitry which is unfamiliar to many engineers who would otherwise be interested in control. The medium of communication can be in a language as simple as FORTRAN.

SELF-CHECKING, ALARM AND PRINTOUT CAPABILITY

The logic and computation capability can be used extensively in the areas of reasonableness checking and diagnostics, whereas the same functions would have been prohibitively costly in analog systems.

CALIBRATION, SCALING AND DOCUMENTATION

Adjustment of analog systems is somewhat complicated by the necessity to size resistor and capacitor variables to properly fit within the analog signal range of the hardware. Once adjusted, the process of knowing quantitatively the settings of the control system can be cumbersome, often requiring the measurement of resistor and capacitor values. The digital machine, on the other hand, presents relatively minor scaling problems and offers a complete record of the control law from a printout of the program and control constants. Further, once programmed, the settings are not subject to inadvertent changes such as occur with analog systems that become victims of knob twiddlers.

COMPLEXITY OF THE CONTROL FUNCTION VS. RELIABILITY

Sophistication of a control function, if implemented with analog hardware, invariably involves more components, and the idea is often discarded because complexity can have adverse effects on reliability. In the digital machine, complexity, if needed, is implemented in the software and has little effect on reliability.

ECONOMY OF TIME SHARING

All of the features listed above are in the area of the digital machine having the capability of better, more complex and more easily implemented control, features which are especially important in justifying D.D.C. for plant controls. Others, especially in the chemical industry, have stressed the time-sharing feature of D.D.C. whereby a large number of single loop controllers could be eliminated with one computer, thereby affecting cost savings and reduction in size of control rooms and associated instrumentation. In such applications, the digital machine was not called upon to perform control functions particularly different or more complicated than the analog equipment it replaces. We believe that this aspect of cost reduction and consolidation is secondary in the power plant area and unlike in the chemical industry, the benefits to be derived are mainly in the area of better control with attendant effects on plant reliability and availability.

STRUCTURING THE SYSTEM

The use of a digital computer in control must invariably require interfacing with analog equipment. Any complete system will be a combination of digital and analog components, and evidently an infinite number of such combinations can exist. In arriving at logical recommendations for structuring the combination, one must take a close look at various aspects of the control task in question. These aspects fall into the following categories:

1. Pertinent control concepts and their application
2. Interface considerations
3. Reliability and redundancy considerations

CONTROL CONCEPTS

A review of some control concepts as applied in the boiler control area is helpful in visualizing the role that the digital computer can best play. Reference is made to the discussion of control concepts in Chapter II, where the use of proportional feedforward control, feedback and cascade control was presented.

There was an intermediate stage in the industry where, because of reliability considerations, a type of hybrid analog digital control structure was suggested. The structure was advocated with the intent that, in the event of computer failure, an underlying minimal structure of proportional analog control would remain in operation providing some adequate degree of control.

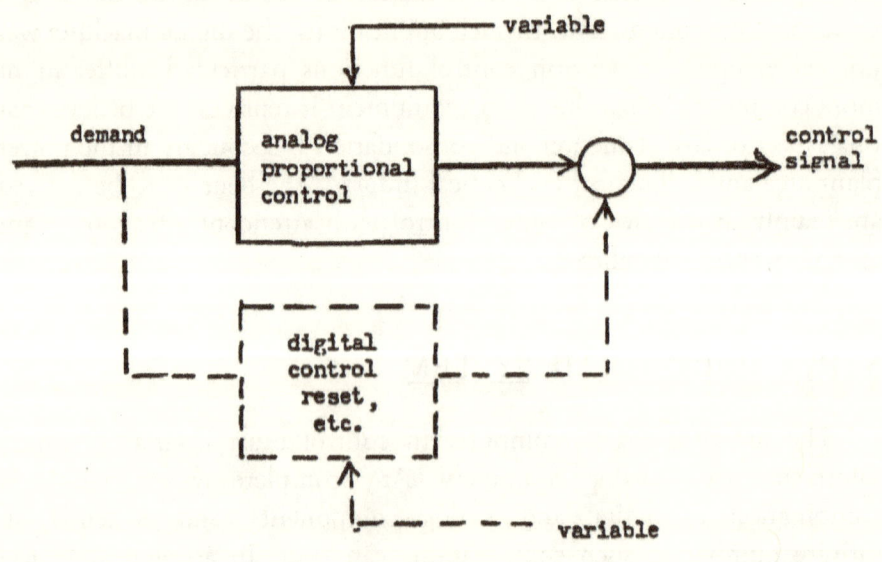

FIGURE 11.1

This concept has not gained favor because of the complexity of interfacing dual control actions (analog and digital) and the inadequacy of proportional control as described in Chapter II.

The principle of cascade control is widely used in boiler controls. Some limited degree of cross coupling between loops also does occur. In thinking through another division in roles between analog and digital control equipment, it is natural to consider the possibility of having the higher level (usually slower) of the cascade control as digital with the lower level or subloop level as analog.

In the early days of computers with slow scanning rates and long cycle times, there was some merit to having fast subloops kept as analog. The scanning rates and compute times in today's generation of computers are so fast that these two items are no longer a consideration as far as boiler controls are concerned. We find that one of the main areas of application of the superior logic and computational capability of the digital computer is at the subloop level where a great many problems of nonlinearity occur.

An example of desirability of preserving D.D.C. down to the subloop level is given in Fig. 11.2.

The block diagram in Fig. 11.2 shows the flow subloop for a once-through boiler. The flow demand signal generated by another cascade loop is cleared through a low limiter that prevents the demand signal from dropping below 30%. A controller (pure fast reset in this case) looks at the error between demand and flow feedback and develops a signal to the speed governor of the boiler feed pump turbine through a function generator.

On Fig. 11.2 are shown the very nonlinear relationships between demand for pump speed and pump flow. If these relationships were fixed, some reasonable compensation could be accomplished with an analog function generator. However, due to the fact that the supply pressure to the feed pump turbine can vary significantly (steam source, either an auxiliary boiler or an extraction from the turbine crossover pipe), the fixed function generator can at best be a mediocre compromise.

Another problem in this flow loop is the requirement to prevent the flow dropping below minimum. An alarm and time delay trip is provided to protect the unit from operating at flow rates below minimum. In the configuration of Fig. 11.2 there is no intelligence to the controller to indicate the rate at which the demand signal is approaching the minimum and the particular amount by which the flow is away from minimum. The combination of these two items of information could be used to advantage with some nonlinear logic to slow down control action in anticipation of the demand signal hitting the low limit, thereby preventing an undershoot in flow.

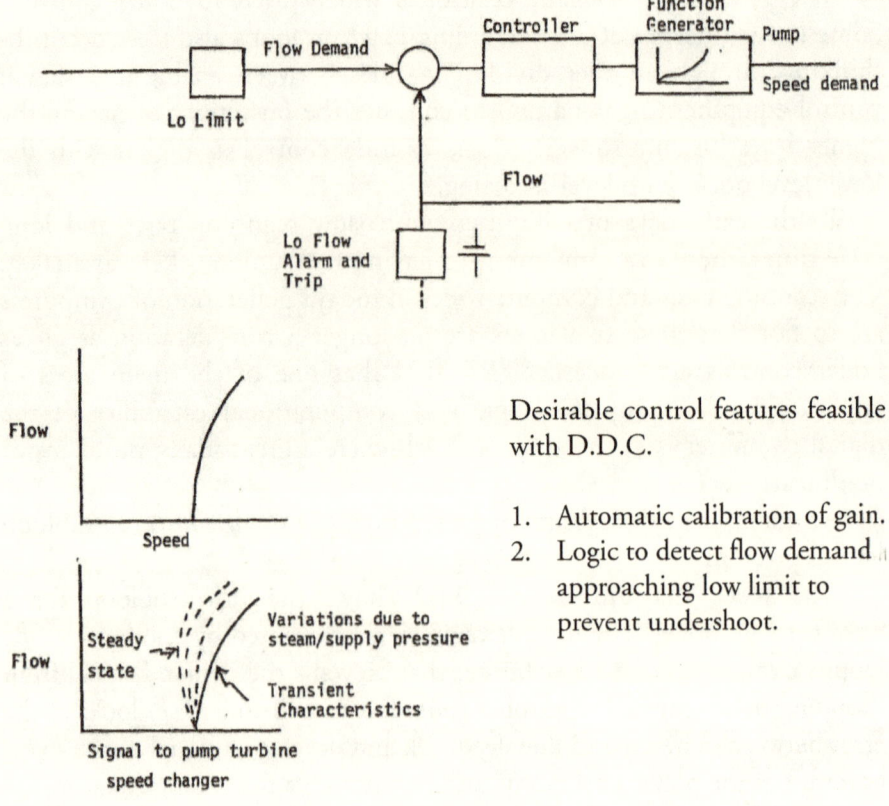

Desirable control features feasible with D.D.C.

1. Automatic calibration of gain.
2. Logic to detect flow demand approaching low limit to prevent undershoot.

FIGURE 11.2

Both the problem of adaptive loop gain compensation from continuous monitoring of the process characteristics as well as the logic required for smooth control of flow in and out of the minimum flow region are natural applications of D.D.C. The complexity of the analog configuration required to do the equivalent makes it impractical to attempt these desirable functions with analog hardware.

Summarizing from this review of pertinent control concepts that are applicable in the area of boiler controls, we conclude that:

1. The most reasonable and straightforward manner of deriving the potential benefits of D.D.C. is to implement down to and including the subloop level.
2. There is no sound basis for advocating a hybrid system where control modes are mixed, some being done digitally and others, notably the so-called proportional mode, being done analog.

3. In the days where computer costs were high, the use of a minimal analog control structure designed to yield acceptable control in the event of computer outage, was advocated. CPU and memory costs today make it natural to provide redundancy with duplicate computer equipment rather than with analog hardware. Fig. 11.3 shows a structure of D.D.C. with limited analog backup.

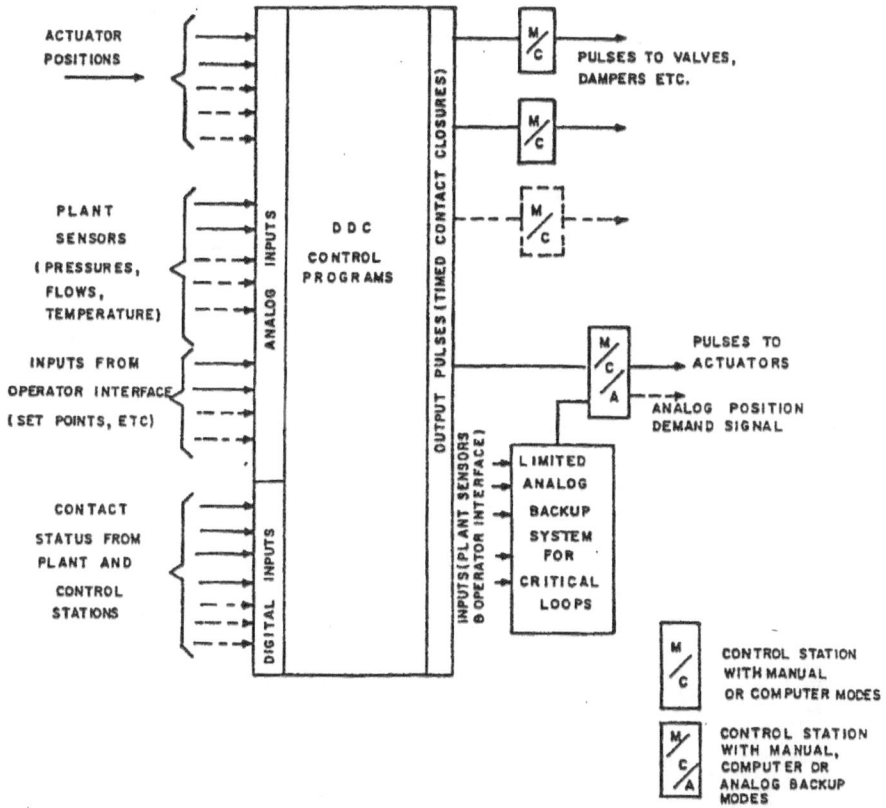

STRUCTURE OF BOILER TURBINE CONTROL SYSTEM

FIGURE 11.3

INTERFACE CONSIDERATIONS

The basic elements of an analog boiler control system are shown in Fig. 11.4(a). The process inputs; the system cabinet with all computation and logic modules; the operator control panels with indicator recorders,

set point stations master manual/auto stations and final element manual/
auto stations; and finally the actuators themselves.

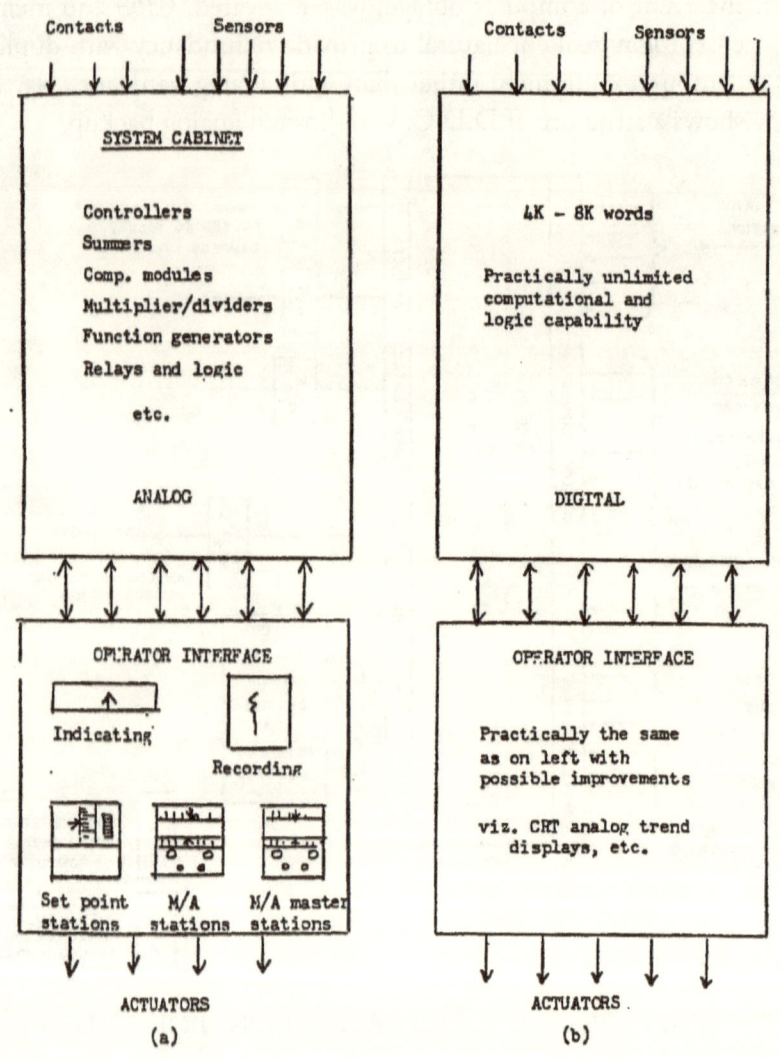

FIGURE 11.4

   Disregarding reliability and redundancy considerations, one can easily
visualize a D.D.C. system, as shown in Fig. 11.4(b), where the prime area
of replacement of analog functions are those performed in the system
cabinet. It is in this area that the logic and computational capability of a
minicomputer can be classed as orders of magnitude larger than the most
complex analog configuration imaginable.

Preservation of the operator panel is essential, and we visualize an evolutionary upgrading of information communication links with more use of CRT displays. Basically, the difference between systems of Fig. 11.4(a) and 11.4(b) can be recognized as replacement of analog electronics in the system cabinet by digital electronics. Significant differences between these configurations lie in the area of reliability and will be discussed in the next section.

An important interface is that between computer and the actuators. The importance of bumpless transfer between manual and computer control has led to the almost universal acceptance of D.D.C. as being outputed with pulses to an integrate-and-hold device external to the computer.

Two basic types are in common use, one (Fig. 11.5(a)) where the integrate and hold device is an analog holding amplifier, such as a self-synchronizing manual/auto station. The computer can modify the output of these analog signals by imparting raise or lower pulses of varying magnitudes and/or durations. The analog signal in turn controls an actuator either through an I/P converter and pneumatic positioning system, or through an electrical actuator and its closed loop positioning system.

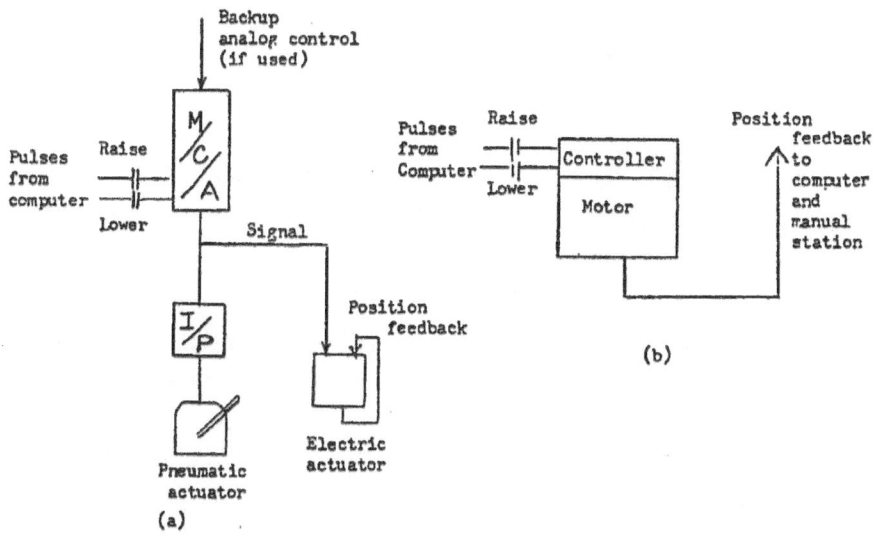

FIGURE 11.5

When electrical actuators are used, a strong argument exists in favor of using the electric motor as the memory device, and introducing D.D.C. pulses, Fig. 11.5(b) directly to the motor controller. This eliminates some intermediate links and may have possible reliability advantages. However, from a program implementation point of view, all these schemes are very similar.

RELIABILITY AND REDUNDANCY

In discussing aspects of reliability, it is well to review the degree of reliability that customers are used to with present analog systems.

Fig. 11.6 describes the nature of typical analog controls. They are made up of several subsystems which can be operated separately on partial automatic or totally with all of them under automatic integrated cascade control. The point is that total control, although desirable, is not mandatory all the time, and it is commonplace to operate under partial control modes wherein, due to some plant equipment or control malfunction, sections or subsystems are under manual control or partial control.

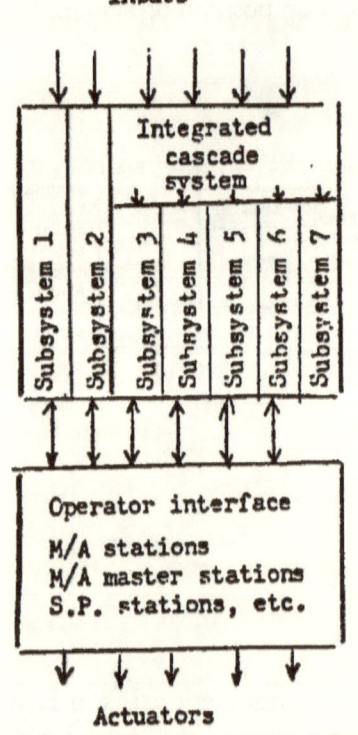

Analog

Component failure permits
operation or partial control   +

Failure analysis too complex and
malfunction possibilities numerous   –

Digital

Possibility of complete outage
leaving all controls on manual   –

Capability of detecting false
inputs by reasonableness checks   +

System design planned
for complete outage of computer   +

FIGURE 11.6

In regard to the reliability of such systems, we can make the following observations. In favor of the analog approach, we conclude that failure of some component permits operation in various degrees of partial control, whereas for the digital counterpart, failure of the computer means all controls on manual except insofar as provided for by redundant backup controls. As stated before, with the present state of the art redundancy can be provided at relatively small cost with complete duplicate computer systems.

In the case of the analog system, one must emphasize that there are an infinite number of modes of failure possible. Failure analysis component by component is almost impossible. To avoid major upsets due to failure of critical sensors, some limited redundancy is sometimes provided and another line of defense is available in such safeguards as fuel-air cutback controllers, deviation monitors, etc. By and large, reliance is placed on the operator to detect malfunctions and avoid trips and/or damage to equipment.

In aspects of self checking and diagnostics, the digital computer has a tremendous capability. For instance, it can check reasonableness of input information in a number of ways. Use of this self checking capability coupled with a system design which is planned for complete digital computer outage with provision of limited analog backup or complete throwover to a redundant computer gives every promise of a superior system both from a performance as well as reliability standpoint.

A valid question raised in the design of the initial few D.D.C. installations, concerned the desirability of a separate computer for control versus one giant plant information processing and control computer. In the early days of plant computer technology, the most economical way of proving the feasibility and gaining experience with D.D.C. was to try it as an incremental function on a large plant computer system. Today it is evident that the function is best done on a small separate dedicated computer. The reasons are:

1. Functional decentralization, much as administrative decentralization is necessary to get the job done in reasonable time and with reasonable effort. There are points beyond which the increased complexity and cost of centralizing and cooordinating a large number of functions far outweigh savings from sharing of common hardware. The overhead monitor

programs necessary for D.D.C. could be greatly reduced in a smaller computer dedicated to control only. The economy of scale has disappeared in today's generation of computers.

2. Reliability considerations favor the separate small computer from the standpoint of decreased numbers of components and decreased exposure to handling by personnel concerned with other functions, such as monitoring, performance calculations, etc. The cost of providing complete redundancy is small relative to that of doing the same on a large monitoring and results computer.

3. Acceptance of digital control is easier when it is recognized that it involves little more than the replacement of a box of electronics (special purpose analog computer) by another box of electronics (digital computer) to functionally do the same thing, although better. Added functions could be economically incorporated in this computer provided they fall in the category of control and dynamics, such as turbine startup controls, turbine stress calculations, etc.

DIGITAL CONTROL ALGORITHMS

Implementation of control laws with a digital computer can be done in an infinite number of ways. Many of these give essentially equivalent performance, and one should be careful not to spend a career analyzing trivial or marginal differences between algorithms. One of the great advantages of the digital computer is the flexibility with which all types of control and logic calculations can be incorporated to suit the particular need. Such different control modes as the use of limits and deadbands that may be fixed or functions of some variable, plus switching on or off of integral or other control action depending on input conditions are purely a function of the process requirements and the ingenuity of the control engineer.

Control knowhow is nevertheless well rooted in analog practices, and it is logical to evolve from these practices developing digital equivalents to well known analog control modes. For reference purposes some frequently used algorithms and some terms that are becoming commonly accepted are described below.

CONTROL MODES

Two most commonly used forms of three mode control are shown in Fig. 11.7(a) and (b). The parallel configuration of Fig. 11.7(a) has noninteracting control modes in the time domain, i.e., proportional integral or derivative gains can be individually set without interacting with each other.

When the overall control is expressed as one transfer function made up of zeros and poles, we note that the adjustment of individual terms of the time-domain function results in changes of the poles and zeros of the control function. For this reason, this form is known as interacting in the frequency domain. One should note that the parallel form of Fig. 11.7(a), when expressed into series form, can give rise to complex zeros which at times may be advantageous.

The form of Fig. 11.7(b). called the series form, is that normally encountered with analog 3 mode controllers using one operational amplifier. The form of the transfer function for a type 3 mode controller is

$$\frac{K_1(1+T_i s)(1+T_D s)}{s\left(1+\dfrac{T_D}{10}s\right)}$$

and, the adjustments normally available are noninteracting in the frequency domain, i.e., proportional band $\dfrac{100}{K_P}$ adjusts the gain of the overall function. Reset time T_i is generally expressed as repeats per min $\dfrac{60}{T_i}$ and can be adjusted without affecting the other parameters of the transfer function, and likewise rate time T_D can be adjusted independent of the other parameters. Note that in this form there is no possibility of complex zeros.

The considerable amount of frequency response thinking, guide rules, etc., that have evolved over the years make it desirable to have control algorithms that are noninteracting in the frequency domain. In other words, the control engineer is more at home with locating such things as lead breaks (T_D and T_I) or corner frequencies $\left(\dfrac{1}{T_D}\right)$, $\dfrac{1}{T_I}$ and gain

K_p as independent parameters than to work with the basic proportional reset and rate gains of the parallel system. As more control experience and guide rules are developed in the time domain, these preferences may change. Digital techniques permit, of course, a wide range of control laws which can be implemented with algorithms and algebraic Z form manipulations, as explained in many texts and in Ref. (11.1).

Some of the variations that will find application are control modes written such that derivative and/or proportional action is bypassed on set point changes, provision for limiting the change in output per computation step and provision for automatic resetting of accumulators to prevent integral windup.

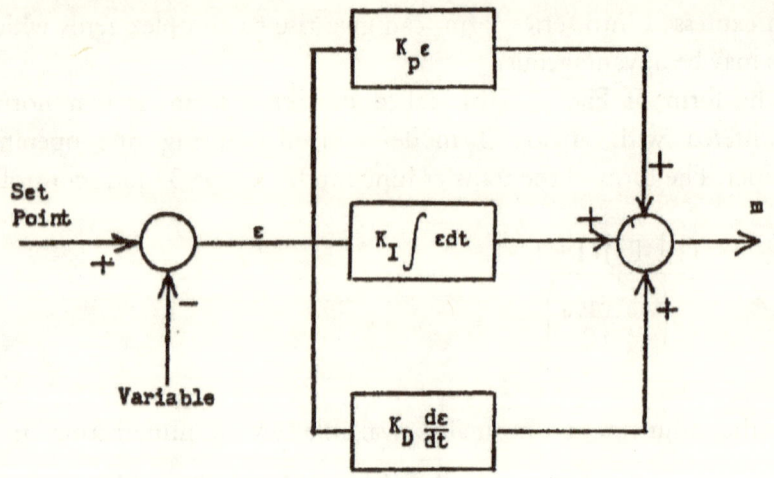

Parallel Configuration of 3-mode Controller

(a)

Series Configuration of 3-Mode Controller

(b)

FIGURE 11.7

POSITION ALGORITHMS

The position algorithm is usually used internally within the computer to form demand signals in cascade loops. It is given as

$$P_n = K_p e + K_D \frac{\Delta e}{\Delta t} + K_I \sum_0^n e \Delta t + P_m$$

where e = error signal and P_m is the initial value. P_n is the actual position or demand out of a 3 mode digital controller.

The formation of $\frac{\Delta e}{\Delta t}$ is usually done using some filtering techniques since differentiation is a noisy process. If a plain backward difference computation between present and last previous sample is not satisfactory from a noise standpoint, one way of filtering over more samples can be attempted as follows:

Let e_n, e_{n-1}, e_{n-2}, e_{n-3} be the last 4 samples of error. Their average value would be

$$e^* = \frac{e_n + e_{n-1} + e_{n-2} + e_{n-3}}{4}$$

and

$$\frac{\Delta e}{\Delta t} = \frac{\left[\dfrac{e_n - e^*}{1.5\Delta t} + \dfrac{e_{n-1} - e^*}{0.5\Delta t} + \dfrac{e^* - e_{n-2}}{0.5\Delta t} + \dfrac{e^* - e_{n-3}}{1.5\Delta t} \right]}{4}$$

$$= \frac{1}{6\Delta t} \left[e_n - e_{n-3} + 3e_{n-1} - 3e_{n-2} \right]$$

VELOCITY ALGORITHMS

In outputting to the plant, the velocity mode algorithm is used. Here the integration is done external to the computer, and the control algorithm must determine merely a change in the desired control signal. The output function is obtained by differencing the position algorithm.

Some interesting variations of this algorithm can be used employing nonlinear logic to improve the response and stability of control.

For instance, in forming the control effort at any given sample time

$$Q_n = K_p(e_n - e_{n-1}) + K_I e_n \Delta t$$

we can note the sign of the proportional and integral terms. When the variable is moving away from set point, both terms have the same sign, whereas when it is returning to set point, they have opposite signs. This property can be used to improve control stability and response. For instance, when the variable is some distance away from set point, logic can be set up to only accept proportional term values that have the same sign as the integral term, whereas when the error falls within a prescribed band, proportional terms are included regardless of their sign and their gain can be increased.

This discussion can by no means be comprehensive. It is included here to merely indicate the wide scope of possible control actions that can be implemented. New schemes and variations of old ones will be applied as needed in the particular application.

D.D.C. COMPUTATIONS IN BOILER CONTROL

Controller algorithms are only a portion of the control computation task. It is convenient to package the software with standard subroutines describing certain controller functions. However, a great deal of computation will use standard algebraic and logic operations available in FORTRAN.

The basic control computation cycle will consist of:

1. Scanning of inputs and storing the latest sample in memory.
2. Reasonableness checks on input data.
3. Control calculations to establish control action.
4. Outputting of control action every control cycle which may or may not be coincident with the scanning cycle.

Control calculations will involve a number of limiting, runback and logic functions. They can generally be described in block diagram form as indicated in Figs. 11.8 to 11.13 which apply to the control computations for a once-through boiler already described in Chapter IX.

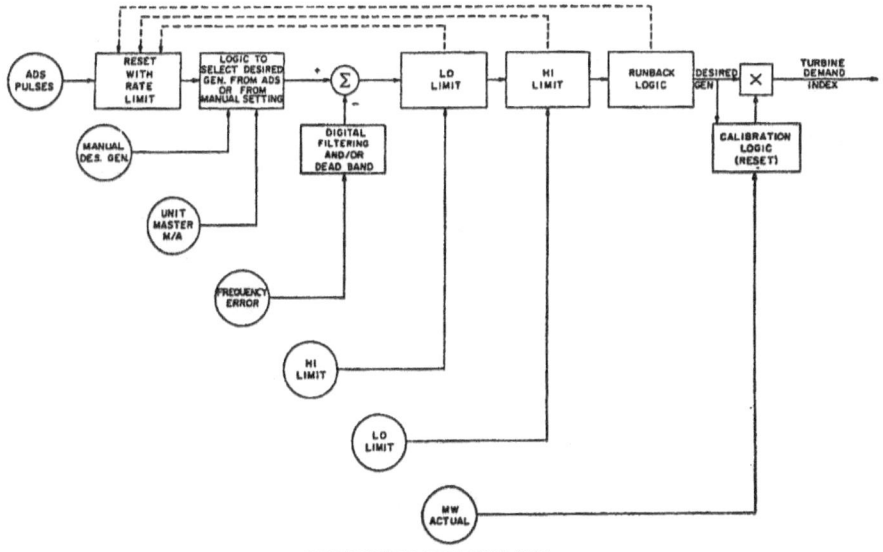

BOILER-TURBINE DDC—MASTER DEMAND SIGNAL

FIGURE 11.8

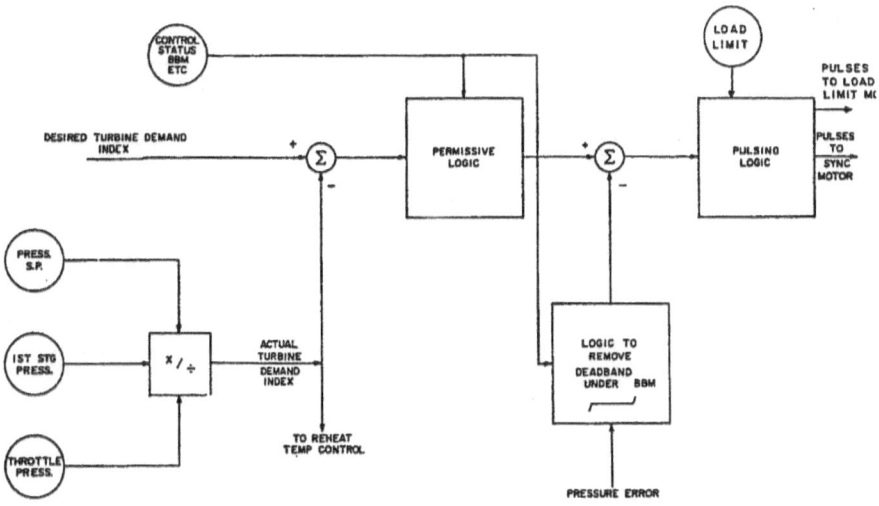

BOILER-TURBINE D. D. C.—TURBINE CONTROL

FIGURE 11.9

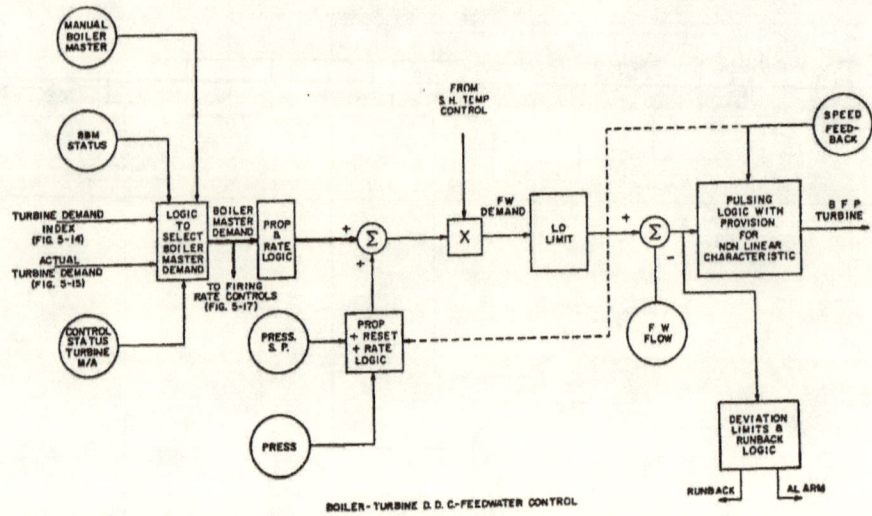

FIGURE 11.10

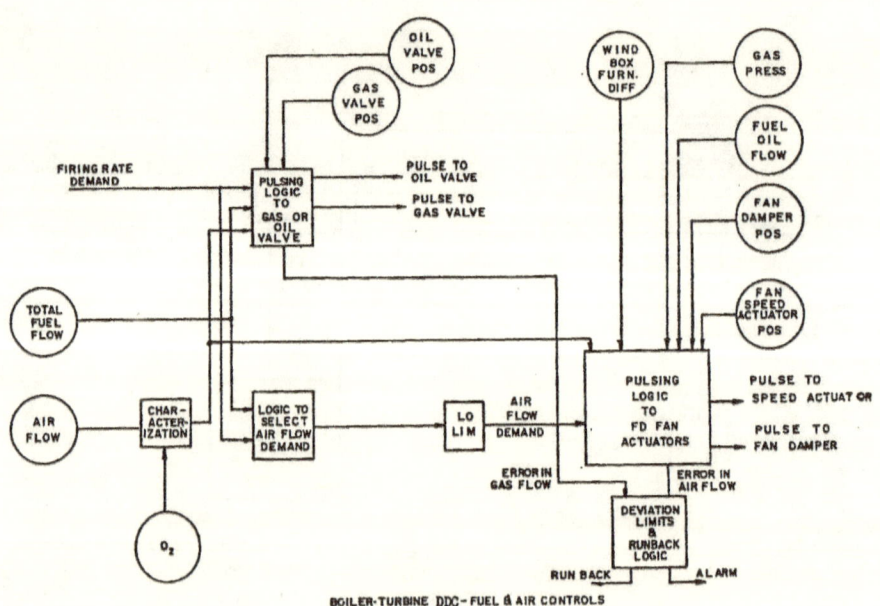

FIGURE 11.11

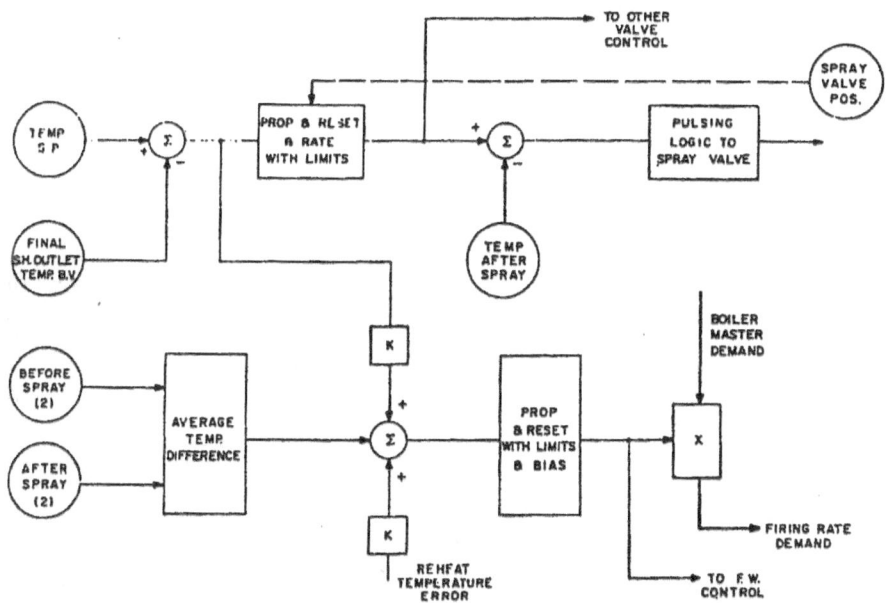

BOILER-TURBINE D.D.C.-SUPERHEAT TEMPERATURE CONTROL

FIGURE 11.12

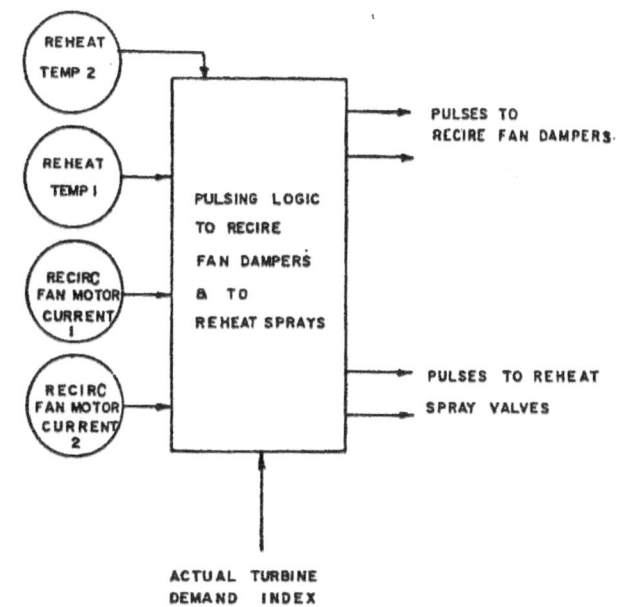

BOILER TURBINE D.D.C.—REHEAT TEMPERATURE CONTROL

FIGURE 11.13

REFERENCES

11.1 Uram, Robert, and Giras, Theodore, "Computer Control Philosophy for a Steam Generator", Instrumentation in the Power Industry Vol. 13, 1970, ISA.

11.2 Mills, R.J., "Tuning Digital Controllers From Time Domain Test Data", ISA Power Instrumentation Symposium, 1972.

11.3 de Mello, F.P., "Z Transform Techniques for the Analysis of Systems", AIEE Conference paper CP 60-1291, Fall General Meeting, Chicago, IL, September, 1960.

11.4 Hill, D.E., Kronstain, L.J., "Versatile Operator Oriented D.D.C. System", ISA paper 73-515, International Conference & Exhibit, October 15-18, 1973, Houston, TX.

11.5 Davis, G.E., Jr., "Hybrid Digital-Analog Power Plant Control", ISA Instrumentation in the Power Industry, Vol. 13, 1970, pp. 51-61.

11.6 Strohmeyer, C., & Girais, T.C., "Combined Digital-Analog Control Approach for Sibley Station Unit No. 3", ASME-IEEE, Joint Power Generation Conference, September 24-28, 1967.

11.7 Clark, D.R., Bilski, R.S., "Power Plant D.D.C.—Where Has It Been And Where Is It Going", ISA Paper 75-537 Industry Oriented Conference and Exhibit, Milwaukee, WI., October, 1975.

11.8 Rocca, J.V., "User's Viewpoint—Automation of a Power Plant", paper presented at Foxboro Company Power Systems Seminar, Denver, Colorado, October, 1975.

CHAPTER XI

MODELING FROM FIRST PRINCIPLES

The basic process dynamics of heat transfer and fluid flow is described by fundamental relations of mass balance, energy balance pressure drop, heat transfer, heat storage as well as fluid properties (steam tables for water and steam, and gas laws for the air and combustion products).

These relations are nonlinear algebraic equations and nonlinear partial differential equations.

For the process of fuid flow in a heated tube, the equations of conditions along the spacial dimension X and as a function of time t are:

Mass Balance

$$\frac{\partial \dot{m}}{\partial X} = v \frac{\partial e}{\partial t}$$

Energy Balance

$$-\frac{\partial (\dot{m}h)}{\partial X} = v \frac{\partial}{\partial t}\left[\rho h - \frac{P}{J}\right] = \dot{q}$$

Pressure Drop (neglecting inertia effects)

$$\frac{\partial P}{\partial X} = f \frac{(\dot{m})|\dot{m}|}{\rho}$$

Heat Transfer From Metal to Fluid

$$\dot{q} = K(\dot{m})^{0.8}(T_m - T)$$

Metal Heat Storage

$$T_m = \frac{1}{MC}\int(\dot{q}_g - \dot{q})dt$$

Fluid Properties (Steam Tables)

$$P = f_1(P, h)$$

$$T = f_2(P, h)$$

where

m	=	flow rate at (x, t)
$\dot{q}$	=	rate of heat input to fluid at (x, t)
v	=	volume per unit length
p	=	pressure at (x, t)
h	=	enthalpy at (x, t)
ρ	=	density at (x, t)
T	=	temperature at (x, t)
T_m	=	metal temperature at (x, t)
f	=	friction coefficient
MC	=	metal thermal capacitance per unit length

One essential step for the solution of these partial differential equations is to convert the continuous space dimension into discrete steps. Finite volume elements are visualized and the equations are used to define inlet and outlet conditions in these finite volume elements.

The other step which was virtually mandatory before the advent of digital computation was to convert the nonlinear equations into linear small perturbation equations about an operating point. Although this linearization approach is still very useful, the full nonlinear approach is quite feasible using digital computers and digital techniques for the solution of nonlinear differential equations.

Examples of modeling techniques will be given to show some of the basic effects of importance.

DRUM BOILER MODELS

Drum type boilers lend themselves better to linearized treatment, because steam conditions stay within fairly limited ranges in given heating sections. Saturation is pegged at the drum which forms a convenient point separating the process between subcooled and boiling regions, and superheated parts. Fig. 12.1 shows the general division of groups of equations which adequately represent the pressure, temperature and drum level phenomena. The coupling between groups of equations is shown, as are also the basic input and output variables.

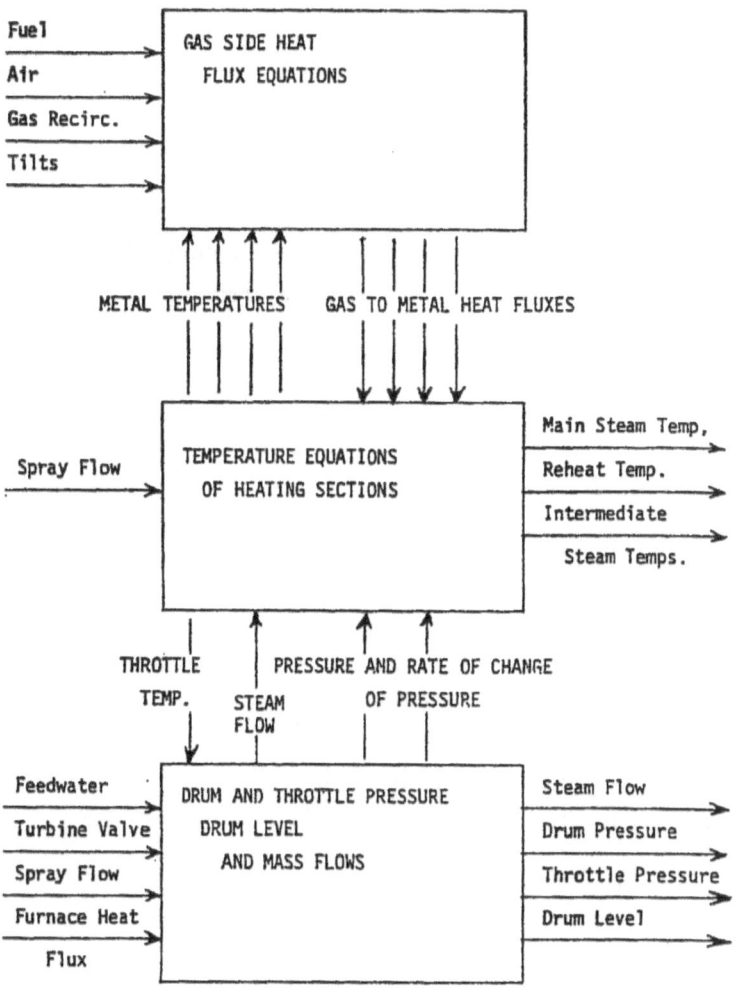

FIGURE 12.1

PRESSURE, FLOWS AND LEVEL MODEL

Fig. 12.2 shows the physical system considered in the development of equations to solve for pressure, flows and level.

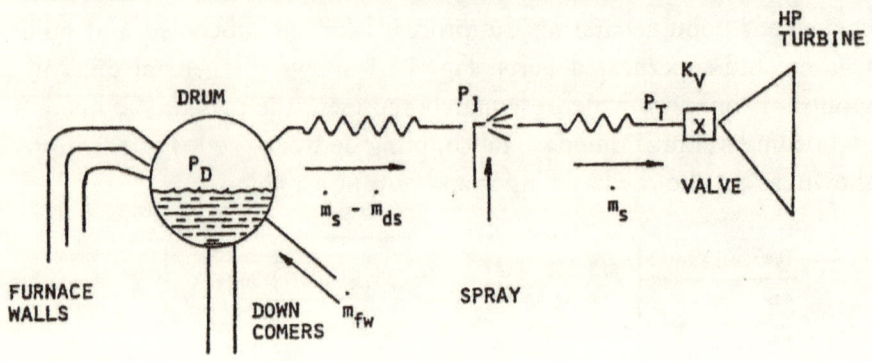

FIGURE 12.2

The boiler storage, including superheater volume effects, is lumped at the drum, while the superheater pressure drop effects are simulated as equivalent orifices.

The pressure mass flow equations are:

$$P_D - P_I = f_1(m_s - m_{ds})^2 \tag{12.1}$$

$$P_I - P_T = f_2(m_s)^2 \tag{12.2}$$

$$\dot{m}_s = \frac{K_V P_T}{\sqrt{T_T + 460}} \tag{12.3}$$

Linearizing equations 12.1 to 12.3, and solving for the dependent variables ΔP_T, ΔP_I, and Δm_s as function of ΔK_V, Δm_{ds} and ΔP_D

$$\Delta P_T = -\left[\frac{P_{To}(K_1 K_2)}{D}\right]\Delta K_V + \left[\frac{1}{D}\right]\Delta P_D + \left[\frac{K_1}{D}\right]\Delta \dot{m}_{ds} + \left[\frac{K_4(K_1 + K_2)}{D}\right]\Delta T_T \tag{12.4}$$

$$\Delta P_I = -\left[\frac{P_{To}K_1}{D}\right]\Delta K_V + \left[\frac{1+K_2K_2}{D}\right]\Delta P_P + \left[\frac{K_1(1+K_2K_3)}{D}\right]\Delta\dot{m}_{ds} + \left[\frac{K_1K_2}{D}\right]\Delta T_T \quad (12.5)$$

$$\Delta\dot{m}_S = \left[\frac{P_{To}}{D}\right]\Delta K_V + \left[\frac{K_3}{D}\right]\Delta P_P + \left[\frac{K_1K_3}{D}\right]\Delta\dot{m}_{ds} - \left[\frac{K_4}{D}\right]\Delta T_T \quad (12.6)$$

where subscript "o" denotes steady state operating values, and

$$K_1 = \frac{2(P_{Do} - P_{To})}{(\dot{m}_{so} - \dot{m}_{dso})}$$

$$K_2 = \frac{2(P_{To} - P_{To})}{\dot{m}_{so}}$$

$$K_3 = \frac{\dot{m}_{so}}{P_{To}}$$

$$K_4 = \frac{\dot{m}_{so}}{2(T_{To} + 460)}$$

$$D = 1 + K_1K_3 + K_2K_3$$

SPRAY FLOW EQUATIONS

Spray flow is proportional to spray valve flow area K_S and the square root of pressure drop across the valve, i.e.,

$$\dot{m}_{ds} = K_S\sqrt{(P_e - P_I)} \quad (12.7)$$

where P_e = economizer pressure = $P_D + f_e(m_{fw})^2$ (12.8)

and "fe" is the economizer pressure drop coefficient.

Linearizing 12.7 and 12.8

$$\Delta \dot{m}_{ds} = K_{So} \frac{1}{2} \left(P_{eo} - P_{Io} \right) \left(\Delta P_e - \Delta P_I \right)$$

$$\Delta \dot{m}_{ds} = \frac{K_{So}}{2} \left(P_{eo} - P_{Io} \right) \left(\Delta P_D + 2 \text{ fo } \dot{m}_{fwo} \Delta \dot{m}_{fw} - \Delta P_I \right) \qquad (12.9)$$

BOILER PRESSURE EFFECTS

Fig. 12.3 shows the process of steam generation with all quantities defined.

Equations can be written for the down-comers, water walls, and drum as follows:

Down-comers

The flow rate is considered constant. The inlet enthalpy is obtained from a mixing equation of feedwater flow with saturated liquid from the drum, since feedwater is assumed to be directed to the down-comers.

$$m_r h_r = (m_r - m_{fw}) hf + m_{fw} h_{fw}$$

i.e., $$h_r = \frac{(\dot{m}_r - \dot{m}_{fw}) h_f + \dot{m}_{fw} h_{fw}}{\dot{m}_r} \qquad (12.10)$$

The enthalpy of the flow entering the water walls h_r' is h_r delayed by the transport time through the down-comers.

i.e., $$h_r' = h_r \varepsilon^{-ST} \qquad (12.11)$$

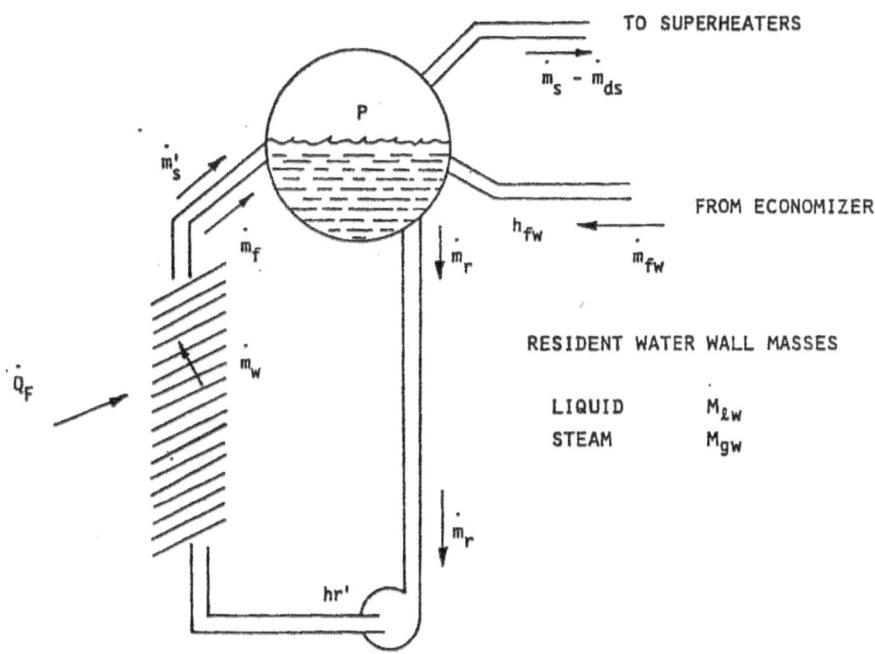

$\dot{Q}_F$ = HEAT FLUX FROM FURNACE TO WATER WALLS

$\dot{Q}_W$ = HEAT FLUX FROM METAL TO FLUID IN WATER WALLS

m_{fw} = FEEDWATER FLOW

m_r = DOWN COMER FLOW

m_w = STEAM GENERATION IN WATER WALLS

m_f = SATURATED, LIQUID FLOW OUT OF WATER WALLS INTO DRUM

m'_s = STEAM FLOW OUT OF WATER WALLS

m'_s-m_{ds} = STEAM FLOW OUT OF DRUM INTO SUPER HEATERS

FIGURE 12.3

where τ = transport time = $\dfrac{V_{DC}\rho_L}{\dot{m}_r}$

V_{DC} = Volume of down-comers

ρ_L = Density of water

Water Walls and Drum

Fig. 12.4 defines the water wall and drum subsystem.

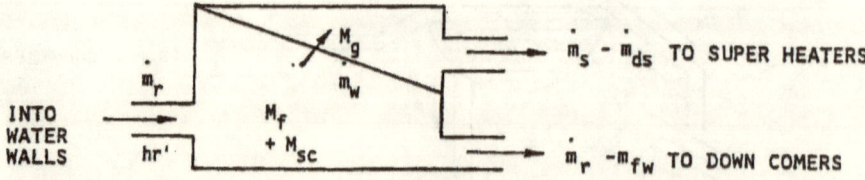

INTO
WATER
WALLS

$\dot{m}_r$

hr'

M_g

$\dot{m}_w$

M_f
+ M_{sc}

$\dot{m}_s - \dot{m}_{ds}$ TO SUPER HEATERS

$\dot{m}_r - \dot{m}_{fw}$ TO DOWN COMERS

FIGURE 12.4

Conceptually the process is described by inlet flow of subcooled fluid $\dot{m}_r$ at enthalpy h_r' and exit flows of saturated steam m_s. The resident mass of subcooled liquid is denoted M_{sc}, of saturated liquid M_f and of steam as M_g. m_w is the steam generation rate. Equations for this process are:

Mass Balance

$$\dot{m}_{fw} - (\dot{m}_s - \dot{m}_{ds}) = \frac{d\,Mf}{dt} + \frac{d}{dt}M_g \qquad (12.12)$$

Volume Balance

$$d/_{dt}(M_g v_g) + d/_{dt}(M_f v_f) = 0 \qquad (12.13)$$

Energy Balance

$$\dot{Q} + \dot{m}_r h_r' - (\dot{m}_r - \dot{m}_{fw})h_f - (\dot{m}_s - \dot{m}_{ds})h_g \frac{d}{dt}M_g h_g + M_f h_f + (M_{SC})\left(\frac{h_f + h_r'}{2}\right) - 144\,V\frac{P}{J} \qquad (12.14)$$

where M_g is resident mass of steam
        M_f is resident mass of saturated liquid
        M_{sc} is resident mass of subcooled liquid
        V is volume of water walls and drum

Equations 12.12, 12.13, and 12.14 can be expressed in terms of the variables (rates of change) $d\,M_f/_{dt}$, $d\,M_g/_{dt}$ and $dP/_{dt}$ as:

$$\frac{dM_f}{dt} + dM_g / dt = \dot{m}_{fw} - (\dot{m}_s - \dot{m}_{ds}) \qquad (12.12)$$

$$v_f dM_f / dt + v_f dM_g / dt + \left[M_g \frac{dv_g}{dP} + M_f \frac{dv_f}{dP} \right] \frac{dP}{dt} = 0 \qquad (12.15)$$

$$h_f \frac{dM_f}{dt} + h_g \frac{dM_g}{dt} + \left[M_g \frac{dh_g}{dP} + M_f \frac{dh_f}{dP} + \frac{M_{sc}}{2} \frac{dh_f}{dP} + \left(\frac{h_f + h_r'}{2} \right) \frac{dM_{sc}}{dt} - \frac{144\,V}{J} \right] dP / dt$$

$$= \dot{Q} + \dot{m}_r h_r' - (\dot{m}_s - \dot{m}_{ds}) h_g - (\dot{m}_r - \dot{m}_{fw}) h_f \qquad (12.16)$$

Making the justifiable assumption that $\dfrac{h_f + h_r'}{2} = h_f$, and that M_{1sc} $<<$ M_f the mass of subcooled fluid M_{sc} can be lumped with saturated liquid and equation 12.16 rewritten as

$$h_f \frac{dM'_f}{dt} + h_g \frac{dM_g}{dt} + \left[M_g \frac{dh_g}{dP} + M'_f \frac{dh_f}{dP} - \frac{144\,V}{J} \right] \frac{dP}{dt}$$

$$= \dot{Q} - \dot{m}_r (h_f - h_r') + \dot{m}_{fw} h_f - (\dot{m}_s - \dot{m}_{ds}) h_g \qquad (12.17)$$

where $M'_f = M_{sc} + M_f$.

Equations 12.12, 12.15, and 12.17 may be solved for $dP/_{dt}$, $d\,M_f/_{dt}$, and $d\,M_g/_{dt}$

$$\frac{dP}{dt} = \cfrac{\left| \begin{array}{ccc} 1 & 1 & [\dot{m}_{fw} - (\dot{m}_s - \dot{m}_{ds})] \\[2mm] v_f & v_g & 0 \\[2mm] h_f & h_g & [\dot{Q} - \dot{m}_r(h_f - h_r') + \dot{m}_{fw}h_f - (\dot{m}_s - \dot{m}_{ds})h_g] \end{array} \right|}{\left| \begin{array}{ccc} 1 & 1 & 0 \\[2mm] v_f & v_g & \left[M_g\dfrac{dv_g}{dP} + M'_f\dfrac{dv_f}{dP} \right] \\[4mm] h_f & h_g & \left[M_g\dfrac{dh_g}{dP} + M'_f\dfrac{dh_f}{dP} - \dfrac{144V}{J} \right] \end{array} \right|}$$

or $dP/dt = \dfrac{[\dot{Q} - \dot{m}_r(h_f - h_r') + \dot{m}_{fw}h_f - (\dot{m}_s - \dot{m}_{ds})h_g](v_g - v_f) + (\dot{m}_{fw} - \dot{m}_s + \dot{m}_{ds})(v_f h_g - v_g h_f)}{\Delta}$ (12.18)

where $\Delta = (v_g - v_f)\left[M_g\dfrac{dh_g}{dP} + M'_f\dfrac{dh_f}{dP} - \dfrac{144V}{J} \right] - (h_g - h_f)\left[M_g\dfrac{dv_g}{dP} + M'_f\dfrac{dv_f}{dP} \right]$

$$= v_{gf}\left[M_g\frac{dh_g}{dP} + M'_f\frac{dh_f}{dP} - \frac{144V}{J} \right] - h_{gf}\left[M_g\frac{dv_g}{dP} + M'_f\frac{dv_f}{dP} \right]$$

$$= M_g v_{gf}\left(\frac{dh_g}{dP} - \frac{h_{gf}}{v_{gf}}\frac{dv_g}{dP} \right) + M'_f v_{gf}\left(\frac{dh_f}{dP} - \frac{h_{gf}}{v_{gf}}\frac{dv_f}{dP} \right) - v_{gf}\frac{144V}{J}$$

Similarly,

$$\frac{dM'_f}{dt} = \left| \begin{array}{ccc} [\dot{m}_{fw} - (\dot{m}_s - \dot{m}_{ds})] & 1 & 0 \\[4mm] 0 & v_g & \left[M_g\dfrac{dv_g}{dP} + M'\dfrac{dv_f}{dP} \right] \\[4mm] \dot{Q} - \dot{m}_r(h_f - h_r') + \dot{m}_{fw}h_f - (\dot{m}_s - \dot{m}_{ds})h_g & h_g & \left[M_g\dfrac{dh_g}{dP} + M'\dfrac{dh_f}{dP} - \dfrac{144V}{J} \right] \end{array} \right|$$

and

$$\Delta$$

$$\frac{dM_g}{dt} = \begin{vmatrix} 1 & [\dot{m}_{fw} - (\dot{m} - \dot{m}_{ds})] & 0 \\ v_f & 0 & \left[M_g \dfrac{dv_g}{dP} + M' \dfrac{dv_f}{dP} \right] \\ h_f & \dot{Q} - \dot{m}_r(h_f - h'_r) + \dot{m}_{fw} h_f - (\dot{m}_s - \dot{m}_{ds})h_g & \left[M_g \dfrac{dh_g}{dP} + M' \dfrac{dh_g}{dP} - \dfrac{144V}{J} \right] \end{vmatrix}$$

$$\Delta$$

Equation 12.18 reduces to:

$$\frac{dP}{dt} = \frac{v_{gf}}{\Delta} \left[\dot{Q} - (\dot{m}_s - \dot{m}_{ds}) \frac{v_g}{v_{gf}}(h_{gf}) - \dot{m}_r(H_f - h'_r) + \dot{m}_{fw} \frac{v_f}{v_{gf}} h_{gf} \right] \qquad (12.19)$$

Equation 12.19 can be implemented in its complete nonlinear form as described in the computational block diagram of Fig. 12.5.

Note that all of the factors of multiplication in Fig. 12.5 are determined from saturation properties as function of pressure or from the resident masses M_f and M_g which are also solved from integration of equations defining $\dfrac{dM_f}{dt}$ and $\dfrac{dM_g}{dt}$.

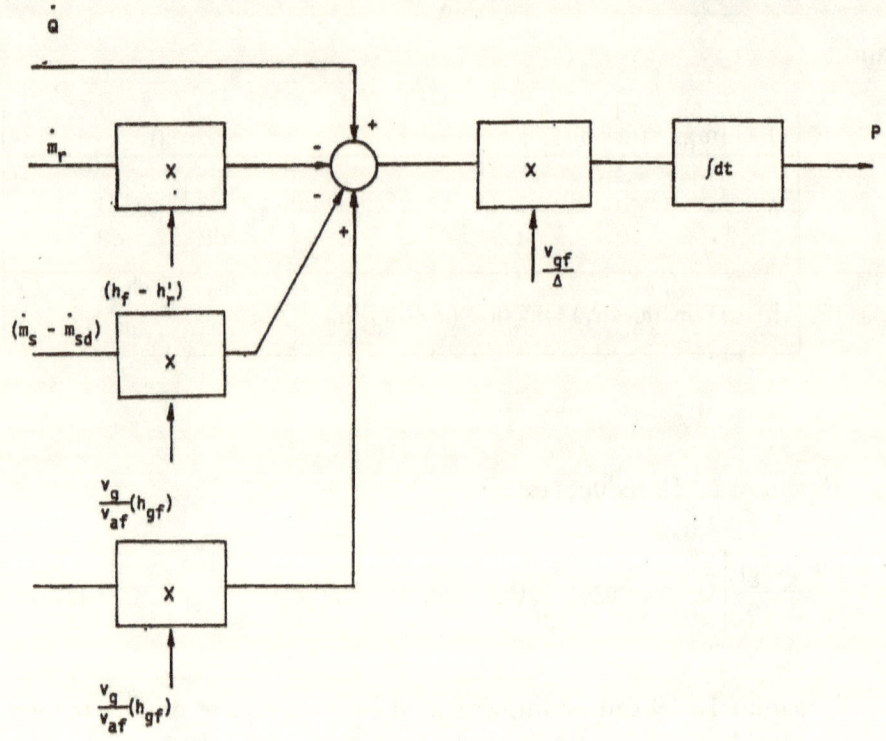

FIGURE 12.5

EFFECT OF SUPERHEATER STORAGE VOLUMES

The compressibility effect of the superheater volumes can be lumped with the water wall and drum storage volume by adding the terms

$$\frac{v_f h_{gf}}{v_{gf}} \sum_i^N v_i \frac{\partial \ell}{\partial}\Big|_{T_i}$$

to the denominator Δ in equation 12.18. These terms correspond to the summation of volumes times compressibility coefficients $\frac{\partial \ell}{\partial}$ corresponding to the pertinent superheat steam conditions.

The heat flux from tubes to fluid $\dot{Q}$ is a function of temperature difference between metal and fluid. Temperature of fluid is tied to saturation temperature at the drum pressure.

Fig. 12.6 shows the solution of $\dot{Q}$ from heat flux in the furnace $\dot{Q}_w$

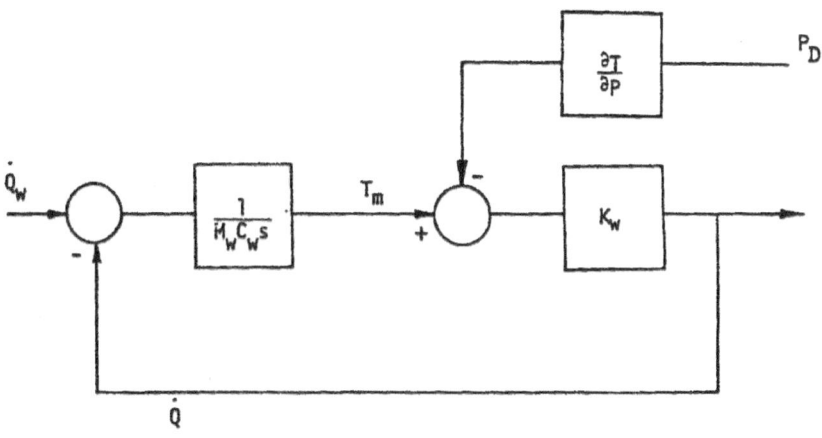

FIGURE 12.6

where M_wC_w is the heat capacitance of water wall tubes and K_w is the thermal film coefficient of heat transfer. The resulting time lag between heat flux in the furnace and heat to the fluid $T_w = \dfrac{K_w}{M_wC_w}$ is in the order of 5 to 7 seconds.

Examination of expression 12.19 shows that the boiler storage factor C defined in the simplified representation of Fig. 3.3 (Chapter III) can be evaluated as:

$$C = \frac{v_{gf}}{v_g h_{gf}}\left[M_{fT}\left(\frac{dh_f}{dP} - \frac{h_{gf}}{v_{gf}}\frac{dv_f}{dP}\right) + M_gT\left(\frac{dh_g}{dP} - \frac{h_{gf}}{v_{gf}} - \frac{dv_g}{dP}\right) - \frac{144}{J}V_T \right.$$

$$\left. + \frac{v_g}{v_{gf}}h_{gf}\sum_i^N V_i \left.\frac{\partial \rho}{\partial P}\right|_{T_i} \right] \tag{12.20}$$

where V_T = Volume of water walls and drum

V_i = Volume of superheater section i

i.e., dP/dt in psi/sec = $\dfrac{1}{C}\Delta\dot{m}_s$ in/ lbs/sec. An example of typical values of these parameters is:

260 MW	2400 psi	1000°F	1000°F boiler

Rated steam flow	480 lbs/sec	
Volume of water walls		2800 cu ft
Volume of drum		950 cu ft
Circulating flow		2000 lbs/sec
Quality of steam leaving		
water walls at full load		0.24

Volumes of superheaters ft³ 140, 750, 750, 320, 340

Average $\dfrac{\partial\rho}{\partial P}\Big|_T$, lb/ft³ $\dfrac{0.6}{100}$, $\dfrac{0.385}{100}$, $\dfrac{0.260}{100}$, $\dfrac{0.185}{100}$, $\dfrac{0.150}{100}$

Assuming uniform steam generation along water wall tube length, and assuming equal velocity of steam and water, the cross section area of steam voids in per unit of total cross section area as function of per unit tube length ℓ;

$$A_s/A_T = \ell\,(v_g\,\dot{m}_s) + (\dot{m}_r - \dot{m}_s)\,(1-\ell)\,v_f$$

where A_s = cross section of tube area occupied by steam
A_T = total cross section of tube area
ℓ = per unit tube length
$\dot{m}_s$ = steaming rate at outlet of water walls
v_g = saturated steam specific volume
v_f = saturated water specific volume
V_w = volume of water walls

The volume of steam voids in the water walls is:

$$V_w \int_0^1 \left[\ell\left(v_g\,\frac{\dot{m}_s}{\dot{m}_r}\right) + v_f\,\frac{(\dot{m}_r - \dot{m}_s)}{\dot{m}_r}(1-\ell)\right]\frac{d\ell}{v_g + v_f}$$

$$= \frac{V_w}{v_g + v_f} \left[v_g \frac{\dot{m}_s}{\dot{m}_r} \frac{\ell^2}{2} + \frac{(\dot{m}_r - \dot{m}_s)}{\dot{m}_r} \ell v_f - \frac{(\dot{m}_r - \dot{m}_s)}{m_r} \frac{\ell_{vf}^2}{2} \right]_0^1$$

$$= \frac{V_w}{v_g + v_f} \left[\frac{(\dot{m}_r - \dot{m}_s)}{\dot{m}_r} v_f + v_g \frac{\dot{m}_s}{\dot{m}_r} - \frac{\dot{m}_r}{2\dot{m}_r} v_f \right]$$

$$= \frac{\left[v_g (1/2) - v_g \, \dot{m}_s / \dot{m}_r \right]}{(v_g + v_\ell)} \tag{12.21}$$

At full load, with drum press = 2600 psi, v_f = .0295, v_g = 0.1213, h_{gf} = 337.2

Volume of steam voids = 2800 $\left| \dfrac{.0295/2 + .1213 \times \dfrac{480}{2000}}{0.1508} \right|$ 814.4 cu ft

Volume occupied by saturated water = 2800 – 814.4 = 2185.6 cu from the

Mass of resident steam in water walls plus drum (assuming water level at 40% of drum volume) = $\dfrac{(814.4 + 06 \times 950)}{.1213}$

M_{gT} = 11413 lbs

Mass of resident saturated water in water walls plus drum $= \dfrac{(2185.6 + 0.4 \times 950)}{.0295}$

M_{fT} = 86970 lbs

Other values from steam properties around 2600 psi:

$$\frac{dh_f}{dP} = \frac{25.6}{200} \quad h_{gf} = 337.6$$

$$\frac{dh_g}{dP} = -\frac{22.8}{200} \quad v_{gf} = .09172$$

$$\frac{dv_f}{dP} = +\frac{.00170}{200} \quad v_g = 0.1211$$

$$\frac{dv_g}{dP} = -\frac{.01874}{200}$$

Calculation of storage factor C due to water walls plus drum

$$C_{WW} = \frac{.0918}{.1213 \times 337.2} \left[86970 \left(\frac{25.6}{200} - \frac{337.2}{.0918} \times \frac{.00170}{200} \right) \right.$$

$$+11413 \left(\frac{22.8}{200} + \frac{337.2}{.0918} \times \frac{.01874}{200} \right)$$

$$\left. -\frac{144}{778}(2800+950) \right] 29.07$$

Due to superheaters

$$C_{sH} = 140 \times \frac{0.6}{100} + 750 \times \frac{0.385}{100} + 750\frac{0.260}{100} + 750\frac{0.185}{100} + 320 \times \frac{0.185}{100} + 340 \times \frac{0.150}{100}$$

$$= 8.167$$

Total storage factor C = 37.24 lb/psi. With 480 lbs/sec as rated flow, the significance of the storage factor is that a one per unit change in steam flow rate would produce a rate of change of drum pressure of 480/37.24 = 13 psi/sec.

CALCULATION OF DRUM LEVEL

Consider the physical system of Fig. 12.7.

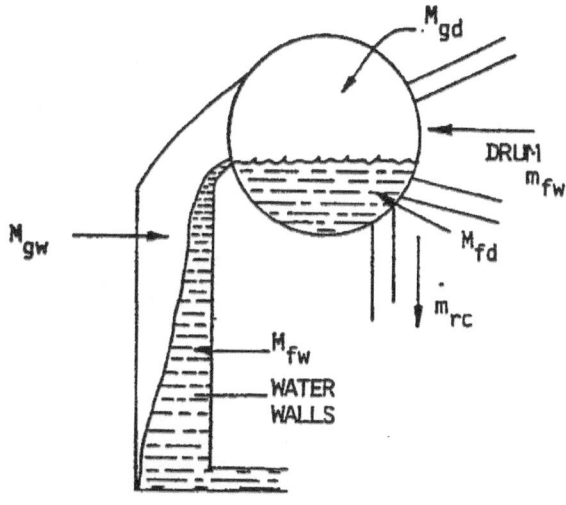

FIGURE 12.7

Here M_{fd} = Mass of liquid in drum
M_{fw} = Mass of liquid in water walls
M_{gd} = Mass of steam in drum
M_{gw} = Mass of steam in water walls

The drum level is calculated from determination of M_{fd} and knowledge of saturated liquid specific volume v_f.

Equations are:

$$M_{fd} = \int (\dot{m}_{fw} - \dot{m}_r + \dot{m}_\ell)dt \qquad (12.22)$$

where $\dot{m}_\ell$ = flow rate of saturated liquid

$$\text{out of water walls} = \dot{m}_r - \dot{m}_w - d\frac{M_{fw}}{dt} \qquad (12.23)$$

Substituting (12.23) in (12.22)

$$M_{fd} = \int (\dot{m}_{fw} - \dot{m}_w)dt - M_{fw} \tag{12.24}$$

Assuming uniform steam generation along the tube length, the relationship between resident water wall liquid and steam generation in the steady state is:

$$M_{fw} = \frac{V_w}{V_{gf}} \left[\frac{v_g/v_f}{(v_g/v_f - 1)\left(\dfrac{\dot{m}_w}{\dot{m}_r}\right)} \, \mathrm{Ln}\left[1 + \left(\frac{v_g}{v_f} - 1\right)\left(\frac{\dot{m}_w}{\dot{m}_r}\right) \right] - 1 \right] \tag{12.25}$$

In the above expression the change in water wall liquid ΔM_{fd} can be related to change in steam rate Δm_w and change in pressure ΔP.

$$\Delta M_{fw} = \left(\frac{\partial M_{fw}}{\partial \dot{m}_{fw}}\right) \Delta \dot{m}_w + \left(\frac{\partial M_{fw}}{\partial \left(\dfrac{v_g}{v_f}\right)}\right) \frac{\partial \dfrac{v_g}{v_f}}{\partial P} \Delta P = A\,\Delta\dot{m}_w + B\,\Delta P \tag{12.26}$$

where A and B are derived by differentiating expression (12.25).

Steam generation m_w is obtained from the energy equation and is expressed as:

$$\Delta \dot{m}_w = \frac{\Delta \dot{Q}}{h_{gf_o}} - \left[M_{gw}\frac{dh_g}{dP} + M_{fw}\frac{dh_f}{dP} - \frac{144}{J}V_w \right]\frac{dP}{dt}$$

$$- \left[\frac{\dot{m}_{so}}{h_{gfo}}\frac{dh_{gf}}{dP} + \frac{\dot{m}_r}{h_{gfo}}\frac{dh_f}{dP} \right]\Delta P + \left(\frac{\dot{m}_r}{h_{gfo}}\right)\Delta h'_r \tag{12.27}$$

A simpler model of drum level which neglects pressure effects on the specific volume of saturated liquid is shown in Fig. 12.8.

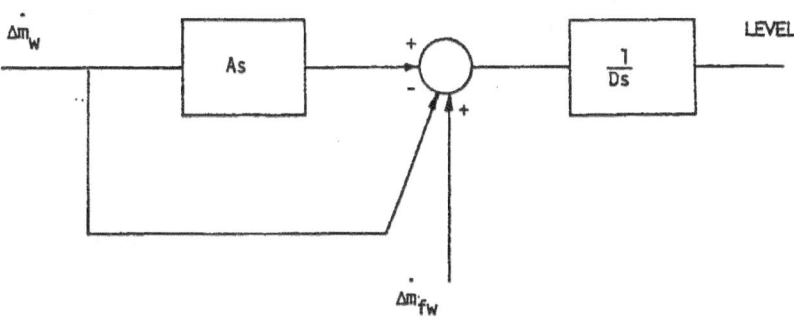

FIGURE 12.8

where A is the change in resident fluid in water walls as function of steaming rate.

$$A = \frac{2 V_w P_g}{\dot{m}_r \left[(2 - X_e) \dfrac{P_g}{P_f} + X_e \right]^2}$$

For a drum 60″ in diameter, 53.5 ft long and a pressure of 2500 psi, circulating flow rate of 2000 lbs/sec and full load steaming rate of 480 lbs/sec.

The quality at full load $= \dfrac{480}{2000} = 0.24$

Water wall volume = 2800 cu from the

$$A = \frac{2 \times 2800 \times 1/.131}{2000[(2 - .24)0.218 + .24]^2} = 55 \text{ sec}$$

D = 775 lbs/in

HEATING SECTION MODELS

The process of flow and heat transfer in long heated tubes is described by partial nonlinear differential equations in space and time. The classical modeling approach is to use finite differences in space and time. Variables are defined at discrete points in space and conditions within a bulk heating section are considered as the average of inlet and outlet conditions. The mass balance energy balance equations are then solved by standard integration methods.

These approaches produce parasitical wringing effects which are due to the inherent assumption of conditions in a heating volume being the average of inlet and outlet values.

A unique modeling approach (Ref. 12.2) avoids these effects by solving the equations within a finite residence time in closed form.

Fig. 12.9 shows a long heated tube broken into finite lengths.

A temperature effects model is derived by treating the continuous tube into lumped volumes separated by orifices (to account for the pressure drop) as indicated in Fig. 12.10.

The mathematical development follows from visualization of the flow as a pulsating phenomenon.

The distributed process of flow and energy transfer is approximated as a lumped parameter representation where the pressure drop is visualized as being concentrated in a number of equivalent volumeless orifices separating lumped volumes where heating occurs.

The continuous flow process is now visualized as a series of pulsations whereby fluid is instantaneously admitted into a volume and instantaneously ejected from that volume to the adjacent one after having resided in the volume for a period equal to the residence time.

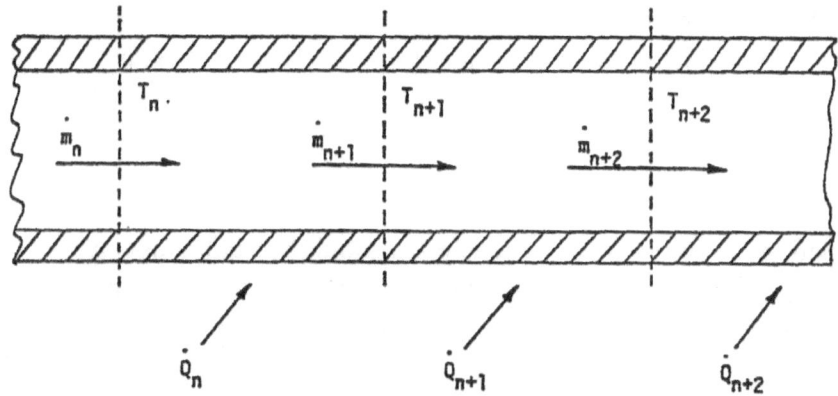

FIGURE 12.9

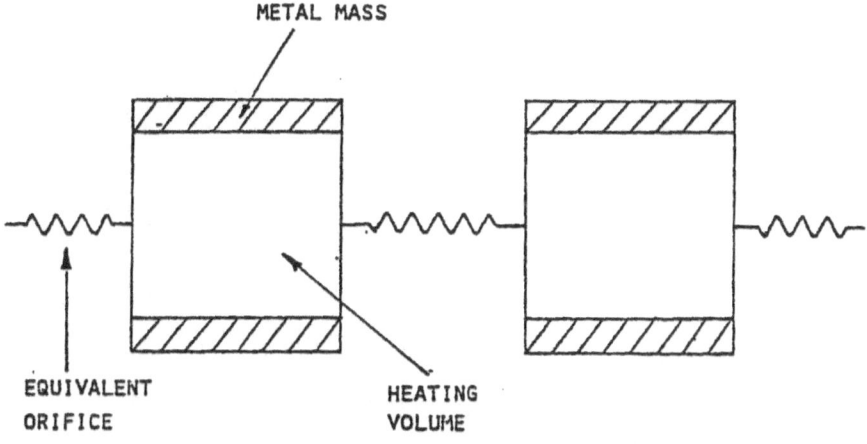

FIGURE 12.10

The derivation of the model follows from assumptions listed below:

1. m lbs introduced instantaneously at the beginning of residence time at temperature T_o and ejected at temperature $T(t_r)$ at end of tr secs.
2. Metal temperature at T_{mo} at beginning of residence time, and $T_m(t_r)$ at end.
3. $\dot{Q} \times t_r$ = total heat input from gas to metal during residence time.
4. Change in pressure with respect to time P is treated as occurring as a discrete step in pressure $\dot{P}t_r$ occurring in zero time immediately preceding injection.

5. Change in pressure with respect to distance is treated as an adiabatic constant enthalpy pressure drop process in series with the volume residence phenomena.

The heating of the trapped fluid during the residence time is treated in closed form with the simultaneous solution of two differential equations defining the metal temperature and the fluid temperature.

These equations solve the fluid temperature and metal temperature at the end of the residence time t_r as function conditions at the beginning of the residence and heat flux during residence.

$$T(t_r) = T_o \left[\frac{1 + Z\varepsilon^{-x}}{1+Z} \right] + T_{mo} \left[\frac{Z(1-\varepsilon^{-x})}{1+Z} \right] + \frac{\dot{Q}t_r}{MC} \left[\frac{Z(1-\varepsilon^{-x})}{1+Z} \right] \qquad (12.28)$$

$$T_m(t_r) = T_o \left[\frac{1-\varepsilon^{-x}}{1+Z} \right] + T_{mo} \left[\frac{Z+\varepsilon^{-x}}{1+Z} \right] + \frac{\dot{Q}t_r}{MC} \left[\frac{Z+\varepsilon^{-x}}{1+Z} \right] \qquad (12.29)$$

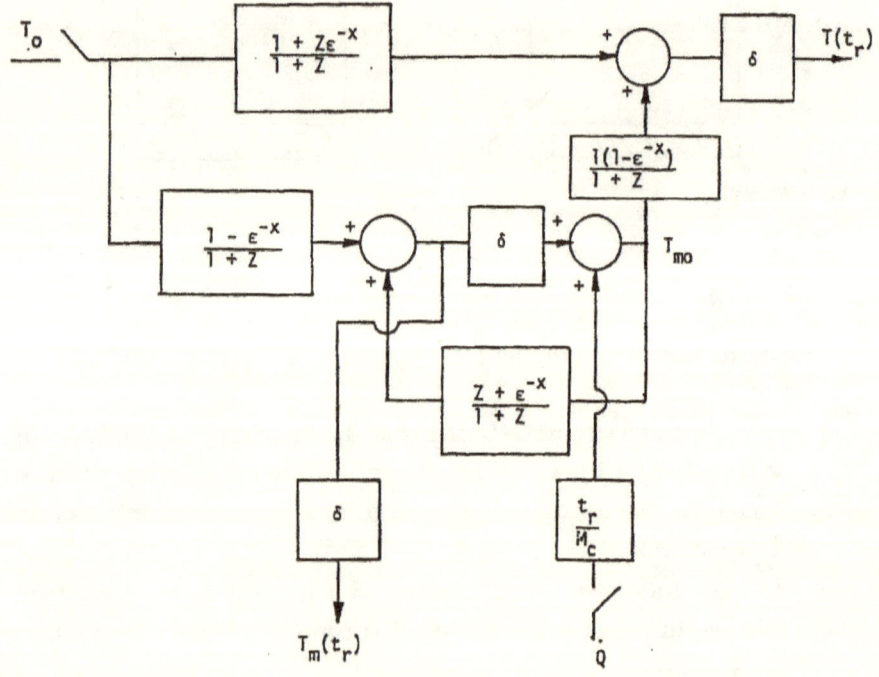

FIGURE 12.11

The block diagram of Fig. 12.11 describes this discrete pulsating process. The heat input $\int_0^{t_r} \dot{Q}dt$ is considered as a pulse $\dot{Q}t_r$ at the beginning of the residence time.

The symbols are:

Z = capacity ratio $\dfrac{MC}{V\rho C_p}$

M = mass of metal
C = specific heat of metal
V = inner volume of section
ρ = density of fluid
C_p = specific heat of fluid

$X = \dfrac{HA}{\dot{m}C_p}$

HA = heat transfer coefficient from tube to fluid
m = fluid flow rate
$\dot{Q}$ = heat flux from gas to metal
δ = delay operator for residence time t_r

Fig. 12.12 is a refinement of Fig. 12.11 where the heat input $\int_0^{t_r} \dot{Q}dt$ is introduced as 2 pulses, $\dfrac{\dot{Q}t_r}{2}$ at beginning and end of t_r interval. Further block diagram reduction yields the model of Fig. 12.13 which is a sampled data model giving an output at every t_r secs. To convert this into a continuous model, use is made of standard smoothing techniques.

Remembering that the continuous function

$$\frac{1}{S} = \frac{t_r}{2}\left|\frac{1+\delta}{1-\delta}\right| = t_r\left|\frac{1}{2}+\delta+\delta^2+\ldots\ldots\right|$$

it follows that

$$\frac{\delta}{1-\delta} = \frac{1}{t_r s} - \frac{1}{2} = \frac{1}{t_r s}\varepsilon^{-s\,t_r/2} = \frac{1}{t_r s\left(1+\dfrac{t_r}{2}s\right)}$$

and

$$\frac{1}{1-\delta} = \frac{1}{t_r s} + \frac{1}{2} = \frac{1}{t_r s} \varepsilon^{-s^1 r/2}$$

The sampled data model of the pulsating flow model can be converted to a continuous model as shown on Figs. 12.14 and 12.15.

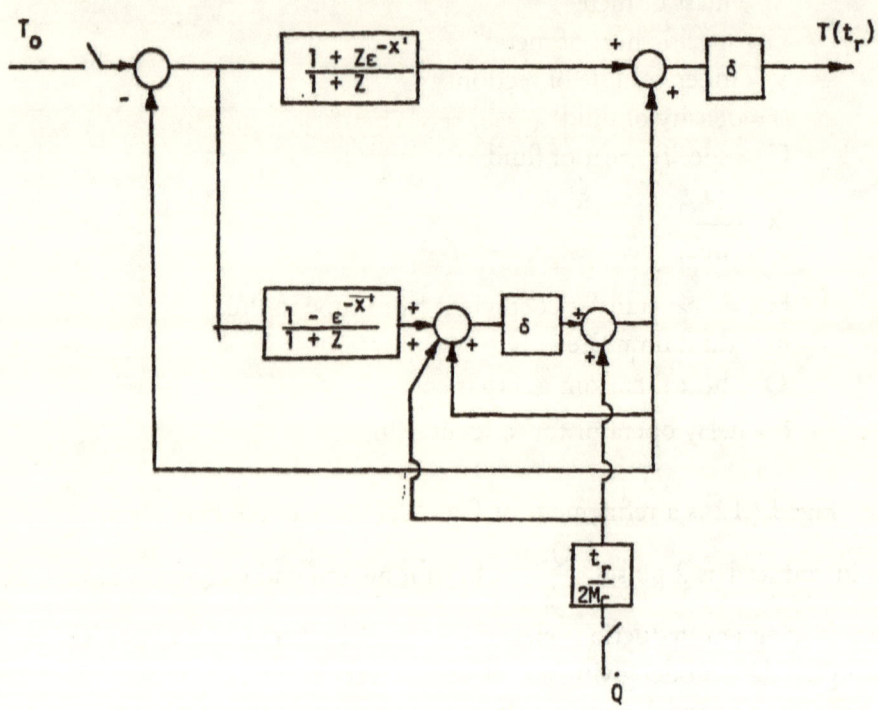

FIGURE 12.12

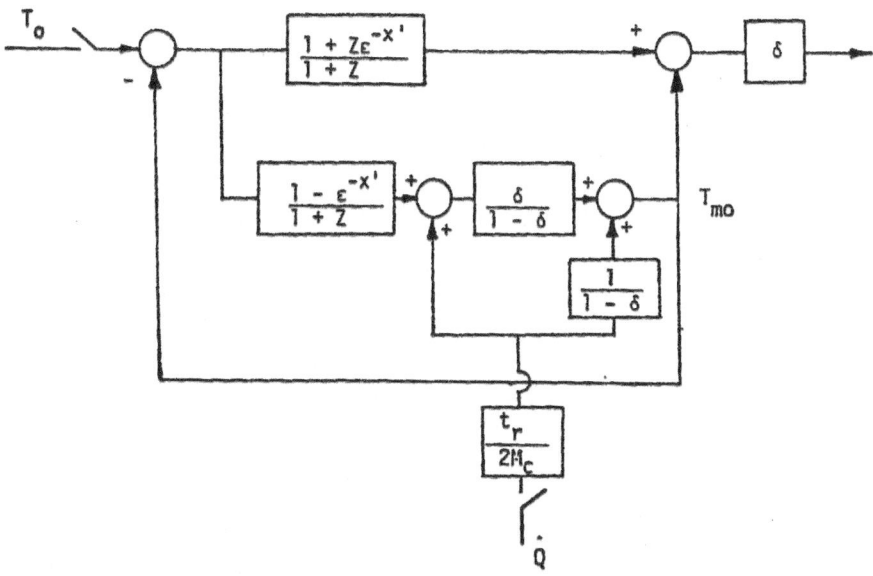

FIGURE 12.13

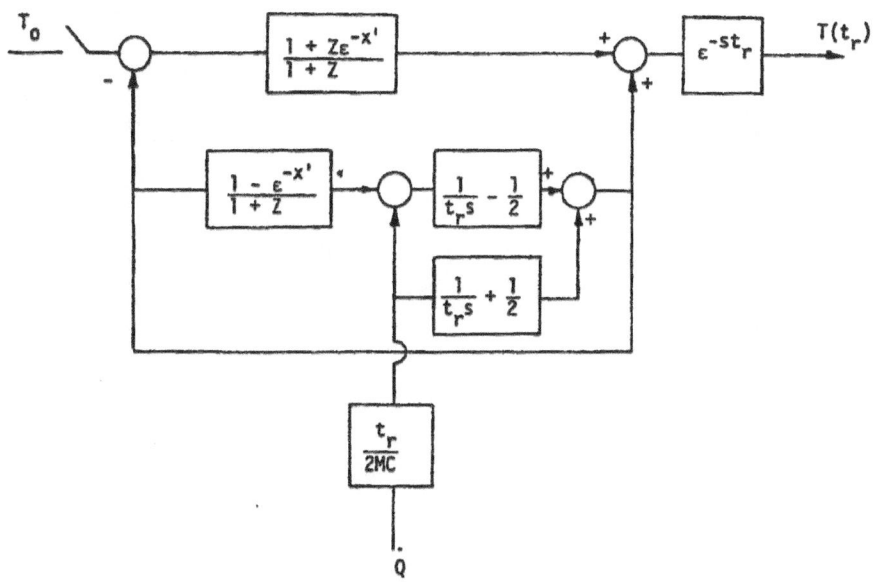

FIGURE 12.14

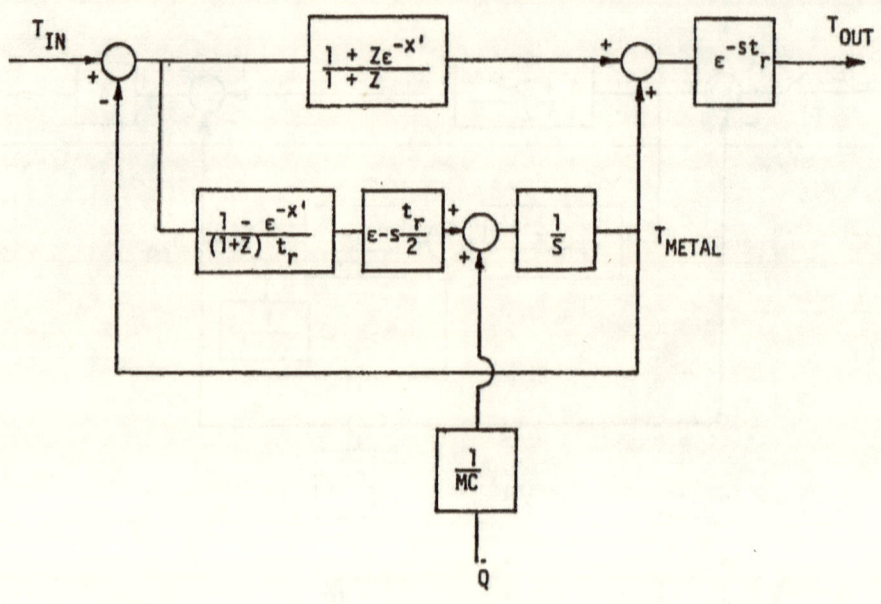

FIGURE 12.15

Finally, smoothing ϵ^{-st_r} to $\dfrac{1}{1+st_r}$, we obtain the model of Fig. 12.16.

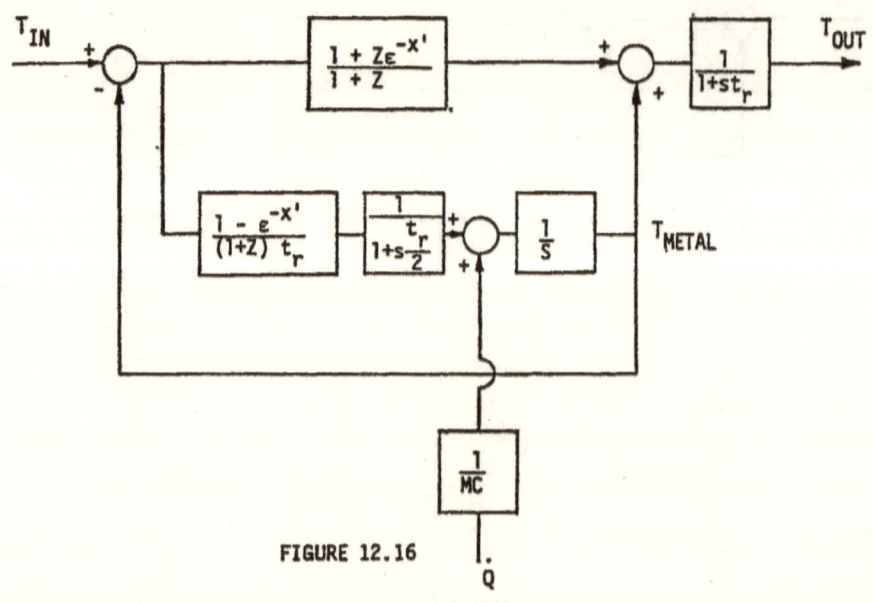

FIGURE 12.16

FIGURE 12.16

The effect of rate of change of pressure is taken into account by modifying the inlet temperature an amount equal to $\dot{P}t_r \frac{\partial T}{\partial P}\big|_s$. This is equivalent to the effect on temperature due to isentropic compression or expansion by the amount of pressure change during the residence interval.

The effect of pressure drop, ΔP, is also accounted for by modifying the inlet temperature an amount $\Delta \frac{\partial T}{\partial P}\big|_h$ which is the temperature change due to a throttling loss ΔP at constant enthalpy.

The complete model is shown in Fig. 12.17. It is apparent from inspection of Fig. 12.17 that the temperature model requires the definition of several nonlinear relations, e.g., C_p, $\frac{\partial T}{\partial P}\big|_h$, $\frac{\partial T}{\partial P}\big|_s$ which are derived from steam properties and are functions of pressure and temperature.

It should be noted that all the coefficients appearing in the temperature model of Fig. 12.17, including t_r, $\varepsilon^{-X'}$, Z, $\frac{\partial T}{\partial P}\big|_s$ can be evaluated at each time interval based on the most recent values of pressure and temperature.

The temperature model for the reheater is essentially the same as for the superheater. The main difference is that due to linear relationship that exists between reheat flow and pressure, and the almost linear relationship between pressure and density, the parameter t_r (residence time) need no longer be a variable to be calculated every time interval, but is taken as a constant.

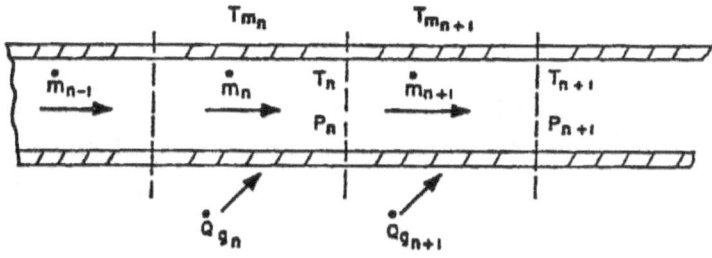

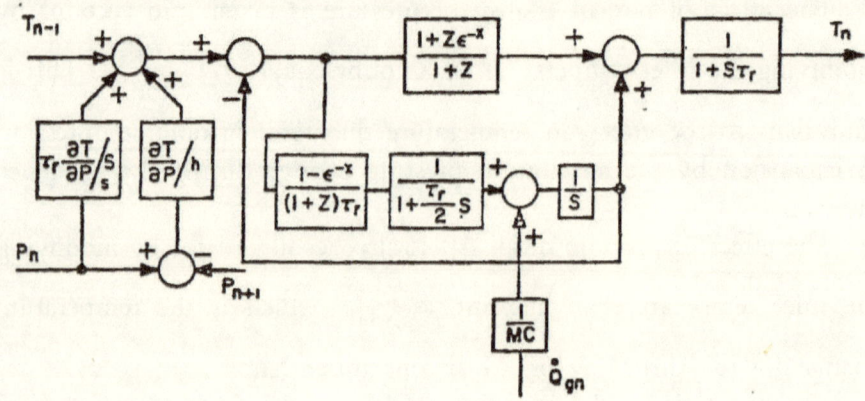

VARIABLES IN n^{th} SECTION

P_n = AVG PRESS.
T_n = OUTLET TEMP
ρ_n = AVG DENS.
V_n = INNER VOLUME
$\dot{m}_n$ = AVG FLOW RATE

τ_{r_n} = RESIDENCE TIME $\dfrac{V_n \rho_n}{\dot{m}_n}$

$\dfrac{\partial T}{\partial P}/s$ = PARTIAL OF TEMPERATURE WITH PRESSURE AT CONSTANT ENTHROPY

$\dfrac{\partial T}{\partial P}/h$ = PARTIAL OF TEMPERATURE WITH PRESSURE AT CONSTANT ENTHALPY

M_n = MASS OF METAL
C_{p_n} = AVG SP HEAT OF FLUID
C_n = SP HEAT OF METAL

Z = CAPACITY RATIO $\dfrac{M_n C_n}{V_n \rho_n C_{p_n}}$

X = TRANSMISSION

FACTOR $\dfrac{HA_n}{\dot{m}_n C_{p_n}}(1+Z)$

HA_n = HEAT TRANSFER COEF
$\dot{Q}_{g_n}$ = HEAT FLUX
T_{m_n} = AVG METAL TEMP
S = LA PLACE OPERATOR

MATHEMATICAL MODEL FOR HEATING SECTION

FIGURE 12.17

The model of Fig. 12.17 can be implemented in complete nonlinear form in digital computers. The linearized version of the model of a heating section solving for small perturbations about an opeating point can be similarly derived and is displayed in Fig. 1.13 of Chapter I. The effect of changes in C_p between inlet and outlet conditions are accounted for.

The important parameter in these temperature models is the transmission factor $X = \dfrac{H\,A}{\dot{m}C_p}$. The value of ε^{-X} represents the fraction of temperature change at the inlet that appears at the outlet at the end of the travel time. Evidently, neglecting travel time effects, the model is basically a first order model. Hence, whenever X is high, say, above 1.5, it is advisable to split the heating section into more than one. The response is then second or higher order depending on the number of sections in series.

The response of outlet temperature to changes in inlet temperature is governed by this transmission factor X. At low flow rates, residence times increase. The heat transfer coefficient HA varies as the 0.8 power of flow. Hence, X tends to increase as flow rate decreases.

Figs. 12.18 to 12.20 show typical response characteristics of a secondary superheater to changes in inlet temperature, changes in heat flux and changes in steam flow.

The superheater is broken into 3 sections and the curves show intermediate temperatures at the end of each section. The results are obtained with the model of Fig. 1.13 (Chapter I).

The cases show the effect of film temperature coefficient HA which may also be expressed as θ_o, the average metal to fluid temperature drop for a given steady state heat flux and flow condition. Figs. 12.18 and 12.19 are for an assumed high film temperature drop θ_o (90° at flow rate of 350 lbs/sec and 78.3° at flow rate of 175 lbs/sec).

Fig. 12.20 is for an assumed low film temperature drop coefficient (10° at flow rate of 350 lbs/sec and 8.7° at flow rate of 175 lbs/sec).

The effect of film coefficient can be seen by comparing the responses in the two sets of cases. For high resistance to heat flow (high θ_o) the outlet temperature responds in part almost instantaneously to changes in inlet temperature.

For low values of film coefficient (low θ_o) the outlet temperature response is lagged considerably as the change gets first absorbed by the metal.

The response in Figs. 12.18 to 12.20 do not include the lag of the transducer (thermocouple). Thermocouples well seated in wells generally exhibit a response characterized by one time constant in the order of 20 to 30 sec.

Film coefficients can usually be obtained from estimates of average temperature drop between metal and fluid in heating sections. If not available, they may be calculated from the physical dimensions of the heating section as indicated in the following section.

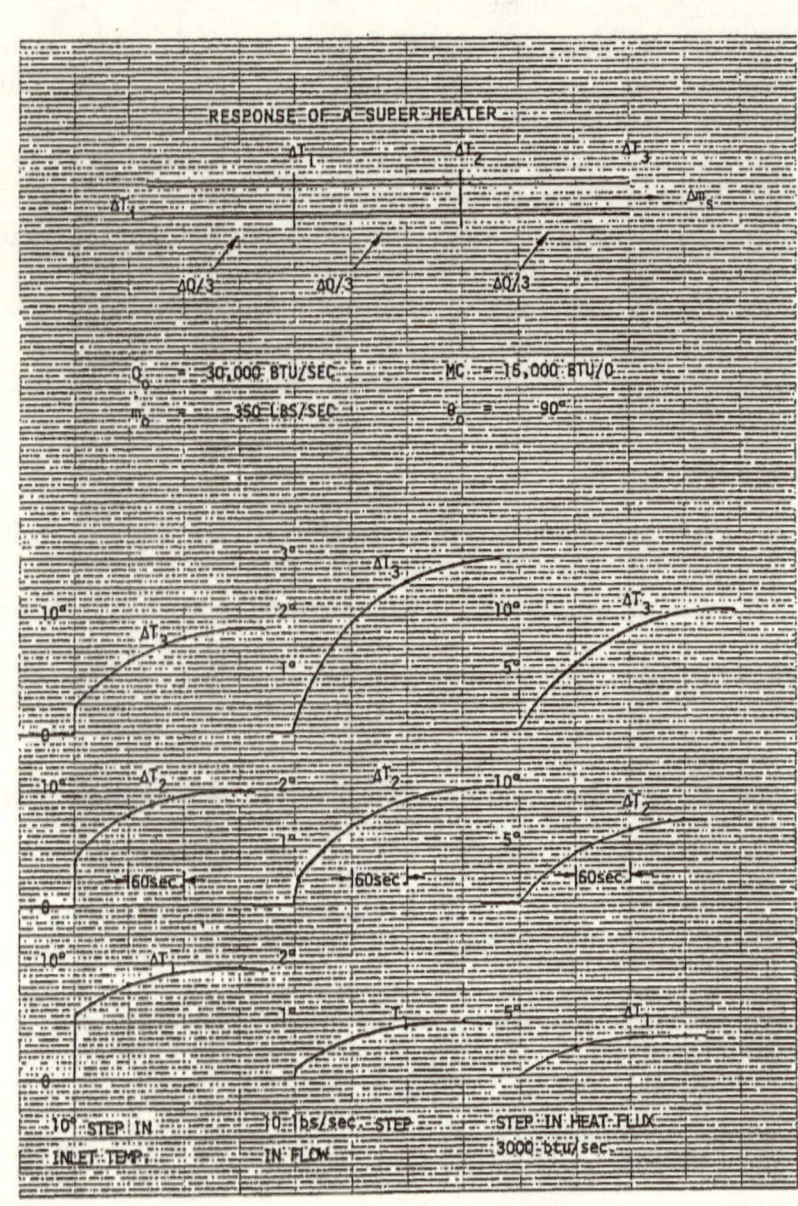

FIGURE 12.18

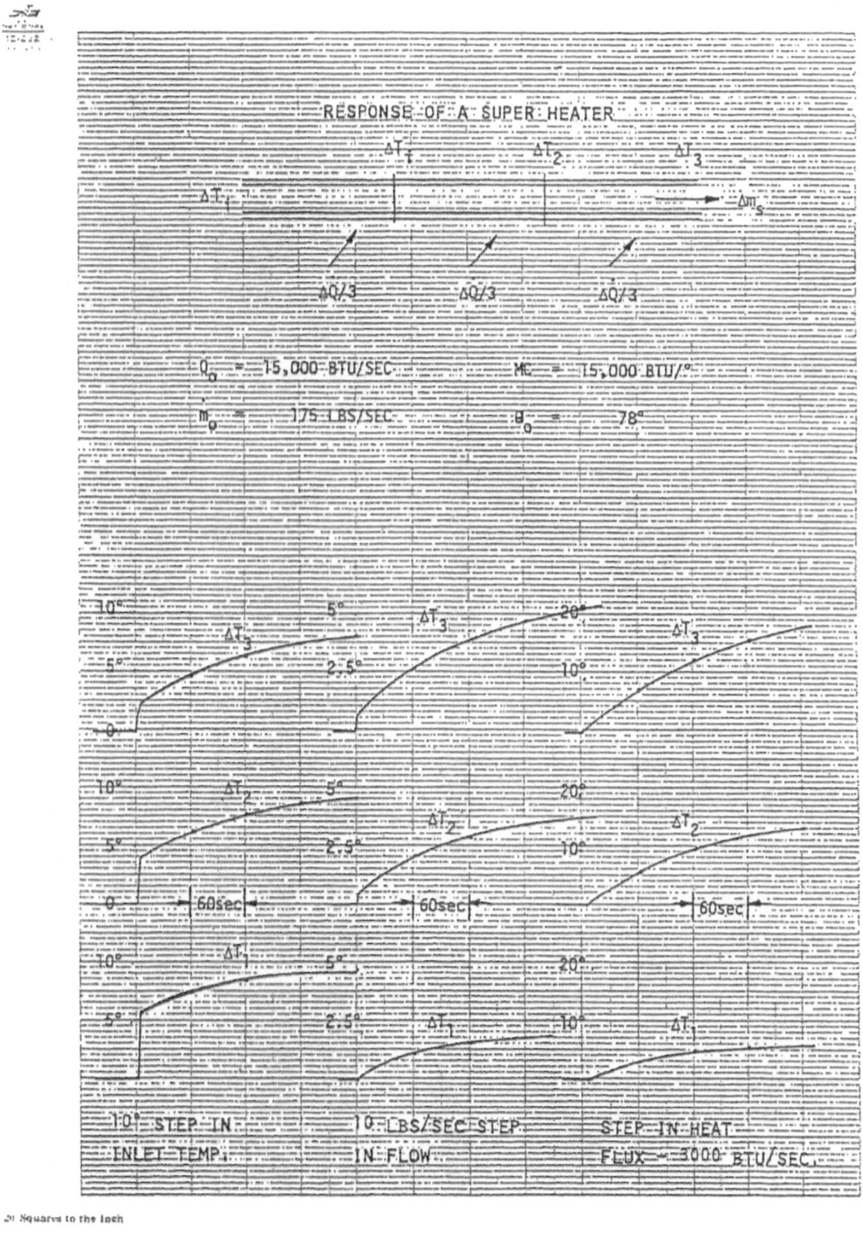
RESPONSE OF A SUPER HEATER
ΔT1
ΔT2
ΔT3
Δm_s
ΔQ/3
ΔQ/3
ΔQ/3
Q_o = 15,000 BTU/SEC
MC = 15,000 BTU/°
m_o = 175 LBS/SEC
θ_o = 78°
10°
5°
0
5°
2.5°
20°
10°
60sec
60sec
60sec
10° STEP IN INLET TEMP.
10 LBS/SEC STEP IN FLOW.
STEP IN HEAT FLUX - 3000 BTU/SEC.
20 Squares to the Inch

FIGURE 12.19

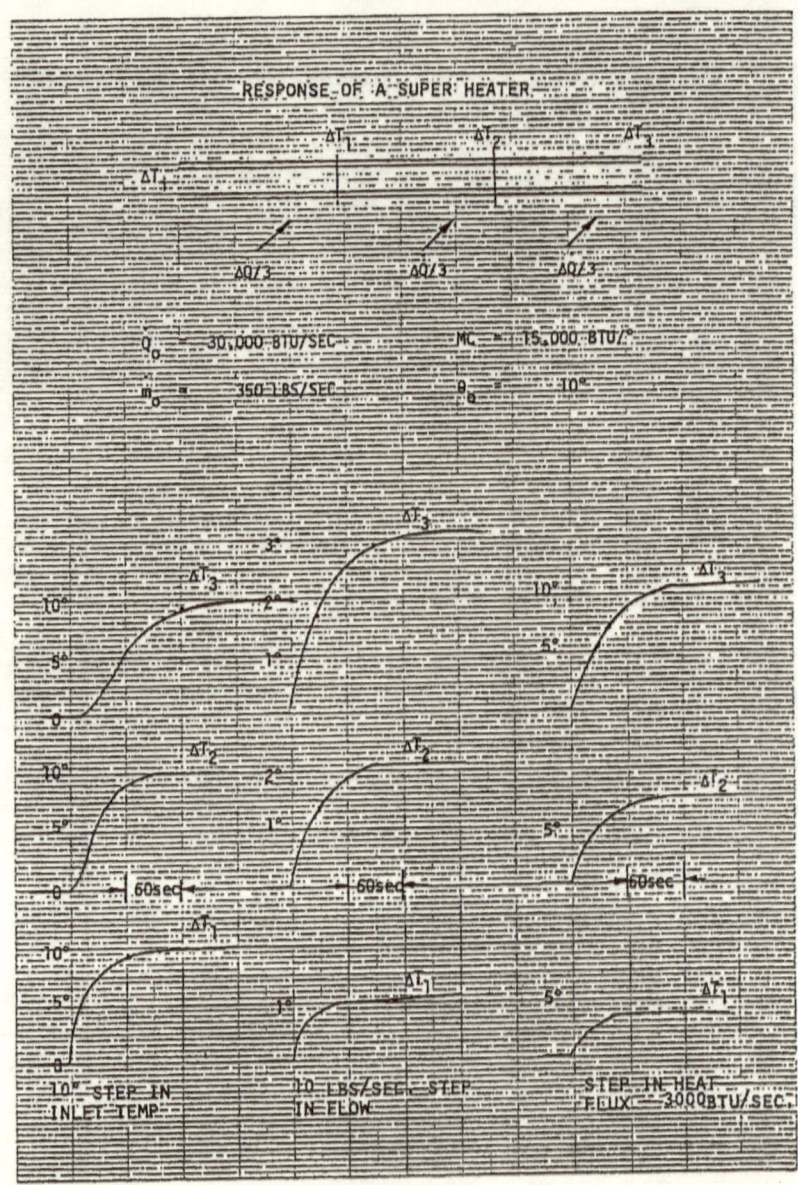

FIGURE 12.20

FILM COEFFICIENTS

Heat transfer from metal to steam is governed by the relation:

$$\frac{h_d}{K} = .023\left(\frac{V_d\rho}{\mu}\right)^{0.8}\left(\frac{C_p\mu}{K}\right)^{04} \tag{12.30}$$

where:

h = film conductance BTU/hr - sq ft - deg F
d = inside diameter of tube - ft
K = thermal conductivity of fluid BTU/hr - sq ft - deg F
V = velocity - ft/hr
ρ = density - #/ft^3
C_p = pecific heat - BTU/lb - deg F
μ = viscosity - lb/ft - hr

The quantity $\left(\dfrac{V_d\rho}{\mu}\right)$ is known as the Reynolds number and $\left(\dfrac{C_p\mu}{K}\right)$ is the Prandlt number. $\left(\dfrac{h_d}{K}\right)$ is known as the Nusselt number.

Fig. 12.21 shows the variation of the Prandlt number for steam as function of pressure and temperature.

Fig. 12.22 is a plot of viscosity of steam as function of pressure and temperature.

In the model of heating sections the temperature transmission factor $\varepsilon^{-\frac{\dot{Q}_o}{\theta_o \dot{m}_o C_p}}$ contains the effect of $\dfrac{\dot{Q}_o}{\theta_o}$ area, may be thought as the thermal admittance of the heating section.

The exponent $X = \dfrac{\dot{Q}_o}{\theta_o \dot{m}_o C_p}$ expressed in terms of the factors in eq. 12.30 is:

$$X = \frac{\dot{Q}_o}{\theta_o \dot{m}_o C_p} = 0.1243\left(\frac{1}{P_r}\right)^{06}\left(\frac{D_i\mu N}{\dot{m}_o}\right)^{0.2}\frac{L}{D_i} \tag{12.31}$$

where D_i = inside diameter of tube, inches
        L   = length of tube-feet
        N   = number of tubes in parallel.
        m_o = steam mass flow rate
        μ   = viscosity
        P_r = Prandlt number

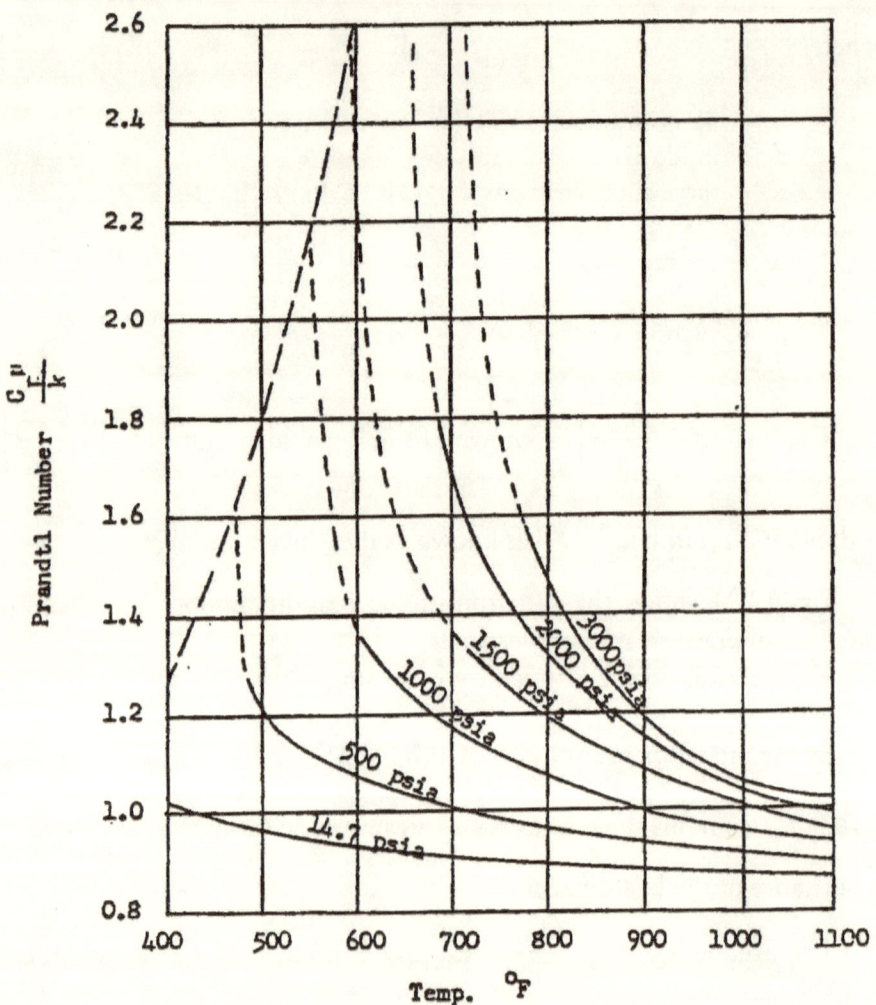

Prandtl Number for Steam

FIGURE 12.21

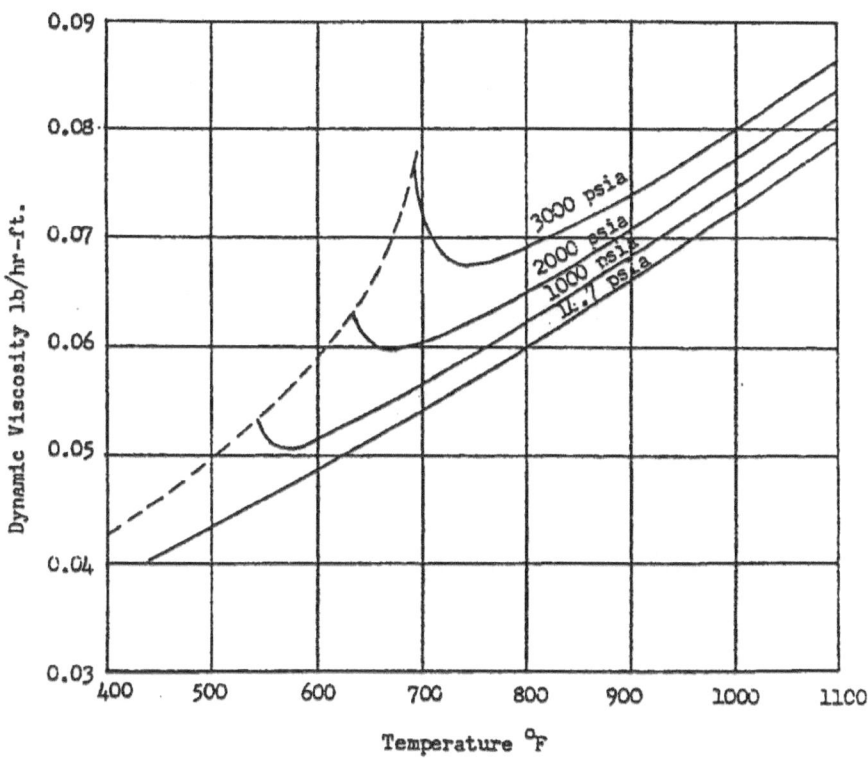

Dynamic Viscosity of Steam

FIGURE 12.22

TURBINE REPRESENTATION

Flow through the high pressure turbine is proportional to turbine valve opening, to throttle pressure and inversely proportional to the square root of absolute steam temperature.

Since temperature deviations are small relative to the absolute temperature, the temperature effect can be linearized and the turbine flow equation can be expressed as:

$$m_s = K_v P \left(1 - (T - T_0) \times 0.342 \times 10^{-3}\right) \qquad (12.32)$$

where K_v is proportional to turbine opening
       P = throttle pressure
       T = throttle temperature
       T_o = rated throttle temperature

Reheat pressure is nearly a linear function of reheat flow and changes from this linear relationship are due to changes in hot reheat temperature.

$$\dot{m}_r = \dot{m}_{ro} \frac{P_R}{P_{Ro}} \sqrt{\frac{T_{Ro}+460}{T_R+460}} \tag{12.33}$$

For a nominal T_{Ro} of 1000°F, linearization of 12.33 yields

$$P_R = \frac{P_{Ro}\dot{m}_r}{\dot{m}_{ro}}(1+3.73\times10^{-4}(T_{HR}-1000)) \tag{12.34}$$

Now reheat flow is proportional to main steam flow and lagged by the reheater charging time T_R (in the order of 6 to 10 sec). Due to extraction to the high pressure heaters, reheat flow is typically 0.82 to 0.85 times main steam flow.

$$\dot{m}_r = \frac{\dot{m}_s(.82)}{(1+S\,T_R)} \tag{12.35}$$

Additional relationships are required to establish input conditions to the reheater that is to define the temperature of the cold reheat steam exhausting from the high pressure turbine.

Heat balances from the turbine manufacturer generally define the expected cold reheat temperature and pressure as function of steam flow at rated main steam pressure and temperature and expected cold reheat pressure.

Variations of cold reheat temperature from this expected value which is defined as $T_{Rc}' = f(m_s)$ would be due to deviations in throttle conditions ΔP_T, ΔT_T and deviations of reheat pressure from nominal for the particular load point ΔP_R.

An example of heat balances for the turbine is given for a 2400 psi, 1000°, 1000° turbine with a full load flow rate of 435 lbs/sec.

m_s	P_T	T_T	T_{RC}	hr	P_{RC}	ηAE	K_v
435	2400	1050	612°	1311.5	430	182.5	0.1812
311	2400	1050	568°	1296	314	198	0.1296
164	2400	1050	523°	1284	172	210	0.0683

where:

m_s = steam flow rate   lbs/sec

P_T = throttle steam pressure   psi

T_T = throttle steam temperature   °F

T_{Rc} = cold reheat steam temperature   °F
       (exhaust of high pressure turbine)

hr = hot reheat enthalpy   BTU/lb

ηAE = efficiency times available energy of steam   BTU/lb
       conditions across high pressure turbine

K_v = valve coefficient   lbs/psi-sec

The variation in reheat temperature from the curve fit obtained from the table above is derived by establishing the variation in cold reheat enthalpy as follows:

$$\Delta h = \left\{ \frac{\partial h}{\partial T}\bigg|_{P_T} \Delta T_T + \frac{\partial h}{\partial P}\bigg|_{T_T} \Delta P_T - \frac{\partial nAE}{\partial P_T}\bigg|_{T_T,P_{Rc}} \Delta P_T - \partial \frac{nAE}{\partial T_T}\bigg|_{P_T,P_{Rc}} \Delta T_T - \frac{\partial nAE}{\partial P_{Rc}}\bigg|_{P_T,T_T} \Delta P_{Rc} \right\} \quad (12.36)$$

The first two terms represent the change in throttle steam enthalpy and the following three terms account for the change in work $\Delta(\eta AE)$ performed by the high pressure turbine due to changes in throttle conditions and reheat pressure.

The various partials in 12.36 are evaluated from the steam tables. Some of these vary with load point and can be expressed either as function of steam flow or valve opening.

HEAT FLUX FROM GAS TO METAL

The analytical determination of heat flux from gas to metal is a very complex problem relying heavily on empirical data. Of all the boiler physical data, the gas path heat flux characteristics are those most subject to error. These characteristics are also subject to change during operation due to slagging, soot blowing, etc.

Basically there are three types of heat transfer effects:

Convection

$$\dot{Q}_c = K_c W_g^{0.75} T_g^{0.5} (T_g - T_m) \tag{12.37}$$

Gas Radiation

$$\dot{Q}_{gr} = K_{gr}(T_g - T_m) \tag{12.38}$$

Luminous Radiation

$$\dot{Q}\ell_r = K\ell_r(T_g^4 - T_m^4)$$
$$= K\ell_r (T_g^2 + T_m^2)(T_g + T_m)(T_g - T_m) \tag{12.39}$$

where W_g = mass gas flow
T_g = average gas temperature
T_m = average metal temperature
K_c = convection coefficient
K_{gr} = gas radiation coefficient
$K_{\ell r}$ = luminous radiation coefficient

For any particular section, the downstream gas temperature $T_{g\ out}$ is related to the upstream temperature by the energy equation

$$T_{g\ out} = T_{g\ in} - \frac{\dot{Q}_g}{W_g C_{gi}} \tag{12.40}$$

where $\dot{Q}_g = \dot{Q}_c + \dot{Q}_{gr} + \dot{Q}_{\ell r}$

Due to the complex nature of combustion and heat absorption process in the furnace walls, the heat flux to these walls is obtained from an energy balance, given the relationship of the furnace exit gas temperature T_{ge} by the boiler manufacturer.

$$T_{ge} = f(\dot{Q}_T, W_g) \tag{12.41}$$

where $\dot{Q}_T$ = fuel burned
W_g = mass flow of fuel and air

In the case of boilers with tilting burners, the relationship would also include burner tilt effects.

The heat absorbed in the furnace can then be derived as:

$$\dot{Q}_F = \dot{Q}_T - \dot{Q}_e + \dot{Q}_A \tag{12.42}$$

where $\dot{Q}_F$ = furnace heat flux

$\dot{Q}_e = W_g T_{ge} C_{pe}$ = furnace exit gas flux

$\dot{Q}_T$ = fuel energy rate

$\dot{Q}_A = C_{pa} T_a W_a$ = inlet air heat input

In the case of drum boilers the furnace walls are lumped for their effect on steaming rate, hence it is not necessary to subdivide the furnace heat flux into portions to the various sections.

In the case of once-through boilers, the subdivision of furnace heat flux into various sections can be done in proportion to the steady state predicted heat absorptions along the furnace sections.

Fig. 12.24 shows the general scheme of solution of gas path equations from basic inputs of fuel flow $\dot{Q}_T$, air flow W_a, and others such as gas recirculation or burner tilts. Heat fluxes for each heating section are solved from knowledge of upstream gas temperature, mass flow rate and average metal temperature which is a state obtained from the heating section model of Fig. 12.16.

Since gas residence times are short relative to the response times of the process, it is usually not necessary to include the time delay of gas travel from one section to the next.

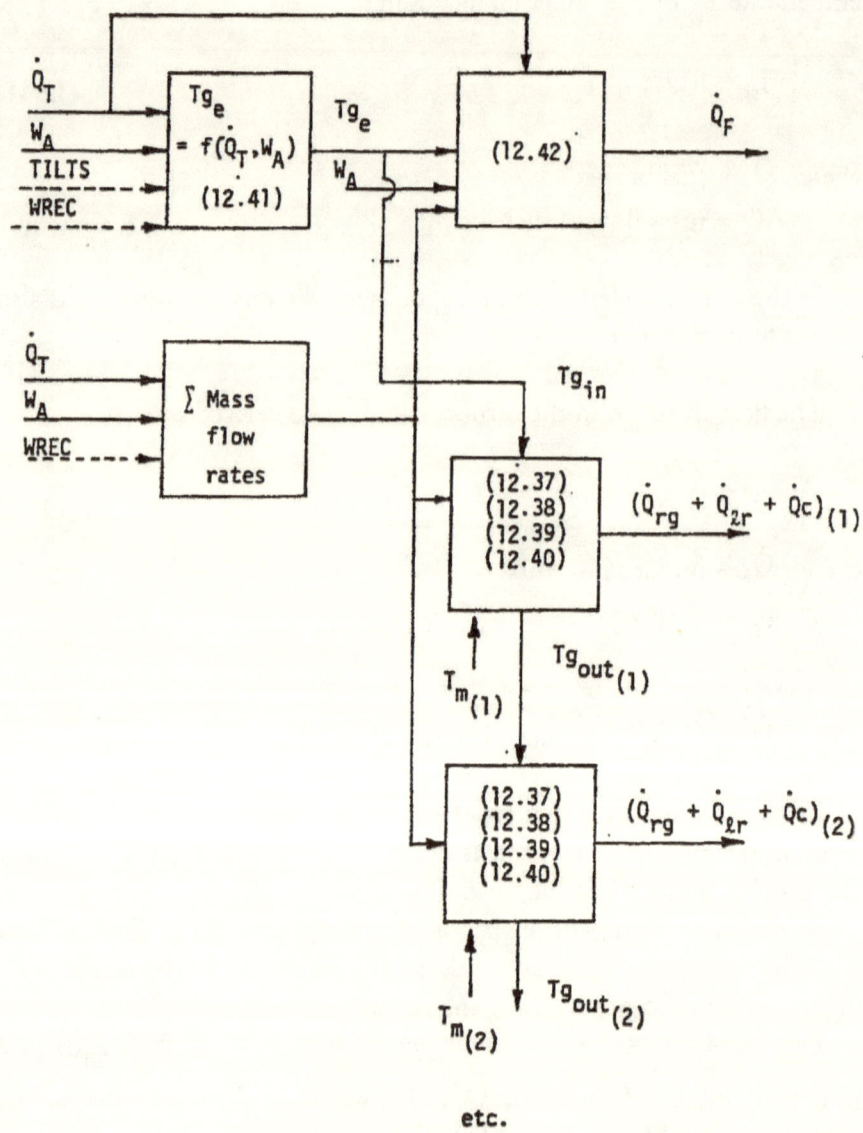

FIGURE 12.24

REFERENCES

12.1 Ennis, M., "Comparison of Dynamic Models of a Superheater", ASME Paper No. 61-WA-171, 1961.

12.2 Ahner, D.J., de Mello, F.P., Dyer, C.E., and Summer, V.C., "Analysis and Design of Controls for a Once-Thru Boiler Through Digital Simulation", ISA Proceedings, 9th National Power Instrumentation Symposium, 1966, p. 11.

12.3 Thompson, F.T., "Dynamic Model of Drum-Type Boiler System", Proceedings of 1965 Power Industry Computer Application Conference, p. 260.

12.4 Schmidt, J.R., Clark, D.R., "Analog Simulation Technique for Modeling Parallel-Flow Heat Exchangers", IEEE Paper 31 PP 67-143, Winter Power Meeting, Feb. 1967.

12.5 Thal-Larsen, H., "Dynamics of Heat Exchangers and Their Models", Journal of Basic Engineering, Trans. ASME, Series D. Vol. 82, 1960, p. 494.

12.6 McDonald, J.P., Kwatny, H.G., "A Mathematical Model for Reheat Boiler-Turbine Generator Systems", IEEE Conference paper 70 CP 221, IEEE Winter Power Meeting, Jan. 1970.

12.7 MacDonald, E.H., and O'Brien, J.T., "Unit Response Testing", IEEE Transactions, PAS-85, July 1966, p. 701.

12.8 Chien, K.L., Ergin, E.I., Ling, C., and Lee, A., "Dynamic Analysis of a Boiler", ASME Transactions, Vol. 80, pp. 1809-1819.

12.9 Daniels, J.H., Enns, M., Hottenstine, R.D., "Dynamic Representation of a Large Boiler-Turbine Unit", ASME Paper No. 61-SA-69, ASME Summer Annual Meeting, 1961.

12.10 Stephens, W.M., de Mello, F.P., Ewart, D.N., "Simulation as a Design Tool for Plant Jack McDonough Boiler Controls", ISA Proceedings, 7th National Power Instrumentation Symposium, 1964, Vol. 7, pp. 35-46.

APPENDIX A

DYNAMIC SYSTEMS, DIFFERENTIAL EQUATIONS—TRANSIENT AND STEADY STATE SOLUTIONS— OPERATIONAL IMPEDANCE

The study of "control and dynamics" requires the use of certain mathematical tools and techniques which have become an essential part of the technology of control. These tools are all related to methods of solution and analysis of systems described by differential equations. It is not the intent here to go through a detailed theoretical development of the pertinent mathematics that form the basis for the various analysis tools. There are numerous texts that may be referenced for this purpose, some of which are in the reference list.[25] The treatment in these appendices will be in the form of a brief review of some basic techniques to supplement and support the material in the main text on "Generation Dynamics and Control."

Dynamic Systems

The behavior of dynamic systems is expressed by differential equations relating the systems' variables. In many cases these equations turn out to be or can be approximated by linear differential equations. When this is the case, classical or closed form solutions can be obtained.

For the general case of non-linear differential equations, solutions must be sought through the use of simulation by analog computation methods or by numerical integration techniques carried out on digital computers.

Although any problem can be solved by these simulation methods, the insight that can be derived from linear system analysis is invaluable as a guide to control system design and performance evaluations.

System Differential Equations

Dynamic systems can be thermal, mechanical, electrical or a combination of all these. In order to stay on familiar ground we will illustrate with an electrical example and limit the discussion to linear differential equations.

Consider the circuit in Fig. A-1.

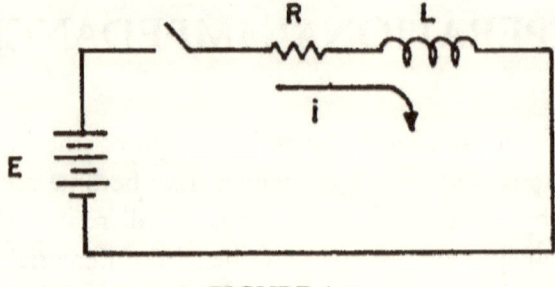

FIGURE A-1

The differential equation is

$$E = iR + L\frac{di}{dt} \tag{A-1}$$

By separating variables, equation A-1 can be put in the form of A-2

$$L \int \frac{di}{E - Ri} = \int dt \tag{A-2}$$

Integration of equation A-2 yields

$$-\frac{L}{R} \ell n(E - Ri) = t + C_1 \tag{A-3}$$

where C_1 is the constant of integration.

Equation A-3 may also be expressed in exponential form as

$$i = E/R + C_2 e^{-(R/L)t} \tag{A-4}$$

where C_2 is derived from constants of integration which in turn are determined from initial conditions in energy storage elements. The current in inductance L at time t = o before the switch is closed is i_o = o.

Substitution of i = o at t = o in equation A-4 yields

$$C_2 = - E/R \tag{A-5}$$

and equation A-4 can be written as

$$i = \frac{E}{R}\left[1 - e^{-(R/L)t}\right] \tag{A-6}$$

plotted in Fig. A-2 as function of time.

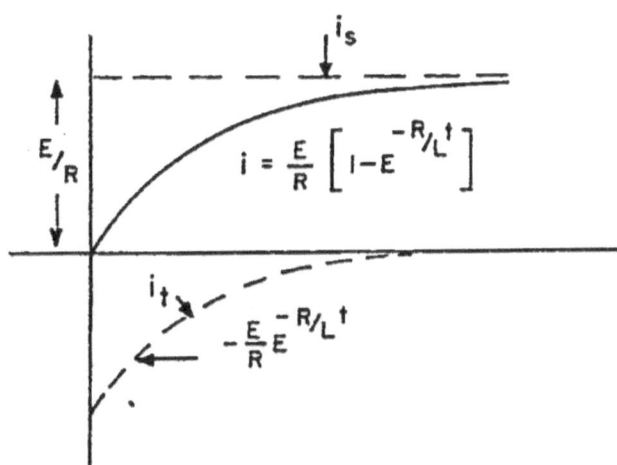

FIGURE A-2

This classical solution can be recognized as containing two components:

(1) The steady state component

$$i_s = E/R \tag{A-7}$$

which has the same form as the applied voltage.
(2) The transient component

$$i_t = -\frac{E}{R}\varepsilon^{-(R/L)t} \tag{A-8}$$

which decays exponentially to zero.

An alternate method of solution for the current in the circuit of Fig. A-1 is to solve separately for the steady state and transient components as follows:

Let $\qquad i = i_s + i_t$ \hfill (A-9)

Substituting equation A-9 in equation A-1

$$E = i_s R + L\frac{di_s}{dt} + Ri_t + L\frac{di_t}{dt} \tag{A-10}$$

Since E is constant di/dt = 0.
By definition also, i_t and $di_t/dt = 0$ in the steady state.

Hence $\qquad i_s = E/R$ \hfill (A-11)

Substituting equation A-11 into equation A-10 yields the relation from which the transient component may be solved, i.e.:

$$Ri_t + L\frac{di_t}{dt} = 0 \tag{A-12}$$

By definition of the transient component it is exponential in nature, and one can express it as

$$i_t = I_{T\varepsilon}^{pt} \tag{A-13}$$

Substituting equation A-13 in equation A-12

$$(R + pL) \, I_T \varepsilon^{pt} = 0 \qquad \text{(A-14)}$$

From equation A-14 the value of p is determined as

$$p = - R/L \qquad \text{(A-15)}$$

which can be noted, is independent of the applied voltage E but merely a function of the circuit parameters.

Substituting equation A-15 in equation A-13 we have

$$i_t = I_T \varepsilon^{-(R/L)t} \qquad \text{(A-16)}$$

The value for I_T is determined from initial conditions, i.e., the value of i at t = o. The total current $i = i_s + i_t$

$$i = E/R + I_T \varepsilon^{-(R/L)t}$$

At t = o

$$i = E/R + I_T = 0$$

whence $I_T = - E/R$.

The resultant expression for current is naturally the same as obtained by the classical solution:

$$i = E/R - E/R \varepsilon^{(-R/L)t}$$

This example was for a system described by a first order differential equation. For the general system of nth order the transient component must be chosen in the form $I_n e^{p_n t}$. The values of p_n are evaluated by setting the coefficient of $I_n e^{p_n t}$ equal to zero. These principles are covered by other more commonly used methods of differential equation solution such as those which use the LaPlace Transform method. We will not pursue the classical method of differential equation solution any further except to introduce the idea of the "characteristic equation" which is basic and which will also be derived with the LaPlace Transform method.

Characteristic Equation

The choice of the exponential form for the transient component of the solution of a linear set of differential equations was guided by the results of the classical solution.

This form of solution has the following interesting properties.

If
$$i = I e^{pt} \qquad \text{(A-17)}$$

then
$$\frac{di}{dt} = Ipe^{pt} \qquad \text{(A-18)}$$

and
$$\frac{d^2 i}{dt^2} = Ip^2 e^{pt} \qquad \text{(A-19)}$$

Also
$$\int i\,dt = \frac{I}{p} e^{pt} \qquad \text{(A-20)}$$

Hence in the equation for the transient solution, if i is substituted by Ie^{pt}, the terms in the equation

$$\frac{d^n}{dt^n}$$

are replaced by p^n and the terms $\int^n (\ )\,dt^n$ are replaced by $1/p^n$. For instance the differential equation

$$a_o \frac{d^n i}{dt^n} + a_1 \frac{d^{n-1}}{dt^{n-1}} i + \ldots a_n i + a_{n+1} \int i\,dt$$

$$+ \ldots a_{n+m} \int^m i\,dt^m = f(t) \qquad \text{(A-21)}$$

with $i = Ie^{pt}$ becomes

$$(a_o p^n + a_1 p^{n-1} + \ldots a_n + \frac{a_{n+m}}{p^m}) Ie^{pt} = f(t) \qquad \text{(A-22)}$$

The polynomial form of the equation formed by substituting derivatives and integrals by the appropriate p and 1/p operators is called the operational form of the equation.

The basic equation which determines the transient modes is independent of the applied forcing function f (t). It is known as the system characteristic equation and in the example above is

$$(a_0 p^n + a_1 p^{n-1} + \ldots a_n + \frac{a_{(n+1)}}{p} + \ldots \frac{a_{(n+m)}}{p^m}) = 0$$

or

$$(a_0 p^{n+m} + a_1 p^{n+m-1} + \ldots a_n p^m + \ldots a_{n+m}) = 0 \qquad \text{(A-23)}$$

The values of p which satisfy equation 23 are the roots of the characteristic equation and are the values that appear in the solution $i = I_n e^p n^t$ determining the transient modes of the system.

The characteristic equation of a system and its roots are fundamental to the evaluation of response and stability of dynamic systems.

Example 1

Fig. A-3 shows a series RLC network connected to a source E(t) by switch S at t = o.

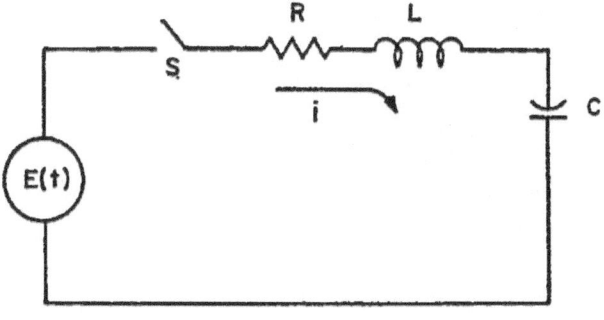

FIGURE A-3

The circuit equation for the time after closure of s is

$$Ri + L\frac{di}{dt} + \frac{1}{C}\int i\,dt = E(t) \tag{A-24}$$

Breaking up the solution into two components (steady state, with same form as E(t) and transient) let us examine the case where E = constant.

The <u>Steady State solution</u> is found from equation A-24 by noting that i_s has the same form as E(t)

i.e. $$\frac{di_s}{dt} = 0 \tag{A-25}$$

Substituting equation A-25 in equation A-24

$$Ri_s + \frac{1}{C}\int i_s\,dt = E \tag{A-26}$$

The only way that i_s can be a constant and satisfy equation A-26 is for $i_s = 0$ and $\frac{1}{C}\int i_s\,dt = E$. The <u>Transient Solution</u> is found by writing the left hand side of equation A-24 in operational form and setting it to zero.

i.e. $$(Lp + R + \frac{1}{Cp})\,i = 0 \tag{A-27}$$

where i is of the form $I_n e^{p_n t}$.

The characteristic equation of equation A-27 is

$$CLp^2 + RCp + 1 = 0 \tag{A-28}$$

which yields the roots

$$P_1 = -\frac{R}{2L} + \frac{1}{2}\sqrt{(\frac{R}{L})^2 - \frac{4}{LC}}$$

and
$$P_2 = -\frac{R}{2L} + \frac{1}{2}\sqrt{(\frac{R}{L})^2 - \frac{4}{LC}}$$

Depending on whether $(R/L)^2$ > or < $(4/LC)$ the roots P_1 and P_2 will be real or complex.

The expression for the transient current is

$$i_t = I_1 e^{P_1 t} + I_2 e^{P_2 t} \qquad \text{(A-29)}$$

To evaluate I_1 and I_2, we note that the system's initial conditions were i = 0 and the voltage across the condenser = 0.

i.e.
$$i = 0 \qquad \text{(A-30)}$$

and
$$\frac{I}{C} \int i \, dt = 0 \qquad \text{(A-31)}$$

Since the steady state component i_s = 0, condition equation A-30 applied to equation A-29 yields

$$I_1 = -I_2 \qquad \text{(A-32)}$$

Also, applying equation A-30 and A-31 to equation A-24 at t = 0^+

$$L\frac{d}{dt}\left(I_1 0^{P_1 t} + I_2 e^{P_2 t}\right)\Big|_{t=0} = E$$

i.e.
$$I_1 p_1 + I_2 p_2 = \frac{E}{L} \qquad \text{(A-33)}$$

Solving equation A-32 and equation A-33

$$I_1 = \frac{E}{L(p_1 - p_2)}, \; I_2 = \frac{E}{L(p_2 - p_1)} \qquad \text{(A-34)}$$

And the total solution for i is

$$i = \frac{E}{L}\left[\frac{e^{p_1 t} - e^{p_2 t}}{p_1 - p_2}\right] \tag{A-35}$$

For the case where p_1 and p_2 are complex conjugate roots $\left(\left(\frac{R}{L}\right)^2 < \frac{4}{LC}\right)$,

i.e.; where $p_1 = -\alpha + j\beta$

and $p_2 = -\alpha - j\beta$ (A-36)

Substituting these values in equation A-35

$$i = \frac{E}{L}e^{-\alpha t}\left[\frac{e^{j\beta t} - e^{-j\beta t}}{2j\beta}\right] \tag{A-37}$$

which can be expressed as, from the definition of $\sin \beta t$

$$i = \frac{E}{L}\frac{e^{-\alpha t}}{\beta}\sin \beta t \tag{A-38}$$

Figure A-4 shows the nature of the current transient.

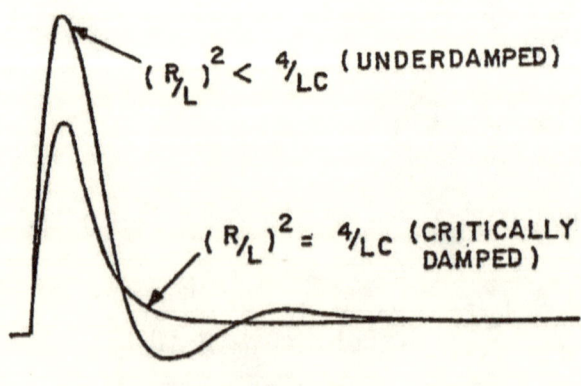

FIGURE A-4

Example 2

Take the same example except that let the applied voltage be a sinusoidal function E = E cos ωt with the switch again closed at t = o.

Again taking up the steady state solution, equation A-24 becomes

$$Ri_s + L\frac{di_s}{dt} + \frac{1}{C}\int i_s dt = \frac{E}{2}(e^{+j\omega t} + e^{-j\omega t})\qquad \text{(A-39)}$$

where $$\frac{e^{+j\omega t} + e^{-j\omega t}}{2}$$

is the exponential form of cos ωt.

Since i_s by definition will be of the same form as the applied voltage, we may further divide i_s into components corresponding to the applied voltage components

$$i_{s_+} = I_+ e^{+j\omega t}\qquad \text{(A-40)}$$

$$i_{s_-} = I_- e^{-j\omega t}\qquad \text{(A-41)}$$

where I_+ is the complex magnitude of i_{s_+} and I_- is the complex magnitude of i_{s_-}. Considering these components individually, from equation A-39

$$R\left[I_+ e^{+j\omega t}\right] + L\frac{d}{dt}\left[I_+ e^{+j\omega t}\right] + j\int I_+ e^{+j\omega t} dt = \frac{E}{2}e^{+j\omega t}$$

or $$RI_+ e^{j\omega t} + LI_+ j\omega e^{+j\omega t} + \frac{I_+}{Cj\omega}e^{j\omega t} = \frac{E}{2}e^{+j\omega t}\qquad \text{(A-42)}$$

Dividing both sides by $e^{+j\omega t}$

$$RI_+ + LI_+ j\omega + \frac{I_+}{Cj\omega} = \frac{E}{2}$$

or $$I_+ = \frac{E}{2\,(R + j(\omega L - \frac{1}{\omega C}))}\qquad \text{(A-43)}$$

A similar derivation for I_- yields

$$I_- = \frac{E}{2\left(R - j\left(\omega L - \dfrac{1}{\omega C}\right)\right)}$$

(A-44)

Equation A-43 can be expressed as

$$I_+ = \frac{E\, e^{-j\theta}z}{2\,Z}$$

(A-45)

where

$$Z = \sqrt{R^2 + \left(\omega L - \frac{1}{\omega C}\right)^2}$$

(A-46)

$$\theta_z = \tan^{-1} \frac{\left(\omega L - \dfrac{1}{\omega C}\right)}{R}$$

(A-47)

Likewise

$$I_- = \frac{E\, e^{+j\theta}z}{2\,Z}$$

(A-48)

The total steady state current $i_s = i_{s_+} + i_{s_-}$. Using equation A-42 and equation A-43 and substituting equation A-47 and equation A-48,

$$i_s = \frac{E}{Z}\left(\frac{e^{j(\omega t - \theta_z)} + e^{-j(\omega t - \theta_z)}}{2}\right)$$

$$i_s = \frac{E}{Z}\cos(\omega t - \theta_z)$$

(A-49)

Z is the impedance of the network and indicates the ratio of voltage to current in the steady state for a sinusoidally varying applied voltage. Equation A-49 is of the same form as the applied voltage $E \cos \omega t$. Its magnitude is E/Z and its phase angle with respect to the applied voltage sinusoid is θ_z.

The concept of operational impedance $Z(p) = R + Lp + \dfrac{1}{Cp}$ is self-evident from equation A-27. By substituting $p = j\omega$ one can derive the impedance to a fixed alternating voltage of frequency ω rads/sec.

These concepts are important in the application of "frequency response" techniques which characterize the system in terms of its behavior as function of the frequency of the exciting function, ω.

Although the example was for an electric circuit, yielding the relationship between current and voltage the method is equally applicable to any variables of a system, be they mechanical, electrical or thermal, as long as they are related by linear differential equations.

APPENDIX B

LAPLACE TRANSFORMS[26,27]

The previous sections have reviewed the classical method of solving linear differential equations. We have seen how the transient solution and steady state solution are derived and have developed the concept of operational impedance and impedance to a constant frequency applied excitation function.

These same results can be derived in a greatly simplified fashion through the use of the direct and inverse LaPlace transform which uses one approach for both the steady state and transient solution. LaPlace transform operational calculus is the cornerstone of control system analysis.

Some Basic Theorems of the LaPlace Transform

A function of time $f(t)$ has a LaPlace transform $F(s)$ where

$$F(s) = \int_0^\infty f(t)\varepsilon^{-st}dt \tag{B-1}$$

The value of the LaPlace transform lies in the fact that a differential equation or expression of the variable "t" transforms into an algebraic equation or expression of the variable "s." This algebraic expression in turn may be operated upon and converted to a form easily recognized in terms of a time function. The process of obtaining the time function from the transform expression is called taking the inverse LaPlace transformation. Mathematical operations which in the time domain involve convolution, convert to simple algebraic multiplications in the s domain. A summary of the important theorems governing the use of the LaPlace transform are:

1. $L[f(t)] = \int_0^\infty f(t)\, \varepsilon^{-st}\, dt$ (B-2)

2. The inverse LaPlace transformation L-1 is defined implicitly by the relation

$L-1\{[f(t)]\} = f(t)\ 0 \le t$ (B-3)

3. If the functions $f(t)$, $f_1(t)$ and $f_2(t)$ have L transforms $F(s)$, $F_1(s)$ and $F_2(s)$ respectively and "a" is a constant of a variable which is independent of t and s, then

$L[a\, f(t)] = a\, F(s)$ (B-4)

and

$L[f_1(t) \pm f_2(t)] = F_1(s) \pm F_2(s)$ (B-5)

Also

$L^{-1}[a\, F(s)] = a\, f(t)\ 0 \le t$ (B-6)

and

$L^{-1}[F_1(s) \pm F_2(s)] = f_1(t) \pm f_2(t)\ 0 \le t$ (B-7)

4. If a function $f(t)$ has the L transform $F(s)$, then

$L\dfrac{df(t)}{dt} = s\, F(s) - f(0+)$ (B-8)

where $f(o+)$ is the value of $f(t)$ at t = o+. It is evident then that

$L\dfrac{d^2 f(t)}{dt^2} = s^2\, F(s) - s\, f(0) - f'(0)$ (B-9)

and

$$L\, f^{(n)}\,(t) = s^n F(s) - \sum_{k=1}^{n} f^{(k-1)}\,(0)\; s^{n-k} \qquad \text{(B-10)}$$

5. If the function $f(t)$ has the transform $F(s)$, its integral
$$f^{(-1)}(t) = \int f(t)dt = \int_0^t f(t)dt + f^{(-1)}(0^+) \text{ has the transform}$$

$$L\left[\int f(t)\,dt\right] = \frac{F(s)}{s} + \frac{f^{(-1)}(0+)}{s} \qquad \text{(B-11)}$$

Similarly,

$$L\left[f^{(-2)}\,(t)\right] = \frac{F(s)}{s^2} + \frac{f^{(-1)}(0)}{s^2} + \frac{f^{(-2)}(0)}{s} \qquad \text{(B-12)}$$

and

$$L\left[f^{(-n)}\,(t)\right] = \frac{F(s)}{s^n} + \sum_{k=1}^{n} \frac{f^{(-k)}(0)}{s^{n-k+1}} \qquad \text{(B-13)}$$

6. The LaPlace transforms of some common functions are as follows:

f(t) $0 \le t$	F(s)
1 or u(t)	$\dfrac{1}{s}$
$e^{-\alpha t}$	$\dfrac{1}{s+\alpha}$
$\dfrac{1}{\beta}\sin \beta t$	$\dfrac{1}{s^2 + \beta^2}$
$\cos \beta t$	$\dfrac{s}{s^2 + \beta^2}$

$\dfrac{1}{\beta} e^{-\alpha} \sin \omega t$	$\dfrac{1}{(s+\alpha)^2 + \beta^2}$
t	$\dfrac{1}{s^2}$
$\dfrac{1}{(n-1)!} t^{n-1}$	$\dfrac{1}{s^n}$
$t\, e^{-\alpha t}$	$\dfrac{1}{(s+\alpha)^2}$
$\dfrac{1}{(n-1)!} t^{n-1} e^{-\alpha t}$	$\dfrac{1}{(s+\alpha)^n}$
$u\,(t-a)$	$\dfrac{1}{s} e^{-as}$
$u\,(t-a) - u\,(t-b)$	$\dfrac{1}{s}(e^{-as} - e^{-bs})$

(B-14)

Unit impulse

$$u_1(t) = \lim_{a \to 0} \frac{u(t) - u(t-a)}{a}$$

The L Transformation

We shall now apply LaPlace transform methods to the solution of differential equations. Take for instance:

$$A\frac{d^2 y}{dt^2} + B\frac{dy}{dt} + Cy = f(t)\ldots, \; y \triangleq y(t) \tag{B-15}$$

in which A, B, and C are known constants.

The unknown y(t) will be called the response function and the known f(t) will be called the driving function. The initial values of the unknown and its first derivative are y(o) and y'(o).

Applying the L transformation to both members of equation B-15

$$L\left[A\frac{d^2y}{dt^2} + B\frac{dy}{dt} + Cy\right] = L\left[f(t)\right] \qquad (B-16)$$

Calling F(s) the L transform of f(t) and $Y(s) \triangleq L\left[y(t)\right]$, the response transform. Then, using equation B-8 and equation B-9

$$L[y'(t)] = sY(s) - y(0)$$

and

$$L[y''(t)] = s^2 Y(s) - y(0) s - y'(0)$$

This discloses the way in which the initial conditions y(o) and y'(o) are incorporated in the solution during the process of transformation.

Equation B-16 becomes

$$AL\left[\frac{d^2y}{dt^2}\right] + BL\left[\frac{dy}{dt}\right] + CL\left[y\right] = L\,f(t)$$

$$A\left[s^2Y(s) - y(0)\,s - y'(0)\right] + B\left[sY(s) - y(0)\right] + CY(s) = F(s)$$

or

$$(A s^2 + B s + C)\, Y(s) = F(s) + y(0)(A s + B) + y'(0)\, A \qquad (B-17)$$

Equation B-17 is called a transform equation. The polynomial coefficient of Y(s)—in this case (As² ± Bs + C)—is called the characteristic function since it completely characterizes the physical system described by the differential equation. Note that this is identical with the system characteristic equation derived in Appendix A, except for the variable "s" instead of the operator "p." The equation formed by setting it to zero is called the characteristic equation of the system. Solving equation B-17 algebraically,

$$Y(s) = \frac{1}{As^2 + Bs + C} \left[F(s) + y(0)(As+B) + y'(0)A \right] \quad \text{(B-18)}$$

This algebraic equation has a form which will be found typical of all transform solutions, viz:

Response transform = System function × Excitation function

The system function in this example is the reciprocal of the characteristic function, but in general it will be a fraction of which the characteristic function is the denominator. It incorporates in one function all the essential knowledge regarding the physical system.

The excitation function includes the driving transform and the initial conditions. It contains all the essential specifications of the excitations applied to the system.

When the form of the driving function f(t) is specified, the algebraic form of Y(s) can be determined and

$$y(t) = L^{-1}\left[Y(s) \right] = L^{-1} \left[\frac{F(s) + y(0)(As+B) + y'(0)A}{As^2 + Bs + C} \right] \quad \text{(B-19)}$$

If Y(s) were an algebraic function of the form of any one of the various transforms listed so far, the inverse could be written immediately by reference to the table. But since Y(s) is a more complicated function than listed, such a direct method of determining the inverse transform fails.

This difficulty may be surmounted by resolving the function into a sum of simpler components whose inverse transforms are readily recognized.

L^{-1} Transformation

Consider the general rational algebraic fraction

$$F(s) = \frac{A(s)}{B(s)} \triangleq = \frac{a_p s^p + a_{p-1} s^{p-1} + \ldots a_1 + a_0}{s^q + b_{q-1} s^{q-1} + \ldots b_1 s + b_0} \quad \text{(B-20)}$$

where $p \leq q$

By solving for the roots of the equation B(s) = 0, and calling these s_1, s_2 ... s_q, the fraction may be expressed as

$$F(s) = \frac{A(s)}{B(s)} = \frac{A(s)}{(s-s_1)(s-s_2)(s-s_3)....(s-s_q)} \qquad \text{(B-21)}$$

and the above may in turn be written as a sum of partial fractions, each partial fraction having for its denominator one of the fa tors of B(s).

There will be "q" of these partial fractions.

i.e.,

$$\frac{A(s)}{B(s)} = \frac{K_1}{(s-s_1)} + \frac{K_2}{(s-s_2)} + \frac{K_3}{(s-s_3)} + ... \frac{K_q}{(s-s_q)} \qquad \text{(B-22)}$$

To evaluate the typical coefficient K_k, multiply both members of equation B-22 by $(s - s_k)$ obtaining

$$\frac{(s-s_k)A(s)}{B(s)} = K_1 \frac{(s-s_k)}{(s-s_1)} + K_2 \frac{(s-s_k)}{(s-s_2)} + ..$$

$$+ K_k + K_q \frac{(s-s_k)}{(s-s_q)} \qquad \text{(B-23)}$$

In the fraction forming the left member of equation B-23, $(s - s_k)$ is a factor of both numerator and denominator and should be divided out. Then letting $s = s_k$, this left member becomes a numer, and in the right member all terms except K_k become zero.

i.e.,
$$K_k = \left[\frac{(s-s_k)A(s)}{B(s)} \right] s = s_k \qquad \text{(B-24)}$$

$$= \frac{A(s_k)}{(s_k-s_1)(s_k-s_2)...(s_k-s_{k-1})(s_k-s_{k+1})...(s_k-s_q)}$$

But
$$(s_k-s_1)(s_k-s_2)...(s_k-s_{k-1})(s_k-s_{k+1})...(s_k-s_q)$$

$$= \left[\frac{d}{ds} B(s) \right]_{s=sk} \triangleq B'(s_k) \tag{B-25}$$

so equation B-24 can be written

$$\frac{A(s)}{B(s)} = \sum_{k=1}^{9} \frac{A(s_k)}{B'(s_k)} \frac{1}{(s-s_k)} \tag{B-26}$$

The actual problem of inverse transformation is now a simple one.

$$L^{-1} \left[\frac{1}{s-s_k} \right] = e^{s_k t}$$

The above holds for $\dfrac{A(s)}{B(s)}$ having first order poles only; i.e., the roots of B(s) being

$$(s + s_1)^n (S + s_2)^m (s + s_3)^l \dots$$

where n, m, l = 1
and $s_1 \neq s_2 \neq s_3 \dots$

Example:

Find the $L^{-1} \left[\dfrac{a_1 s + a_0}{(s+\alpha_1)(s+\alpha_2)(s+\alpha_3)} \right]$

in which α_1, α_2, and α_3, are real numbers, all different.

$$L^{-1} \left[\frac{a_1 s + a_0}{(s+\alpha_1)(s+\alpha_2)(s+\alpha_3)} \right] = K_1 e^{-\alpha_1 t} + K_2 e^{-\alpha_2 t} + K_3 e^{-\alpha_3 t}$$

in which

$$K_1 = \left[\frac{a_1 s + a_0}{(s+\alpha_2)(s+\alpha_3)} \right]_{s=-\alpha_1} = \frac{-a_1 \alpha_1 + a_0}{(-\alpha_1 + \alpha_2)(-\alpha_1 + \alpha_3)}$$

$$K_2 = \left[\frac{a_1 s + a_0}{(s+\alpha_1)(s+\alpha_3)}\right]_{s=-\alpha_2} = \frac{-a_1 a_2 + a_0}{(-\alpha_2 + \alpha_1)(-\alpha_2 + \alpha_3)}$$

$$K_3 = \left[\frac{a_1 s + a_0}{(s+\alpha_1)(s+\alpha_2)}\right]_{s=-\alpha_3} = \frac{-a_1 a_3 + a_0}{(-\alpha_3 + \alpha_1)(-\alpha_3 + \alpha_2)}$$

Special case: One pole lies at the Origin.

In $\dfrac{A(s)}{B(s)}$ of equation B-21 let $s_1 = 0$, then

$$\frac{A(s)}{B(s)} = \frac{A(s)}{s(s-s_2)(s-s_3)(s-s_4)\ldots(s-s_q)} = \frac{A(s)}{s\,B_1(s)}$$

where $B_1(s) = \dfrac{B(s)}{s}$

The form above occurs frequently. It arises, for example, when the excitation function is a constant step and the system function does not have a pole or a zero at $s = 0$.

The final result can be shown to be

$$L\left[\frac{A(s)}{s\,B_1(s)}\right] = \frac{A(0)}{B_1(0)} + \sum_{k=2}^{q} \frac{A(s_k)}{s_k B'(s_k)}\, e^{s_k t} \qquad (B-27)$$

Example:

Find　　　$L^{-1}\{\dfrac{a_1 s + a_0}{s[(s+\alpha)^2 + \beta^2]}\}$

here　　　$A(s) = (a_1 s + a_0)$

$$B_1(s) = \left[(s+\alpha)^2 + \beta^2\right]$$

$B_1'(s) = 2(s + \alpha)$ and $s_2, s_3 = -\alpha \pm j\beta$ and $s_1 = 0$

Using equation B-27,

$$L^{-1} \left\{ \frac{a_1 s + a_0}{s[(s+\alpha)^2 + \beta^2]} \right\}$$

$$= \frac{A(o)}{B_1(o)} + K_2 \varepsilon^{(-\alpha+j\beta)t} + K_3 \varepsilon^{(-\alpha-j\beta)t}$$

where

$$K_2 = \left[\frac{a_1 s + a_0}{2s(s+\alpha)} \right]_{s=-\alpha+j\beta}$$

$$= \frac{a_0 - a_1\alpha + ja_1\beta}{2j\beta(-\alpha+j\beta)}$$

$$= \frac{1}{2\beta\beta_0} \left[(a_0 - a_1\alpha)^2 + a_1^2\beta^2 \right]^{1/2} e^{j(\phi-\frac{\pi}{2})}$$

where $\quad \beta_0^2 = \alpha^2 + \beta^2$

and $\quad \phi = \left[\tan^{-1} \frac{a_1\beta}{a_0 - a_1\alpha} - \tan^{-1} \frac{\beta}{-\alpha} \right]$

Similarly $\quad K_3 = \left[\frac{a_1 s + a_0}{2s(s+\alpha)} \right]_{s=-\alpha j\beta}$

$$= \overline{K}_2 \text{ (conjugate } K_2)$$

Coefficients K_2 and K_3 are conjugate complex numbers.
The final result can be written

$$L^{-1} \left[\frac{a_1 s + a_0}{s\left[(s+\alpha)^2 + \beta^2\right]} \right]$$

$$= \frac{a_0}{\beta_0^2} + \frac{1}{\beta\beta_0} \left[(a_0 - a_1\alpha)^2 + a_1^2\beta^2 \right]^{1/2} e^{-\alpha t} \sin(\beta t + \phi) \qquad \text{(B-28)}$$

A convenient rule to remember in obtaining the L^{-1} of a function where one pair of roots are $[(s + \alpha)^2 + \beta^2]$ is as follows:

The time function corresponding to the roots $[(s + \alpha)^2 + \beta^2]$ in

$$\frac{A(s)}{C(s)[(s+\alpha)^2 + \beta^2]} \quad \text{is } K e^{-\alpha t} \sin(\beta t + \phi) \qquad \text{(B-29)}$$

where

$$K = \frac{|A(-\alpha + j\beta)|}{\beta |C(-\alpha + j\beta)|} \qquad \text{(B-30)}$$

and $\qquad \phi = $ angle of $A(-\alpha + j\beta)$ minus angle of $C(-\alpha + j\beta)$ $\qquad$ (B-31)

x ———————— x

Similarly for a function where one pair of roots are $(s^2 + \omega^2)$ the time function component corresponding to these roots in the function

$$\frac{A(s)}{C(s)(s^2 + \omega^2)} \qquad \text{(B-32)}$$

can be obtained as

$$K \sin(\omega t + \phi) \qquad \text{(B-33)}$$

where

$$K = \frac{|A(j\omega)|}{\omega |C(j\omega)|} \qquad \text{(B-34)}$$

and $\phi = $ angle of $A(j\omega)$ minus angle of $C(j\omega)$ $\qquad$ (B-35)

Multiple Order Poles

Consider the function F(s) which has poles of higher order. (s_1 occurs m_1 times, s_2 occurs m_2 times, etc.)

$$F(s) = \frac{A(s)}{B(s)} = \frac{A(s)}{(s - s_1)^{m_1} (s - s_2)^{m_2} .. (s - s_n)^{m_n}} \qquad (B\text{-}36)$$

The fraction $\dfrac{A(s)}{B(s)}$ may be resolved into a sum of partial fractions. For each pole s_k of multiplicity m_k there are m_k partial fractions

$$\frac{K_{k1}}{(s - s_k)^{m_k}}, \frac{K_{k2}}{(s - s_k)^{m_{k-1}}}, \ldots \frac{K_{km_k}}{(s - s_k)}$$

in which the K's are constants yet to be determined.

Thus the expansion of A(s)/B(s) is

$$\frac{A(s)}{B(s)} = \frac{K_{11}}{(s - s_1)^{m_1}} + \frac{K_{12}}{(s - s_1)^{m_1 - 1}} + \ldots \frac{K_{1j}}{(s - s_1)^{m_1 - j + 1}} + \ldots \frac{K_{1m_1}}{(s - s_1)}$$

$$+ \ldots$$

$$+ \frac{K_{k1}}{(s - s_k)^{m_k}} + \frac{K_{k2}}{(s - s_k)^{m_k - 1}} + \ldots \frac{K_{kj}}{(s - s_k)^{m_k - j + 1}} + \ldots \frac{K_{km_k}}{s - s_k}$$

$$+ \ldots$$

To evaluate the K_k coefficients, first multiply both members of the equation above by $(s - s_k)^{m_k}$ obtaining

$$\frac{(s - s_k)^{m_k} A(s)}{B(s)} = K_{k1} + K_{k2} (s - s_k) + K_{k3} (s - s_k)^2 + \ldots + K_{km_k} (s - s_k)^{m_k - 1}$$

$$+ (s - s_k)^{m_k} \left[\frac{K_{11}}{(s - s_1)^{m_1}} + \ldots + \frac{K_{n_{m_n}}}{s - s_n} \right]$$

In the left member $(s - s_k)^{n_k}$ cancels out with that factor which is also a part of B(s). Letting $s = s_k$, this left member becomes a number which should correspond to K_{k1} of the right hand side since all other terms would be zero.

In order to obtain the other coefficients, we note that by differentiating both sides with respect to s, the following expression results:

$$\frac{d}{ds}(s - s_k)^{m_k} \frac{A(s)}{B(s)} = K_{k2} + 2K_{k3}(s - s_k) + \ldots + (M_k - 1)K_{k_{m_k}}(s - s_k)^{m_{k-2}}$$

$$+ \frac{d}{ds}(s - s_k) \left[\frac{K_{11}}{(s - s_1)^{m_1}} + \ldots + \frac{K_{n_{m_n}}}{(s - s_n)} \right]$$

Letting $s = s_k$, we note that K_{k2} is equal to the number resulting from the evaluation of

$$\left[\frac{d}{ds}(s - s_k)^{m_k} \frac{A(s)}{B(s)} \right]_{s = s_k} \tag{B-37}$$

Similarly for the other terms

$$K_{k3} = \frac{1}{2!} \frac{d^2}{ds^2}(s - s_k)^{m_k} \frac{A(s)}{B(s)} \Bigg]_{s = s_k} \tag{B-38}$$

and

$$K_{kj} = \frac{1}{j - 1!}(s - s_k)^{m_k} \frac{A(s)}{B(s)} \Bigg]_{s = s_k} \tag{B-39}$$

Example:

Find

$$L^{-1}\left[\frac{a_2s^2 + a_1 s + a_0}{(s+\alpha)^3 s^2}\right] \tag{B-40}$$

$$L^{-1}\left[\frac{a_2s^2 + a_1 s + a_0}{(s+\alpha)^3 s^2}\right] = L^{-1}\left[\frac{K_{11}}{(s+\alpha)^3} + \frac{K_{12}}{(s+\alpha)^2} + \frac{K_{13}}{(s+\alpha)} + \frac{K_{21}}{s^2} + \frac{K_{22}}{s} + \right] \tag{B-41}$$

$$= (\frac{K_{11}}{2!} t^2 + K_{12}t + K_{13})e^{-\alpha t} + K_{21} t + K_{22} \tag{B-42}$$

where

$$K_{11} = \left[\frac{a_2s^2 + a_1 s + a_0}{s^2}\right]_{s = -\alpha} = \frac{a_2\alpha^2 - a_1\alpha + a_0}{\alpha^2} \tag{B-43}$$

$$K_{12} = \left[\frac{d}{ds}\frac{a_2^2 + a_1 s + s_0}{s^2}\right]_{s = -\alpha} = \frac{-a_1\alpha + 2a_0}{\alpha^3} \tag{B-44}$$

$$K_{13} = \frac{1}{2!}\left[\frac{d^2}{ds^2}\frac{a_2s^2 + a_1 s + a_0}{s^2}\right]_{s = -\alpha} = \frac{-a_1\alpha + 2a_0}{\alpha^4} \tag{B-45}$$

$$K_{21} = \left[\frac{a_2s^2 + a_1 a + a_0}{(s + \alpha)^3}\right]_{s = 0} = \frac{a_0}{\alpha^3} \tag{B-46}$$

$$K_{22} = \left[\frac{d}{ds}\frac{a_2s^2 + a_1 s + a_0}{(s +\alpha)^3}\right]_{s = 0} = \frac{a_1\alpha - 3a_0}{\alpha^4} \tag{B-47}$$

Let us complete this section by taking the same examples as in Appendix A.

Consider the circuit of Fig. B-1.

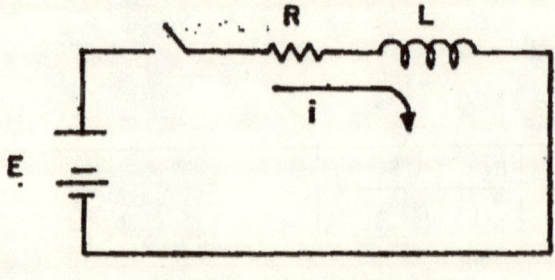

FIGURE B-1

The differential equation for condition after closing of the switch at t = 0‡ is

$$E = iR + L \frac{di}{dt} \tag{B-48}$$

Taking the L transform of both sides of equation B-48

$$E(s) = Ri(s) + Lsi(s) - Li(o) \tag{B-49}$$

where E(s) denotes L E(t)
and i(s) denotes L i(t)
and i(o) = initial condition of i at t = o

Since E is a constant, its L transform is E/S (see table of transforms equation B-14). Also for this case i(o) = 0.
Hence equation B-49 becomes

$$\frac{E}{s} = i(s) \left[R + Ls \right] \tag{B-50}$$

Solving for

$$i(s) = \frac{E}{s[R + Ls]} \tag{B-51}$$

Equation B-51 is the L transform solution of the current.
To obtain the time domain solution of current we must perform the inverse transform of equation B-51.

i.e.,
$$i(t) = L^{-1} \, i(s) = L^{-1} \, \frac{E}{Ls(s + \frac{R}{L})} \tag{B-52}$$

Using the rules of partial fraction expansion (equation B-22 to equation B-24).

$$i(t) = L^{-1} \left[\frac{K_1}{s} + \frac{K_2}{(s + \frac{R}{L})} \right] \tag{B-53}$$

where

$$K_1 = \frac{E}{(R + Ls)} \bigg|_{s=0} = \frac{E}{R}$$

$$K_2 = \frac{E}{Ls} \bigg|_{s=-\frac{R}{L}} = -\frac{E}{R}$$

Using the LaPlace transform tables equation B-14 to obtain the inverse of equation B-53.

$$i(t) = K_1 + K_2 e^{-\frac{R}{L} t}$$

$$= \frac{E}{R} - \frac{E}{R} e^{-\frac{R}{L} t} \tag{B-54}$$

which is the familiar form of the exponential rise in curren: in the inductive circuit of Fig. B-1.

Of particular interest is the exponential term $\varepsilon^{-R/L\,t}$ which reveals the decay of the transient component.

The coefficient L/R has the dimensions of seconds and is known as the "time constant" of the circuit. This time constant is defined as the time in seconds for the transient term to be reduced to $e^{-1} = 0.359$ of its initial value. Another useful interpretation of the time constant is the time that would be required for the transient to disappear completely if its rate continued at its initial value. (Fig. B-2).

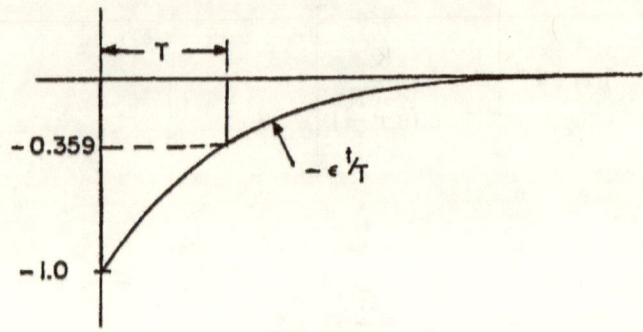

FIGURE B-2

Take now the case treated in Appendix A of the RLC circuit with the sinusoidal excitation voltage.

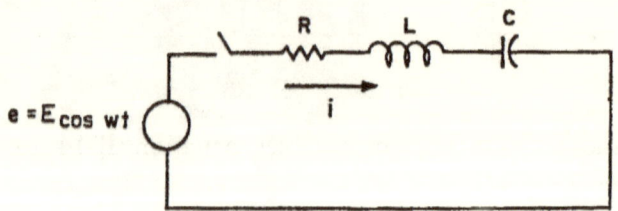

Again the circuit voltage drop equation is

$$Ri + L\frac{di}{dt} + \frac{1}{C}\int idt = E\cos\omega t \qquad (B-55)$$

Taking the L transform of both sides of equation B-55

$$\left(R + L_s + \frac{1}{Cs}\right) i(s) - Li(o) + \frac{V_{co}}{s} = \frac{Es}{s^2 + \beta^2} \qquad (B-56)$$

where i(o) = initial current at t = o

and V_{-co} = initial voltage across the capacitor $\left. \dfrac{1}{C} \int idt \right|_{t=0}$

For the case where these initial conditions are zero, equation B-56 can be expressed as

$$i(s) = \frac{1}{(R + Ls + \dfrac{1}{Cs})} \; \frac{Es}{(S^2 + b^2)} \tag{B-57}$$

Note that equation B-57 is in the form

$$\text{Response transform } i(s) = \left[\text{System function} \; \frac{1}{(R + Ls + \dfrac{1}{Cs})} \right]$$

$$x \left[\text{Excitation function} \; \frac{Es}{(s^2 + \omega^2)} \right]$$

Expressing equation B-57 in terms of poles and zeros

$$i(s) = \frac{EC s^2}{(1 + RCs + LCs^2)(s^2 + \omega^2)} \tag{B-58}$$

$$= \frac{EC s^2}{LC(s + \dfrac{R}{2L} - \dfrac{1}{2}\sqrt{(\dfrac{R}{L})^2 - \dfrac{4}{LC}})(s + \dfrac{R}{2L} + \dfrac{1}{2}\sqrt{(\dfrac{R}{L})^2 - \dfrac{4}{LC}})(s^2 + \omega^2)}$$

where the system poles are the roots of $(1 + RCs + LCs^2)$ which are the same as the roots of the characteristic equation A-26 $(1 + RCp + LCp^2)$ in Appendix A.

The time expression for i(t) is obtained by taking the L^{-1} of equation B-58 using the rules in equations B-29 to B-35 and expressing equation B-58 as

$$i(s) = \frac{EC\,s^2}{LC[(s+\alpha)^2 + \beta^2]\,(s^2+\omega^2)}$$

$$i(t) = K_1\epsilon^{-at}\sin(\beta t + \phi_1) + K_2 \sin(\omega t + \phi_2) \qquad \text{(B-59)}$$

where

$$K_1 = \left. \frac{\left|EC\,s^2\right|}{LC\beta\,\left|s^2 + \omega^2\right|} \right|_{s=-a+j\beta} \qquad \text{See B-29 to B-31}$$

and

$$= \frac{EC\,\left|\left[a^2 - 2ja\beta - \beta^2\right]\right|}{LC\beta\,\left|\left[a^2 - 2ja\beta - \beta^2 + \omega^2\right]\right|}$$

$$= \frac{EC\,\left[(a^2-\beta^2)^2 + 4a^2\beta^2\right]^{1/2}}{LC\,\beta\,\left[(a^2 - \beta^2 + \omega^2)^2 + 4a^2\beta^2\right]^{1/2}}$$

and

$$\phi_1 = \tan^{-1}\frac{-2a\beta}{a^2-\beta^2} - \tan^{-1}\frac{-2a\beta}{a^2-\beta^2+\omega^2}$$

$$K_2 = \left. \frac{\left|EC\,s^2\right|}{\omega LC\,\left|((s+a)^2 + \beta^2)\right|} \right|_{s=j\omega} \qquad \text{See B-32 to B-35}$$

$$= \frac{EC\omega^2}{\omega LC\left[(a^2+\beta^2-\omega^2)^2 + 4a^2\omega^2\right]^{1/2}}$$

and

$$\phi_2 = \pi - \tan^{-1} \frac{2a\omega}{a^2 + \beta^2 - \omega^2}$$

Note that equation B-59 has the total solution, steady state (second term) and transient (first term) obtained by a straight forward routine use of the direct and inverse LaPlace transform.

APPENDIX C

TRANSFER FUNCTIONS, BLOCK DIAGRAMS[25,30]

Recall that in the discussion of equation B-18, Appendix B, the form of the equation was stated as:

Response transform = System function × Excitation function

As will be shown below, another name for the System function is the "System transfer function."

A transfer function is an operational expression describing the incremental functional relationship between two variables. An example will illustrate how a functional relationship or equation relating two variables is expressed in transfer function form. The term "incremental" implies that we are only concerned with changes from a quiescent point, hence initial conditions are assumed zero.

Take the equation

$$e(t) = iR + L\frac{di(t)}{dt} \tag{C-1}$$

describing the relations between current and applied voltage in an R-L circuit.

This equation may be expressed in LaPlace transform form, assuming zero initial conditions, as:

$$e(s) = i(s)R + Lsi(s)$$

$$e_s(s) = i(s)\,[R + Ls] \tag{C-2}$$

By algebraic manipulation we may express equation C-2 as a transfer function between e(s) and i(s)

$$\frac{i(s)}{e(s)} = \frac{1}{R+Ls} \tag{C-3}$$

Figure C-1 shows the schematic representation of equation C-3 in transfer function block diagram form

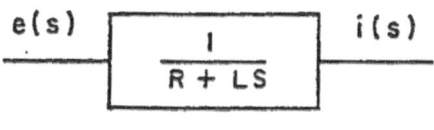

FIGURE C-1

A block diagram is a schematic representation of mathematical relationships or equations between variables. Block diagrams are widely used in the area of control. There are basic block diagram relationships which are useful for reducing the number of branches of block diagrams. These relationships accomplish the same thing as the elimination of variables by substitution, in an array of simultaneous equations. These algorithms that can be used for block diagram reduction are analogous to the formulas that we use in combination of impedances or star-delta, series and parallel transformations so familiar in reduction of networks.

A frequent configuration is the feedback arrangement of Fig. C-2.

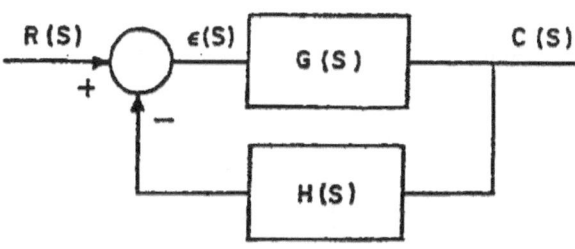

FIGURE C-2

which describes the following relationships

$$C(s) = \varepsilon(s)\, G(s) \tag{C-4}$$

$$\varepsilon(s) = R(s) - C(s)\, H(s) \tag{C-5}$$

where G(s) and H(s) are transfer functions.

Note the symbol <<<IMAGE>>> is a summing or difference junction forming the sum or difference of two or more variables.

Eliminating $\varepsilon(s)$ by substitution of equation C-5 in equation C-4

$$C(s) = [R(s) - C(s)\,H(s)]\,G(s)$$

or $$C(s)\,[1 + G(s)\,H(s)] = R(s)\,G(s)$$

i.e., $$\frac{C(s)}{R(s)} = \frac{G(s)}{1+G(s)\,H(s)} \tag{C-6}$$

It is useful to remember relationship equation C-6 for purposes of block diagram reduction. This relationship defines the closed loop transfer function between the variables C(s) and R(s) expressible in a single block as shown in Fig. C-3.

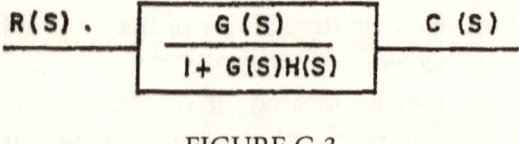

FIGURE C-3

Transfer functions in series can be combined by simple multiplication as shown in Fig. C-4.

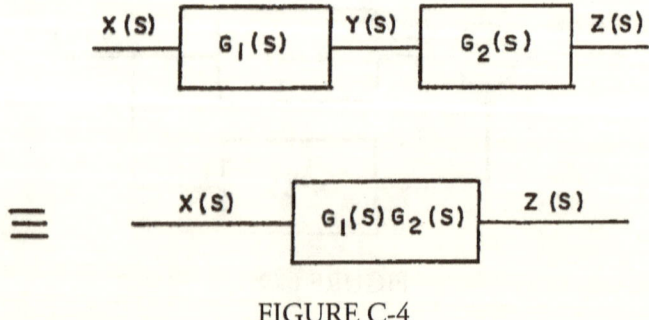

FIGURE C-4

Transfer functions in parallel may likewise be combined by addition as indicated in Fig. C-5.

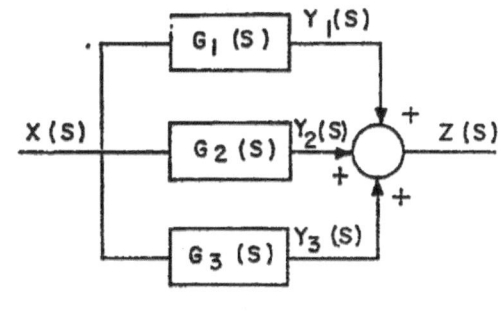

FIGURE C-5

Final Value and Initial Value

Some special properties of the LaPlace transform are of use in various design and analysis short cuts. One of these is the "Final value theorem" which states that if $X(s)$ is the LaPlace transform of $x(t)$ then the numerical value of x at time $t = \infty$ is given by the value of $sX(s)|_{s \to 0}$

i.e., $x(t)|_{t \to \infty} = sX(s)|_{s \to 0}$

provided that $sX(s)$ does not have poles on the axis of imaginaries or on the right hand s plane (I.e., $X(t)$ is a stable function settling to a finite value at $t \to \infty$).

This theorem is of particular use in evaluating the final value of the response of a transfer function to a unit step input.

Since the unit step input is $\dfrac{1}{s}$, the final value can be obtained by merely substituting $s = 0$ in the terms of the transfer function; e.g., the final value of $i(t)$, i.e., i at $t = \infty$ following a unit step input in $e(t)$ to be evaluated from transfer function of equation C-3 is $i(t = \infty) = \dfrac{1}{r + Ls}\bigg|_{s=0} = \dfrac{1}{R}$.

264 F. PAUL DE MELLO

A complementary theorem to the "Final value theorem" is the "Initial value theorem" which states that if X(s) is the LaPlace transform of x(t), then the numerical value of x at t = 0 is given by the limiting value of sX(s) as s goes to infinity.

i.e.,
$$x(t)\big|_{t\to 0} = sX(s)\big|_{s\to\infty}$$

Again for the case of a transfer function G(s), its output to a unit step function is $\dfrac{G(s)}{s}$ and the initial value of the output for the case of exciting it with a unit step function is

$$s\frac{G(s)}{s}\bigg|_{s=\infty} = G(s)\big|_{s=\infty}$$

E.G., in the transfer function equation C-3, the initial response i(t) to a unit step in e(t) is $\dfrac{1}{R+Ls}\bigg|_{s=\infty} = 0$.

APPENDIX D

ANALOG COMPUTERS—STATE SPACE—NUMERICAL METHODS OF[29] DIFFERENTIAL EQUATION SOLUTION

The advent of the analog computer or the electronic differential analyzer in the late 40's represented probably the greatest single breakthrough in the technology of dynamic analysis. This device permits the solution of the system equations by simulation and liberates the engineer from having to wrestle with the mathematics which, except for a limited number of fairly simple cases become entirely unwieldy. Values of voltages correspond through appropriate scaling to values of the problem variables. That is the voltages in the analog computer are analogous to the problem variables, hence thename "Analog Computer."

The basic building block of the analog computer is the high gain operational amplifier, Fig. D-1.

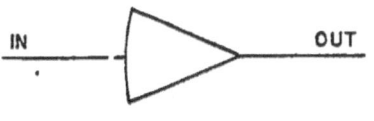

FIGURE D-1

This is a drift stabilized D.C. amplifier with a very high gain (between 50 and 300 × 10^6) over the frequency range of interest, high input impedance and a phase shift of 180°, i.e., the output voltage is the negative of the input voltage.

The basic law which permits use of the operational amplifier as a device performing summing, integrating or other functions is Kirchoff's second law on the summation of currents into a node.

Consider the operational amplifier in the configuration of Fig. D-2.

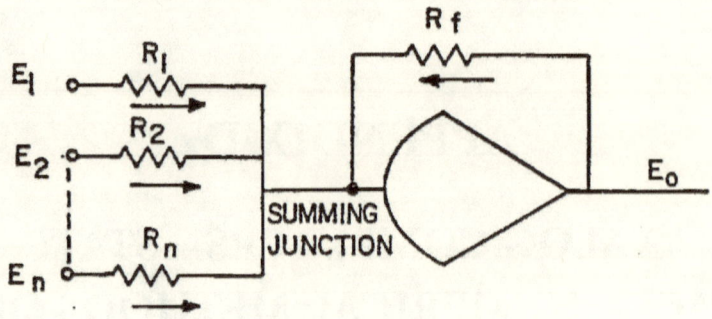

FIGURE D-2

Since the amplifier has a gain of almost infinity and the amplifier input impedance is very high, the summation of currents into the summing junction must equal zero from which fact we derive

$$E_o = -R_f \left[\frac{E_1}{R_1} + \frac{E_2}{R_2} + \dots \frac{E_n}{R_n} \right] \qquad (D\text{-}1)$$

One notes that equation D-1 shows a summing relationship with scale factors of $R_f/R_1, R_f/R_2 \dots R_f/R_n$ on the various inputs $E_1, E_2, \dots E_n$. Note that the amplifier inverts the sign of the inputs.

Examine now the configuration of Fig. D-3.

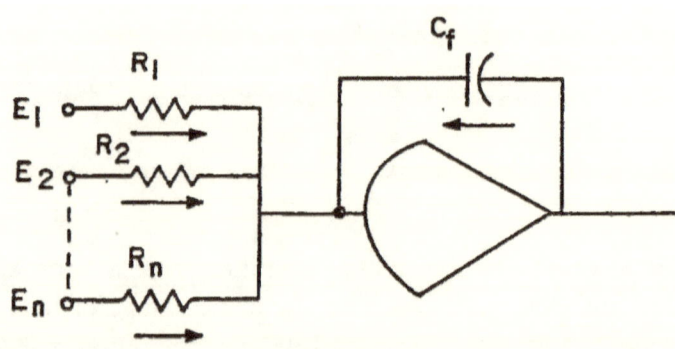

FIGURE D-3

Again Kirchoff's law requires the summation of currents into the summing junction to be zero,

i.e.

$$C_f \frac{dE_o}{dt} + \frac{E_1}{R_1} + \frac{E_2}{R_2} + \ldots \frac{E_n}{R_n} = 0 \tag{D-2}$$

Integrating and rearranging terms

$$E_o = -f\left[\frac{E_1}{(R_1C_f)} + \frac{E_2}{(R_2C_f)} + \frac{E_3}{(R_3C_f)} + \ldots \frac{E_n}{(R_nC_f)}\right] dt \tag{D-3}$$

Equation D-3 shows that the configuration of Fig. D-3 is that of a summing integrator.

The same approach may be applied to an operational amplifier with arbitrary input and feedback impedances and the output voltage can be expressed as

$$E_o = -\sum_{i=1}^{i=n} (\frac{Z_f}{Z_i}) E_i \tag{D-4}$$

Although various combinations of operational amplifiers with especial arrangements of Z_f/Z_i are used on especial purpose analog process controls, for general purpose computation, the analog computer is composed mainly of the following linear elements

1. Summing amplifiers as in Fig. D-2 with fixed ratio of R_f/R_i usually R_f/R_i = 1 or 10. In this configuration the symbol is shown on Fig. D-4

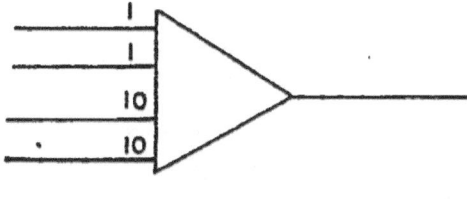

FIGURE D-4

where the feedback is implied and the gain R_f/R_i is indicated on the particular input by a "1" or a "10."

2. Inverters, Fig. D-5, which are no more than summing amplifiers with a gain of "1"

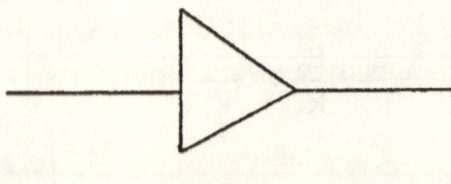

FIGURE D-5

3. Integrators with the configuration of Fig. D-3. By having $C_f = 1$ ufd and $R_i = 100K$ or $1M$, input gains of "10" or "1" are obtained for the factors $\dfrac{1}{R_i C_f}$. The symbol for the integrator is shown on Fig. D-6.

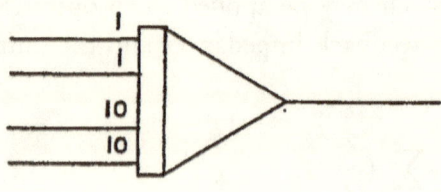

FIGURE D-6

4. Scaling or coefficient potentiometers with relatively low resistance compared with the input resistances R_i of the amplifiers. The symbol used is generally shown on Fig. D-7.

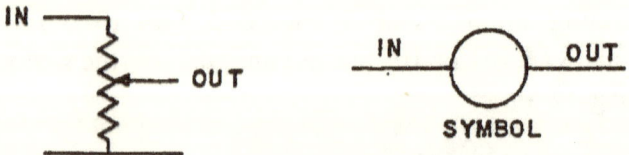

FIGURE D-7

In addition to the above linear elements, there are a variety of non-linear elements such as function generators, multipliers (multiplying two variables) dividers, limiters, etc.

We will illustrate the use of the analog computer technique in the solution of the same RLC circuit problem as was discussed in Appendices A and B.

The basic differential equatio for this circuit is

$$E(t) = Ri + L\frac{di}{dt} + \frac{1}{C} \int idt \qquad \text{(D-5)}$$

The basic rule in solution of differential equations by analog computer techniques is to convert the equation to integral form. The following rules are useful in setting up the equations for ready solution by analog computers.

1. Solve for the highest derivative. In the example of equation D-5, this becomes

$$\frac{di}{dt} = \frac{1}{L}E - \frac{R}{L}i - \frac{1}{CL} \int idt \qquad \text{(D-6)}$$

2. Integrate the highest derivative as many times as the order of the derivative. Equation D-6 becomes

$$i = \int (\frac{1}{L}E - \frac{R}{L}i) dt - \frac{1}{CL} \int \int idtdt$$

$$= \int \left[(\frac{1}{L}E - \frac{R}{L}i) - \frac{1}{CL} \int idt \right] dt \qquad \text{(D-7)}$$

3. Use the form of the resulting equation to connect the analog computer diagram as shown on Fig. D-8. Note that summers and integrators invert the sign.

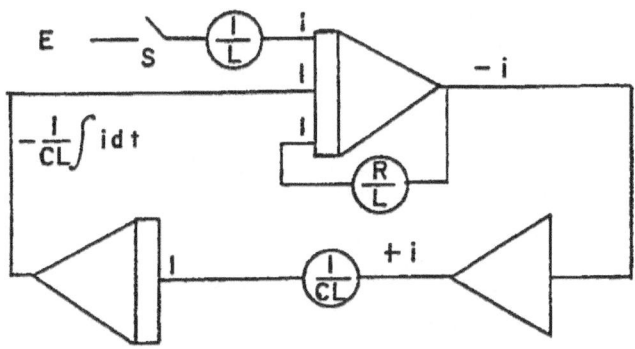

FIGURE D-8

Upon closing the switch S, source E is applied and the solution of the current i will appear as a voltage as function of time out of integrator 1 (as-i)and out of inverter 2 (as+i). This voltage is recorded on a pen recorder yielding a plot of the solution.

Computer Scaling

Both amplitude and time scaling are generally required in the use of analog computers. In any physical system, many different dependent variables can exist, such as temperature, pressure, velocity, force, etc. In the analog simulation only one dependent variable—voltage—is possible. Thus a conversion from the physical variable to the computer variable is necessary. The amplitude scale factor specifies the number of computer volts which are analogous to one physical system unit (degree, psi, ft/sec or amps).

The choice of scale factors is an exercise in common sense from knowledge of the range of the physical variable and the fact that the computer voltage range is specified, usually –100, 0 + 100 volts.

Whereas amplitude scale factors represent a correspondence between computer and physical variable, the time scale indicates the relationship between the computer units of time which correspond to physical system units of time (e.g., 1 second on the computer represents 1 min, or 1 month of the physical system).

Time scaling is accomplished entirely in the gain of integrators. For instance if a 10 to 1 increase in speed is required this can be obtained by increasing the gain on all inputs to integrators by a factor of 10, or sometimes this can be accomplished by changing of integrator feedback capacitors from 1 ufd to 0.1 ufd.

Initial Conditions

Every energy storage element of a physcial system is represented by an integrator or combination of integrators. For instance, in the example of Fig. D-8, the initial condition of current "i_o" is represented by an initial condition on the integrator 1. Likewise an initial condition of voltage across the capacitor C is represented by an initial condition on integrator 3.

Figure D-9 shows a method of setting initial conditions on integrators through establishment of the proper charge across the feedback capacitor.

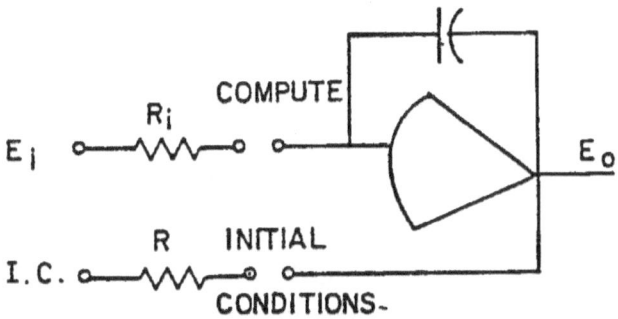

FIGURE D-9

STATE SPACE[29,31]

A very popular formulation of differential equations is in the so called "State Space" form. Basically an "n" th order differential equation is expressed as n, 1st order differential equations. Let us illustrate with examples.

Consider the circuit of Fig. D-10.

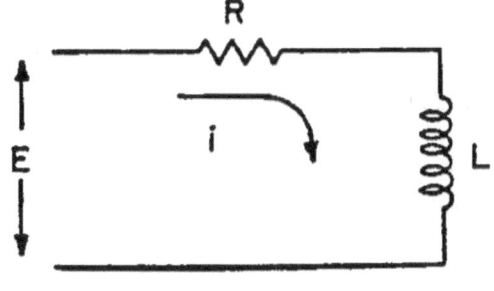

FIGURE D-10

The differential equation for this circuit is

$$L\frac{di}{dt} + Ri = E \qquad (D-8)$$

Now change the nomenclature as follows

  Let i be replaced by X_1

and $\dfrac{di}{dt}$ be replaced by X_1

Using these definitions, equation D-8 can be written

$$\dot{X}_1 = -\frac{R}{L}X_1 + \frac{E}{L} \qquad\qquad\qquad\text{(D-9)}$$

Equation D-9 is a state equation and X_1 is a state variable.

  Consider now the case of the 2nd order differential equation describing the circuit of Fig. D-11.

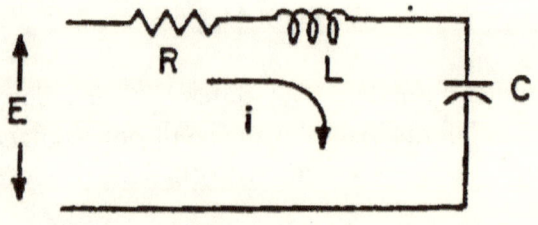

FIGURE D-11

The equation is

$$Ri + L\frac{di}{dt} + \frac{1}{C} \int idt = E \qquad\qquad\qquad\text{(D-10)}$$

Defining $X_1 = i$ and $X_2 = \int idt$ gives by definition

$$\dot{X}_1 = \frac{di}{dt} \text{ and } \dot{X}_2 = i$$

Equation D-10 can now be written as two equations:

$$\dot{X}_1 = -\frac{R}{L}X_1 - \frac{1}{LC} X_2 + \frac{E}{L} \qquad\qquad\qquad\text{(D-11)}$$

$$\dot{X}_2 = X_1$$

As shown by equation D-11, the second order differential equation D-10 can be broken down into two first order equations. Similarly an Nth order system can be represented by N first order equations, which can be put into a very convenient matrix form.

For instance, equation D-11 can be expressed as

$$
\begin{bmatrix} \dot{X}_1 \\ \dot{X}_2 \end{bmatrix} = \begin{bmatrix} \dfrac{R}{L} & -\dfrac{1}{LC} \\ 1 & 0 \end{bmatrix} \begin{bmatrix} X_1 \\ X_2 \end{bmatrix} \begin{bmatrix} E/L \\ 0 \end{bmatrix} \tag{D-12}
$$

Vector   Matrix   Vector   Vector

In this type of formulation one recognizes that the equations are ready for implementation on an analog computer, the $\dot{X}$ vector being the inputs to integrators and the X vector representing outputs to integrators.

Figure D-12 shows the corresponding analog computer diagram derived from matrix formulation equation D-12. Note that this is completely equivalent, as it should be, to the analog computer diagram in Fig. D-8.

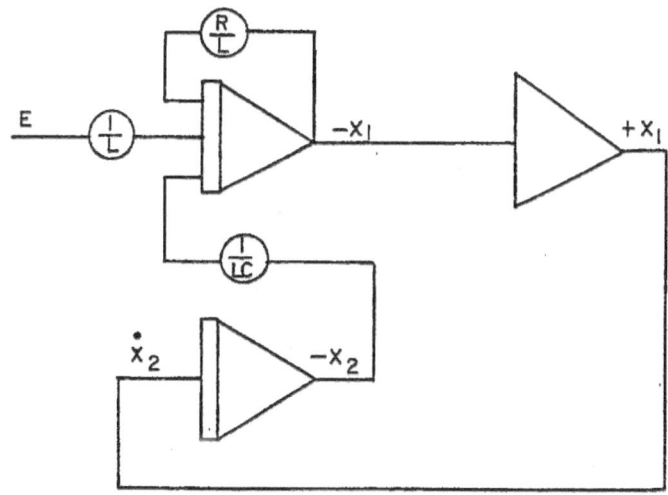

FIGURE D-12

Solution of the State Equation

The state space formulation yields a matrix equation of the form

$$\dot{X}_1 = AX + By \tag{D-13}$$

where, $\dot{X}$, X, B and y are vectors
and A is a square matrix
    The solution of D-13 is

$$X(t) = e^{At} X(o) + \int_0^t e^{A(t-T)} By \, dT \tag{D-14}$$

The term e^{At} in equation D-14 is called the State Transition Matrix. In conjunction with the forcing function y, it determines the time trajectory of the system from its present state $X(o)$ to one in the future.

    A, B and y in general may be functions of time. For the case where they are constant, we can integrate equation D-14 and obtain:

$$X(t) = e^{At} X(o) + A^{-1} \left[e^{At} - I \right] BY \tag{D-15}$$

As an example solve equation D-9 using equation D-15.

Here $A = - R/L$
              $BY = E/L$

$$\therefore X(t) = e^{-R/L \, t} X(o) - L/R \, (e^{-R/L \, t} - 1) \, E/L$$

Which, for $X(o) = 0$, yields the familiar exponential rise in the inductive circuit

$$i(t) = X(t) = \frac{E}{R} (1 - e^{-R/Lt}) \tag{D-16}$$

The first order example yields a very simple solution. For systems of higher order, the solution becomes complicated by the need to evaluate e^{At} where A is a matrix. e^{At} may be expanded into its series form

$$e^{At} = I + \frac{At}{1!} + \frac{A^2t^2}{2!} + \ldots \qquad \text{(D-17)}$$

The larger the value of t, the more terms of the expansion equation D-17 must be included for the required accuracy.

Another approach is to use one or two terms of equation D-17 and limit the time step so that proper accuracy is obtained. The solution uses a recursive algorithm in that it repeats the process of transition from one state to another, where the initial state at the new time step is taken as the final state at the end of the last time step.

For instance if we use only the first two terms of equation D-17,

$$e^{At} = I + At \qquad \text{(D-18)}$$

Substituting equation D-18 into equation D-15 yields

$$X(t) = (I + At) X(o) + BYt \qquad \text{(D-19)}$$

Since equation D-19 is valid for a small time interval, we must use the recursive relationship

$$X(t + \Delta) = (I + A\Delta) X(t) + BY\Delta \qquad \text{(D-20)}$$

where Δ is the time increment. In general equation D-20 can be simplified to the form

$$X(t + \Delta) = A'X(t) + B'y \qquad \text{(D-21)}$$

where A' is an NxN transition matrix and $B'y$ is an N dimension column vector and $X(t + \Delta)$ is likewise an N dimension column vector.

The recursive relationship equation D-21 brings to mind the general area of numerical methods of solving differential equations by using difference equation approximations or numerical integration algorithms.

This is a very vast subject and people have spent careers researching the merits of different algorithms such as Runge Kutta of various orders, Runge Kutta-Gill, Milne, Euler, and others. We will next explore a simple method of differential equation solution which gives good accuracy for most practical problems and will avoid being drawn intothe many fine points which for the most part and from a practical point of view, are unnecessary frills.

A Digital Approach to Differential Equation Solution[25]

A logical step is to try to "digitise" the analog computer technique. This is done with a step by step procedure with the use of integration algorithms.

Before illustrating the method through an example, let us examine the operation of integration and one simple way of approximating this operation with a digital algorithm.

First look at the ideal continuous integrator whose transfer function is represented by 1/s. When inputed with a unit impulse, the response of a continuous integrator is shown on the top trace of Fig. D-13.

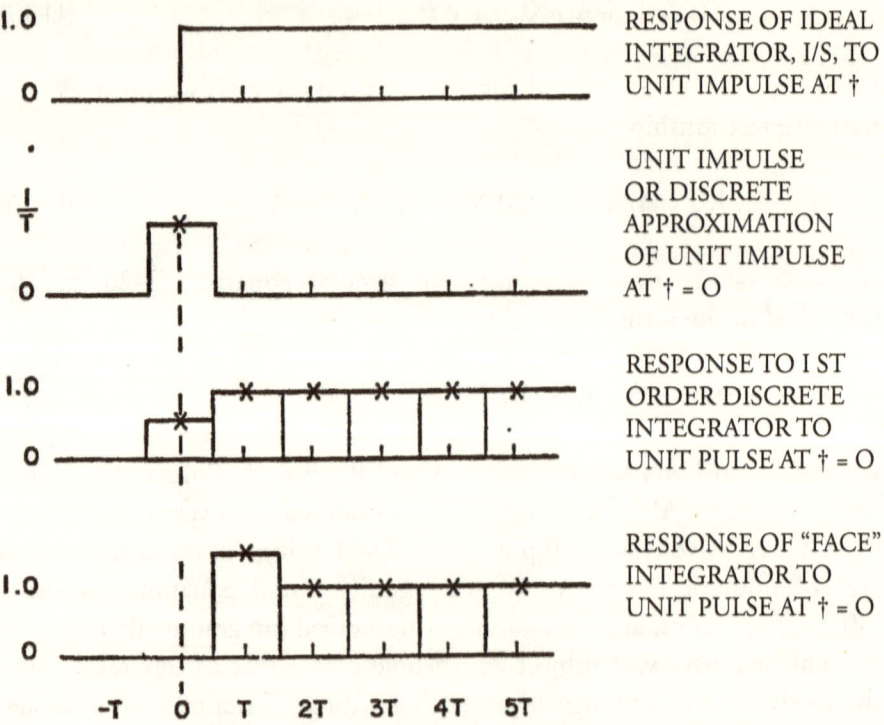

FIGURE D-13—Time-domain relationships illustrating
the response of the FACE integrator.

In the discrete representation, values are defined at times t = 0, t = T, t = 2T . . . etc., and these values may be interpreted as the average value of the corresponding continuous function over the intervals about which the times t = 0, t = ϖ, t = 2T are centered.

On the basis of this interpretation of the digital sequence of numbers, the series that approximates the response of the continuous integrator is given by 3rd trace on Fig. D-13, i.e., the sequence of numbers 1/2, 1, 1, 1 . . . at times t = 0, t = T, t = 2T, etc. This is bascially Tustin's algorithm for integration.

The significant point here is that for this ideal first order integrator, there is an instantaneous undelayed output of 1/2 at time zero when the input is an unit pulse centered around time zero.

Looking now at the analog computer diagram of Fig. D-8 one notes that implementation of this type of algorithm in lieu of the integrators, would require iterations since, because of the simultaneity of output and input, the input cannot be defined independent of the output. In order to avoid this interdependence of input on output at the same instant of time, the next best thing is to delay the output one time interval, breaking up the simultaneous dependence between input and output. When this is done, in order to preserve areas, the best thing that can be done is to tack on the missing 1/2 at time = 0, on to the output at the next time interval as shown on the last trace of Fig. D-13. A block diagram of this simple one-pass digital integration algorithm is shown on Fig. D-14.

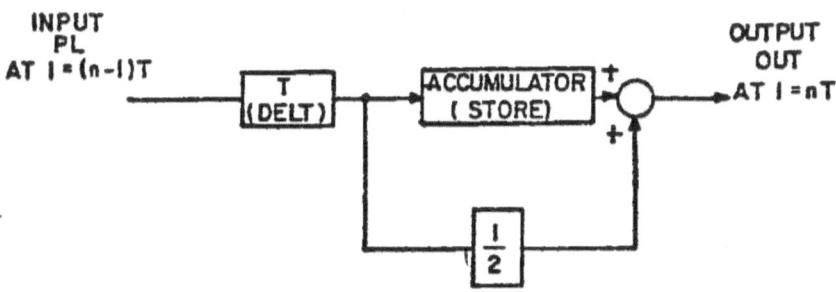

FIGURE D-14

One can now visualize the process of differential equation solution by digital means as a straightforward recursive process of:

1. Defining inputs to integrators at times t_n from the outputs of these integrators (assumed fixed) and from other system inputs at time t_n.
2. Determining the outputs of the integrators at time $t_{(n+1)}$ from the inputs at time t_n in step (1) following the simple algorithm of Fig. D-14.

Note that the output of the integrators can also be viewed as the "states" as explained in the previous section.

APPENDIX E

FEEDBACK CONTROL SYSTEM CONCEPTS[25, 30]

The term "Feedback Control System" is applied to a control system which compares a quantity to be controlled with a reference or desired value and operates on the error between these to bring the controlled quantity towards the desired value.

The closeness with which the controlled quantity is brought towards the desired value is a function of the type of control system. Regulators or type "0" servomechanisms require some finite error in the steady state. Type 1 servomechanism or controllers with "reset" (in process control terminology) have integration in the controller and wipe out the error to zero in the steady state.

One of the greatest aids in understanding control systems is the block diagram. This was introduced in Appendix C. A very common configuration of a feedback control is shown on Fig. E-1.

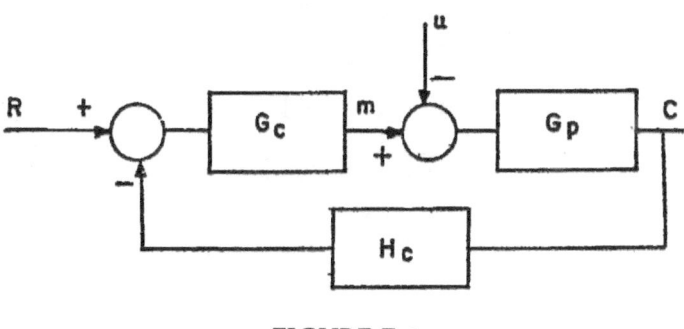

FIGURE E-1

Here G_p is the process transfer function
          G_c is the controller forward function
          H_c is the feedback function
          R   is the reference quantity or set point
          C   is the controlled variable
          m   is the input to the process manipulated by the controls
          u   is the load or disturbance to the process.

This configuration may be understood better through an example. Take the case of a level control system:

G(p)   could be the transfer function K/s of a tank, c being the level of resident fluid,
m   the input controlled, flow and
u   being the uncontrolled (disturbance) drawdown from the tank
R   is the reference level, or set point
H   could be the sensor transfer function, could be a simple lag $\dfrac{1}{1+sT}$.
G_c is the controller function which could be a simple proportional gain K_p or some more complex function.

Using the expressions on block diagram reduction of Appendix C, we can write down

$$C = \frac{G_c G_p}{1+G_c G_p H} R + \frac{G_p}{1+G_c G_p H} u \qquad \text{(E-1)}$$

The effect of closing the loop can be appreciated from equation E-1.

For instance the effect of a disturbance u on the system output C would have been $G_p u$ with the loop open. This has been reduced by a factor by $\dfrac{1}{1+G_c G_p H}$ closing the control loop. If for instance $G_c G_p H$ had a steady state value of 9, the effect of the disturbance will be reduced 10 to 1. If $G_c G_p H$ were to have an integration, (equivalent to having infinite gain in the steady state) the effect of the disturbance would have been zero in the steady state.

Another important point which may be drawn from equation E-1 is the expression for C/R, i.e., the change of the controlled variable in response to a change in desired or reference value.

$$\frac{C}{R} = \frac{G_c G_p}{1 + G_c G_p H} \qquad\qquad (E-2)$$

For very large $G_c G_p$, $C/R \simeq \frac{1}{H}$. This expression shows the importance of accuracy and linearity in the feedback element H, since for systems with large values of $G_c G_p H$ the controlled variable C depends mainly on the feedback element H, and the value of reference R.

Feedback control does not only reduce the effect of disturbances and drifts in the steady state. It also causes the output to respond more rapidly to the command of the references or to counteract dynamically the effects of load disturbances. Here we get involved with the important subject of response and stability of closed loop systems.

Closed Loop and Open Loop Time Response of Simple Systems

The concept of response and stability will be introduced with a couple of simple examples. Take the case of an open circuit generator whose terminal voltage response to a change in field volts is governed by a single time constant as described in Fig. E-2.

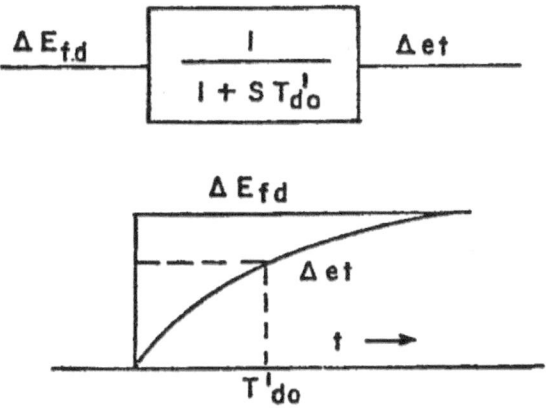

FIGURE E-2

Let us now regulate terminal voltage by means of an idealized regulator and exciter which may be represented by a simple gain as in Fig. E-3.

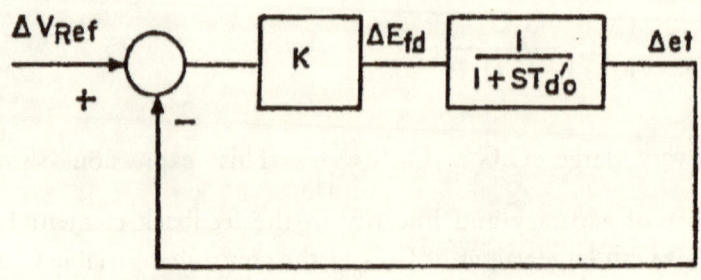

FIGURE E-3

The response of the closed loop

$$\frac{\Delta et}{\Delta V_{ref}} = \frac{G}{1+GH} = \frac{\dfrac{K}{1+sT'_{do}}}{1+\dfrac{K}{1+sT'_{do}}}$$

$$= \frac{K}{1+K+sT'_{do}} = \frac{K}{1+K\left(1+\dfrac{sT'_{do}}{1+K}\right)} \tag{E-3}$$

For large K we note that equation E-3 has a steady state gain of almost unity but a response time constant of $T'_{do}/(1+K)$ which is $(1+K)$ times faster than the open loop response of Fig. E-2.

Fig. E-4 shows the comparison of the open and closed loop performances

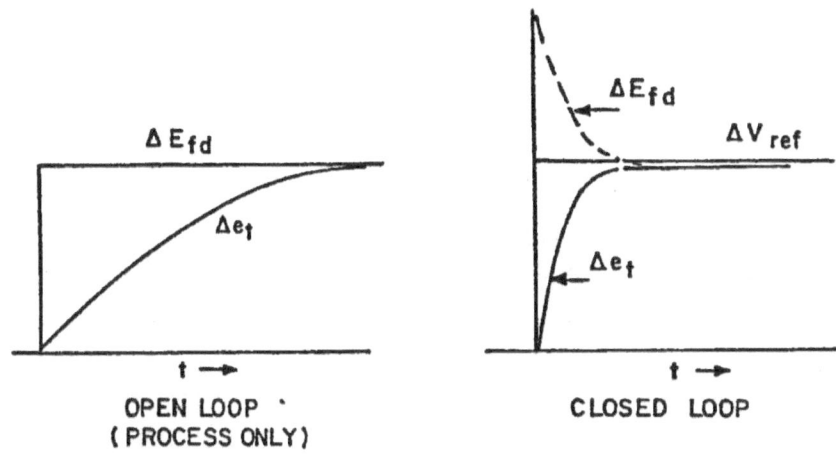

FIGURE E-4

Note that the response of the closed loop is considerably faster than that of the open loop. This is due to the forcing action of the field in response to the high gain operating on the error.

On an idealized system containing only one time constant (first order system) such as in Fig. E-3 there is theoretically no limit to the value of the gain K that can be applied. In practice most systems exhibit more than one time constant, i.e.; are of higher than first order, and a limit to the gain of the closed loop and hence a limit to closed loop's speed of response is reached due to stability considerations.

Let us illustrate by assuming the regulator exciter to be described by one time constant as in Fig. E-5.

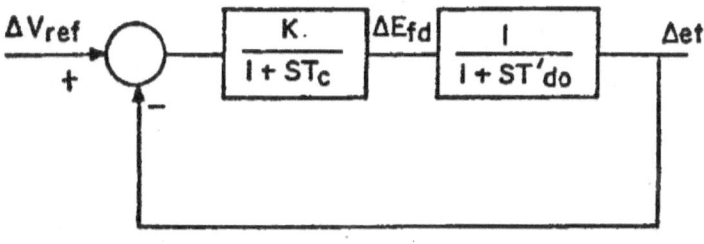

FIGURE E-5

Again solving for $\Delta et / \Delta V_{ref}$ with the use of the $G/1+GH$ formula

$$\frac{\Delta et}{\Delta V_{ref}} = \frac{K}{(1+K)\left[1+s\dfrac{(T^+_\varepsilon T'_{do})}{(1+K)}+s^2\dfrac{T_\varepsilon T'_{do}}{(1+K)}\right]} \tag{E-4}$$

The nature of the closed loop response can be rerived from the roots of the denominator of Equation E-4, which are the roots of the closed loop characteristic equation

$$1 + GH = 0$$

Depending on the value of K, the roots of equation E-4 can be real or complex, the higher the value of K the more oscillatory (less damped) are the roots.

The quadratic form of the denominator can be expressed as

$$1+\frac{2\zeta}{\omega_o}s+\frac{s^2}{\omega_o^2} \tag{E-5}$$

where

$$\frac{2\zeta}{\omega_o} = \frac{T_\varepsilon+T'_{do}}{(1+K)} \tag{E-6}$$

and

$$\frac{1}{\omega_o^2} = \frac{T_\varepsilon+T'_{do}}{(L+K)} \tag{E-7}$$

The roots of equation E-5 are

$$s_1, s_2 = \left[-\zeta \pm j\sqrt{1-\zeta^2}\right]\omega_o \tag{E-8}$$

where ω_o is known as the natural frequency of oscillation and ζ is the damping ratio. A damping ratio $\zeta = 1$, known as critical damping, yields two equal real roots s_1, $s_2 = -\omega_o$. Damping ratio less than 1 yield complex roots while those greater than one yield real (non oscillatory) roots.

In terms of the parameters, K, T_ε and T'_{do}, equation E-6 and equation E-7 yield

$$\zeta = \frac{T_\varepsilon + T'_{do}}{2\sqrt{1-K}\,(T_\varepsilon + T'_{do})} \tag{E-9}$$

$$\omega_o = \sqrt{\frac{1-K}{T_\varepsilon T'_{do}}} \tag{E-10}$$

From equation E-9 and equation E-10 we note that the higher K, the higher the natural frequency and the lower the damping ratio. The second order system of Fig. E-5 cannot become unstable, i.e., cannot have roots with negative values of ζ. However, it can approach exhibiting sustained oscillations as $\zeta \rightarrow 0$ which is unacceptable performance. Systems with characteristic equations of higher orders can easily exhibit instability with increasing loop gains.

The inverse LaPlace transform of expression E-4 multiplied by 1/S (for the input step) yields the time function

$$\Delta et = \frac{K}{(1+K)}\left[1 + \frac{\varepsilon^{-\omega_o t}}{\sqrt{1-\zeta^2}}\sin\left(\sqrt{1-\zeta^2}\,\omega_o t - \tan^{-1}\sqrt{\frac{1-\zeta^2}{-\zeta}}\right)\right] \tag{E-11}$$

Figure E-6 shows the form of the response expression E-11 as function of normalized time "$\omega_o t$" for various values of ζ.

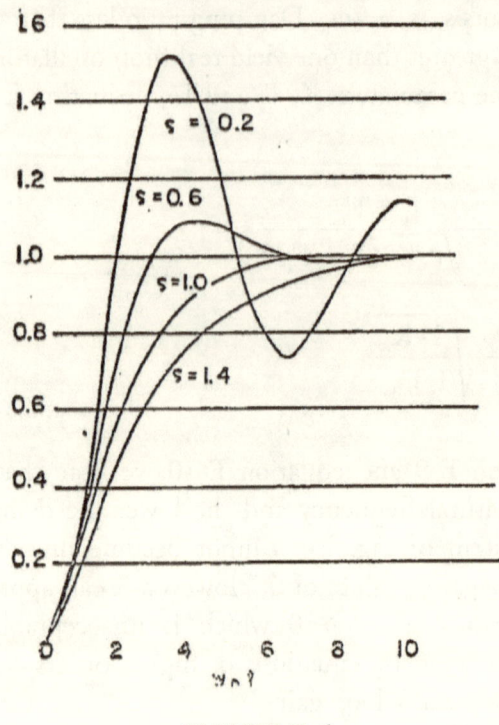

FIGURE E-6

STABILITY OF CLOSED LOOP SYSTEMS

From the above examples it becomes quite clear that the stability of closed loop systems can be investigated from knowledge of the roots of the characteristic equation

$$1 + GH = 0$$

Several methods are available to determine whether or not some of the roots of the characteristic equation lie in the right hand plane (a condition signifying instability). Such methods as Routh's criterion, root locus, etc. have their special application and it is not the intent here to expound further on these methods which may be readily found in the literature. We will merely explore briefly some of the very widely used Frequency Response techniques and Nyquist stability criteria. The table below summarizes briefly the features and names of some of the techniques used for determining stability.

STABILITY CRITERIA CHARACTERISTICS

Method	Answer Obtained	Information Required	Application	Remarks
Routh				Difficult to assess effect of parameter variations
Hurwitz				
Meerov				Involved computation required
Wall	Yes-or-no stability	Closed-loop characteristic polynomial	Analysis	Only computation needed is long division
Schur				Applicable to sampled data systems
Root Locus	Complete system Response	Open-loop transfer function, factored from	Analysis and syntnesis	Can be extended to time delay systems
Myquist Bode	Stability and approximate time response	Open-loop transfer function measured or computed for all frequencies		Application to time delay systems
Dzung	Yes-or-no stability			
Mickallov	Absolute and relative stability	Closed-loop frequency response	Analysis	
Leonhard				

Frequency Response

The concept of the operational impedance and the impedance to a sinusoidally varying excitation function was developed in example 2 of Appendix A.

A transfer function is an operational expression much like that of impedance. The complex number obtained by substitution of $s = j\omega$ in the transfer function gives information on the steady state sinusoidal response of the output of the function to a sinusoidal input excitation of frequency ω. The absolute value of this number corresponds to the magnitude ratio of the output sinusoid to the input sinusoid while the phase angle of the complex number expressed in polar coordinates indicates the angle by which the output sinusoid leads or lags the input sinusoid.

Frequency response techniques use the magnitude and phase characteristics of transfer functions or combinations of transfer functions to derive a great deal of information about the stability and response performance of control systems.

The frequency response characteristics of some common transfer functions encountered in control systems are described in Fig. E-7

Naturally the frequency response of a combination of transfer functions in series is easily obtained by taking the product of these functions, i.e.; the magnitude is the product of the magnitudes of the individual functions and the overall phase angle is equal to the summation of phase angles of the individual transfer functions.

Nyquist Stability Criterion

Recall that the stability of a closed loop system was determined from properties of the characteristic equation $1 + GH = 0$.

The Nyquist criterion is a means used to determine whether or not $1 + GH$ has roots in the right hand half of the s plane. A rigorous derivation of the Nyquist criterion involves use of complex number theorems by Cauchey and examination of the number of cycles of phase rotation of the function $GH(j\omega)$ as $j\omega$ is taken around a closed path from $-j\infty$ to $+j\infty$.

Except for very unusual circumstances Nyquist's criterion applied to practical cases amounts to the following:

> For a closed loop whose characteristic equation is 1 + GH, the stability of the system can be derived by examining the frequency response characteristic of the open loop function GH. This is done by finding the phase angle of GH at the frequency for which the magnitude of GH is 1.0. If the phase angle is 180° the system is borderline unstable. If the phase angle is more than 180° lagging, the system is unstable.

The phase angle of GH at the point where the magnitude of GH is 1 is known as the phase angle at crossover and the frequency ω_c at this point is called the crossover frequency. Many guide rules have been established to relate the shape of the open loop frequency response function to the performance of the closed loop function. One point to remember is that the phase angle at crossover should in general not exceed 130 to 140°. For such cases the closed loop system response will be oscillatory with good damping, the output exhibiting an overshoot of about 25%. The frequency of oscillation of the closed loop is related to and closely approximates the crossover frequency. For smaller phase angles at crossover, in the order of 100° the system in general looks critically damped.

Figure E-8 shows complex plots of typical $G(j\omega)H(j\omega)$ (open loop) functions as ω varies from zero to infinity. Such plots are called Nyquist diagrams. The third locus in Fig. E-8 shows the case of a conditionally stable system—one where either an increase or a decrease in loop gain

TRANSFER FUNCTION
FREQ RESP FUNCTION
TIME RESPONSE TO UNIT STEP
POLES * ZEROS o
FREQUENCY RESPONSE
LOG MAGNITUDE & PHASE (θ) VS LOG FREQUENCY
1/S
1/jω
1/(1+ST)
1/(1+jωT)
TS/(1+ST)
jωT/(1+jωT)
IM
REAL
-1/T
+J
-J
1.0
LOG M
LOG ω
LOG 1/T
0°
-90°
+90°

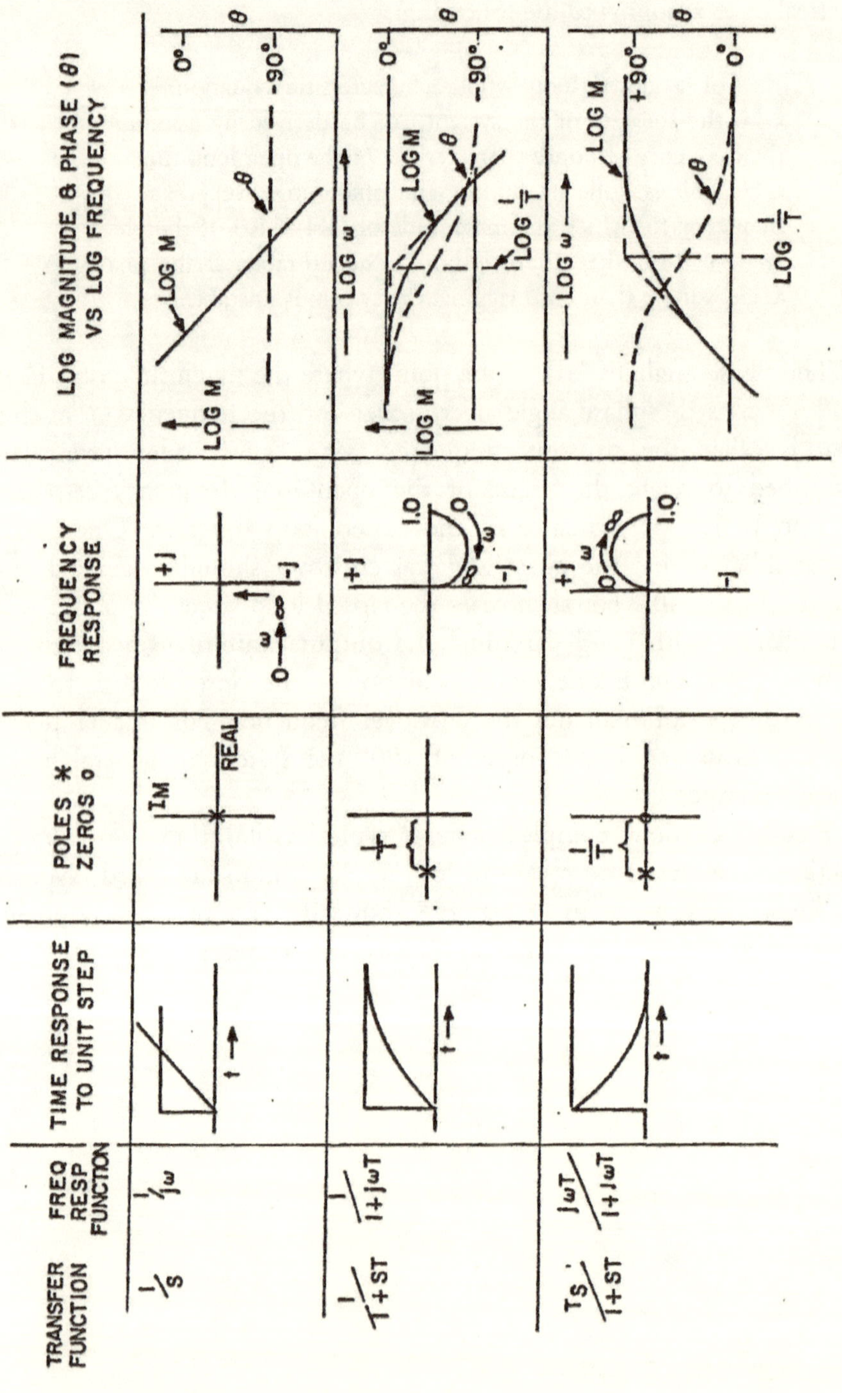

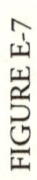

FIGURE E-7

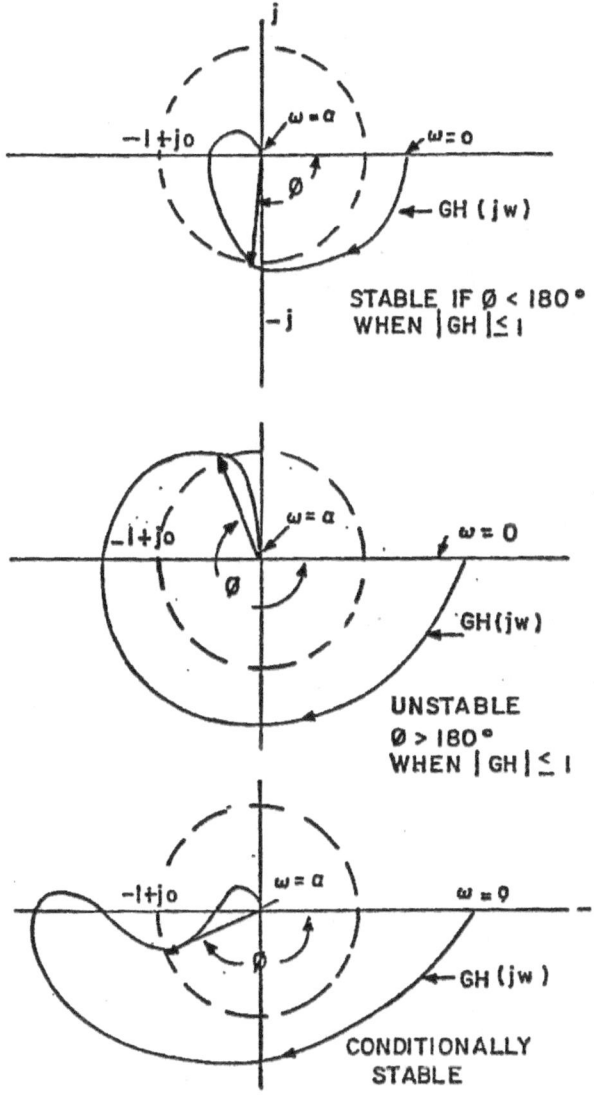

FIGURE E-8

can cause instability. This is in contrast with the usual case where instability is only reached with increasing gain.

Figure E-9 shows typical time responses of the closed loop $\Delta C = \dfrac{G}{1 + GH} \Delta R$ to a step change in reference ΔR.

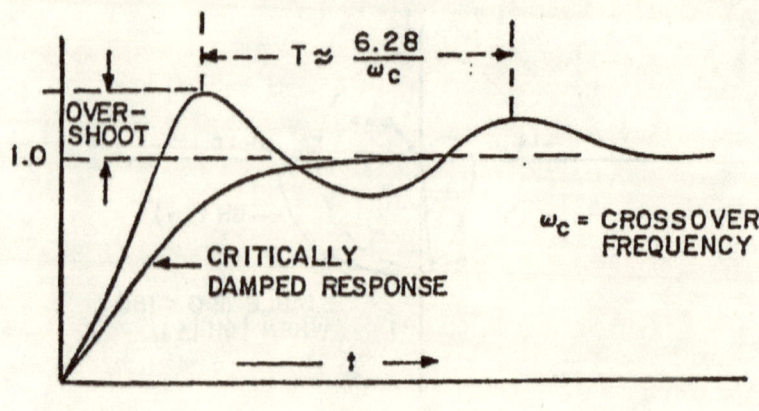

FIGURE E-9

Bode Theorems

From the foregoing discussion, we note that the evaluation of the overall open loop frequency response characteristic $GH(j\omega)$ requires multiplication of transfer functions. The Bode method gives a simple technique for plotting transfer functions in terms of a log of magnitude plot and a phase angle plot.

If two transfer functions are written in polar form

$$A = A_1 e^{j\theta_1} \tag{E-12}$$

and

$$B = B_1 e^{j\theta_2}$$

Then

$$AB = A_1 B_1 e^{j(\theta_1 + \theta_2)} \tag{E-13}$$

That is, to obtain the frequency response of a product of transfer functions, the individual transfer function phases are added and their magnitudes are multiplied. If the magnitudes are expressed as $\log A_1$ and $\log B_1$, then $\log (A_1 B_1) = \log A_1 + \log B_1$. The Bode diagram plots the logarithm of magnitude and phase angle of the open loop function $GH(j\omega)$ as separate functions of frequency.

The Bode plotting technique is based on asymptotic characteristics of transfer functions expressed in factored form. Let us illustrate with examples:

Take the transfer function of a single lag time constant

$$G(s) = \frac{K}{1+sT} \tag{E-14}$$

The frequency response of $G(s) = G(j\omega)$

$$= \frac{K}{1+j\omega T} = \frac{K}{1+j(\frac{\omega}{\omega_0})} \tag{E-15}$$

where $\omega_0 = \frac{1}{T}$

The asymptotes of equation E-15 as $\omega/\omega_0 \to 0$ and as $\omega/\omega_0 \gg 1$ are

$$K \text{ and } -j\frac{K}{(\frac{\omega}{\omega_0})} \tag{E-16}$$

The magnitude of $G(j\omega)$ plotted in log scale versus ω also on a log scale is shown on Fig. E-10 as are also the two asymptotes for $\omega/\omega_0 \to 0$ and $\omega/\omega_0 \gg 1$.

Figure E-10 also shows the plot of phase angle of equation E-15 as function of ω_0. Note that plotted in this form all that is required is the location of the break frequency ω_0 and the shape of the frequency response function is then quickly determined.

A lead function is likewise described by asymptotic straight line approximations of log magnitude versus log ω as also shown on Fig. E-10. The actual function can be quickly determined with the use of templates which give appropriate corrections to the asymptotic straight line approximations a function of the normalized value of ω/ω_0.

Likewise, templates give the phase angle contribution of each lag or lead factor, i.e.; the angle contributed by a pole or zero. Obviously the phase angle of equation E-15 as $\omega/\omega_0 \to 0$ is 0° and it is 90° as $\omega/\omega_0 \to \infty$. At $\omega/\omega_0 = 1$ it is 45°.

Figure E-11 shows typical nomograms for use with the Bode technique.

We will illustrate the use of Bode diagrams in the following example:

Consider the position control system described by the block diagram of Fig. E-12. Find the maximum gain K for which this positioning system will be stable, and also the value of gain K for which the phase angle at crossover is –135°, i.e.; the phase margin is 45°.

Without ever knowing anything about Bode methods, this problem can be solved easily by plotting phase angle and magnitude of "G" with K = 1.0 as function frequency ω. From these plots we can determine the frequencies for which the phase angle is 180° and 135° respectively. Determine also the magnitude |G| at these frequencies. Let these magnitudes be $M_{180°}$ and $M_{135°}$ respectively.

Then the values of gain K required to cause crossover at a phase angle of 180° and 135° respectively are

$$K_{180°} = \frac{1}{M_{180°}} = 12$$

and

$$K_{135°} = \frac{1}{M_{135°}} = 1.92$$

Figure E-13 contains a Bode plot of $\dfrac{1}{s(1+T_1s)(1+T_2s)}$ by use of the asymptotic approximation technique. Phase angle values are marked along the curve from which the value of $M_{135°}$ and $M_{180°}$ can be obtain by interpolation.

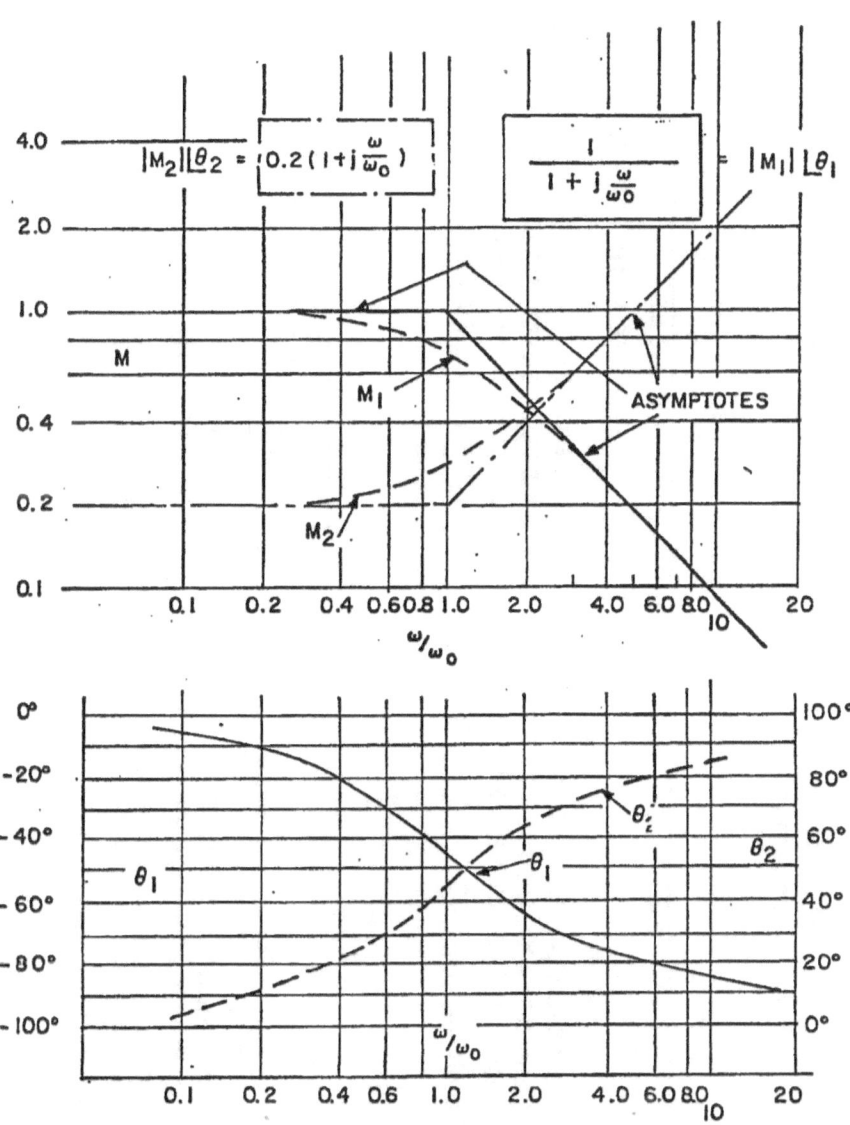

FIGURE E-10

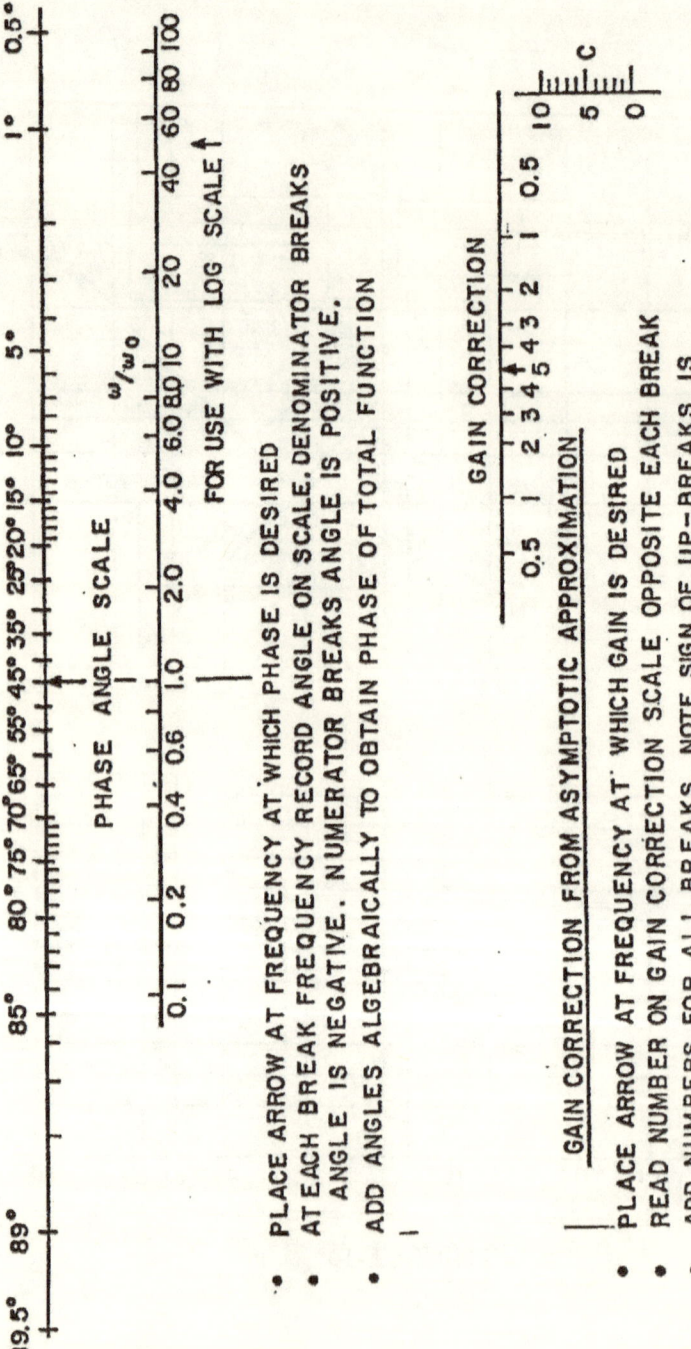

FIGURE E-11

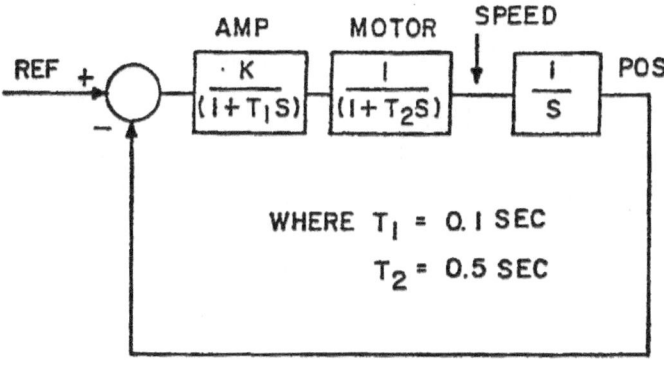

FIGURE E-12

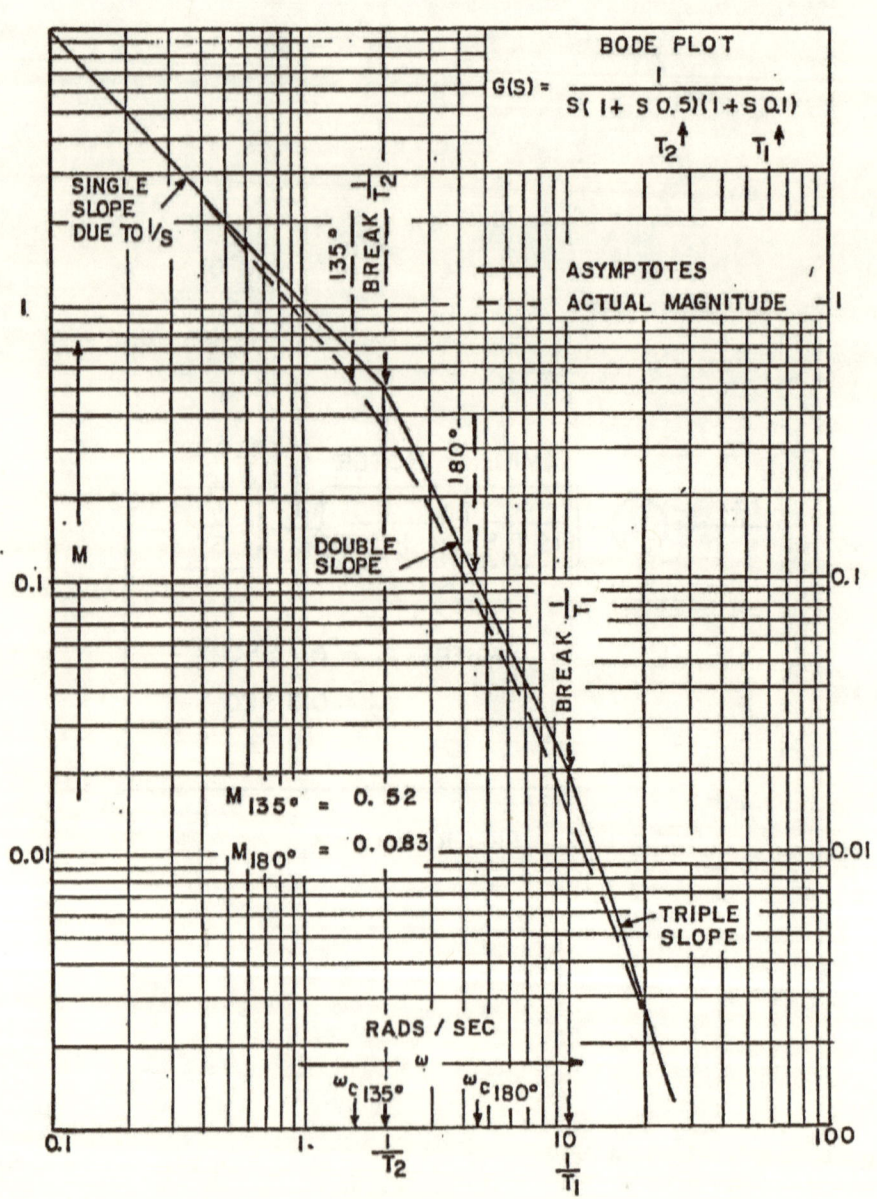

FIGURE E-13

For the value of K which produces borderline stability, the phase angle is 180° and the crossover frequency is ω_c = 4.5 rads/sec. The crossover frequency for the case where phase margin is 45° is ω_c = 1.5 rads/sec.

Control design by frequency response techniques is concerned with the shaping of the frequency response of the controller function to provide the proper phase margin at crossover. Lead/lag functions of the form $(1 + T_1s)/1 + T_2s)$ with $T_1 > T_2$ and lag lead functions with $T_1 < T_2$ are often used, with the location of T_1 and T_2 selected so as to produce the desired effect.

It is apparent that a high or infinite gain at zero frequency is necessary for good steady state and low frequency performance (low, or zero steady state error). The response of the control system is related to the crossover frequency, the higher this frequency the faster the control response, or bandwidth. However, a limiting constraint is the requirement of stability.

The rules of plotting Bode diagrams can be readily derived by elementary reasoning. These can be formulated as follows:

1. Determine the gain and slope at zero frequency.
    a) If all factors contain time constants $(1 + sT)$ both in the numerator as well as the denominator, then the initial gain, at $\omega = 0$, is equal to the steady state gain and the initial slope is zero.
    b) If $(s)^n$ appears in the denominator, the initial gain or magnitude of the function is infinite and the slope is such that the gain is decreased by a factor 10^n for every tenfold increase in frequency. The line can be located by making it pass through the point determined by $\omega = 1$ on the abscissa and overall gain on the abscissa.
    c) If $(s)^n$ appears in the numerator, the initial magnitude ratio is zero and the slope is such that the gain is increased by a factor of 10^n when the frequency is increased tenfold.

2. The initial line is carried to the first break frequency ω_0 = 1/T_1 where T_1 is the longest time constant in the numerator or denominator.

3. At the first break, the change in slope is determined. Since the factor containing this longest time constant will be in the form

$(1 + sT)^n$, the gain will change by a factor 10^n when the frequency changes 10 times.

a) For instance, if the initial slope is zero and $n = 2$ (factor containing the longest time constant in the numerator of the transfer function), then the slope after the break will be such that the gain is increased a hundredfold (10^2) when the frequency changes by a decade. This is called a double break upward.

b) As another example assume that the initial slope is S1; i.e.; the gain increases by ten times per decade increase in frequency and that the longest time constant occurs in the denominator at a break frequency $1/T_1$. Then the slope after the break will be horizontal since the downward slope contributed by the denominator time constant cancels the upward initial slope.

4. The slope thus determined after the first break is continued until the next break which is determined by the next longest time constant in numerator or denominator. The change in slope is determined as before and the process is repeated until all time constants have been accounted for.

Bode theorems relate the slope of the magnitude function to the phase angle. In general for minimum phase functions the phase angle can be approximately determined by the slope of the function.

A single slope, i.e.; magnitude decreasing 10 times for a tenfold increase in frequency carries about 90° phase lag. A slope of two (two decades per decade), i.e.; 100 times decrease in magnitude for every tenfold increase in frequency represents about 180° phase lag and so on.

Frequency Response of the Closed Loop

The closed loop function

$$\frac{G}{1+GH} \tag{E-16}$$

can also be plotted in terms of its gain and phase as function of frequency.

Some easy guide rules can again be used to approximate the shape of the gain versus frequency curve.

Equation E-16 can be approximated under two extreme conditions, i.e.; when GH>>1 and when GH<<1. Under the first condition equation E-16 becomes nearly 1/H and under the second condition equation E-16 is approximately G. This leads to the following set of rules to obtain the approximate shape of the closed loop response.

1. Plot gain curves for G, $\dfrac{1}{H}$ and GH.

2. Follow G when GH < 1, i.e.; follow G if $\dfrac{1}{H} > G$.

3. Change from the G curve to the $\dfrac{1}{H}$ curve when GH > 1, i.e.; follow $\dfrac{1}{H}$ if $\dfrac{1}{H} < G$.

4. The amount of resonant humping at near the transition from one curve to another is a function of the phase angle of GH at crossover, i.e.; at the point where GH = 1.0. Evidently if the phase angle at this point is 180°, the resonant peak would reach infinity. The more oscillatory the system the greater the peak of the closed loop function. Figure E-14 illustrates the various points made above using the example of Fig. E-12 for an arbitrary gain K = 1.5.

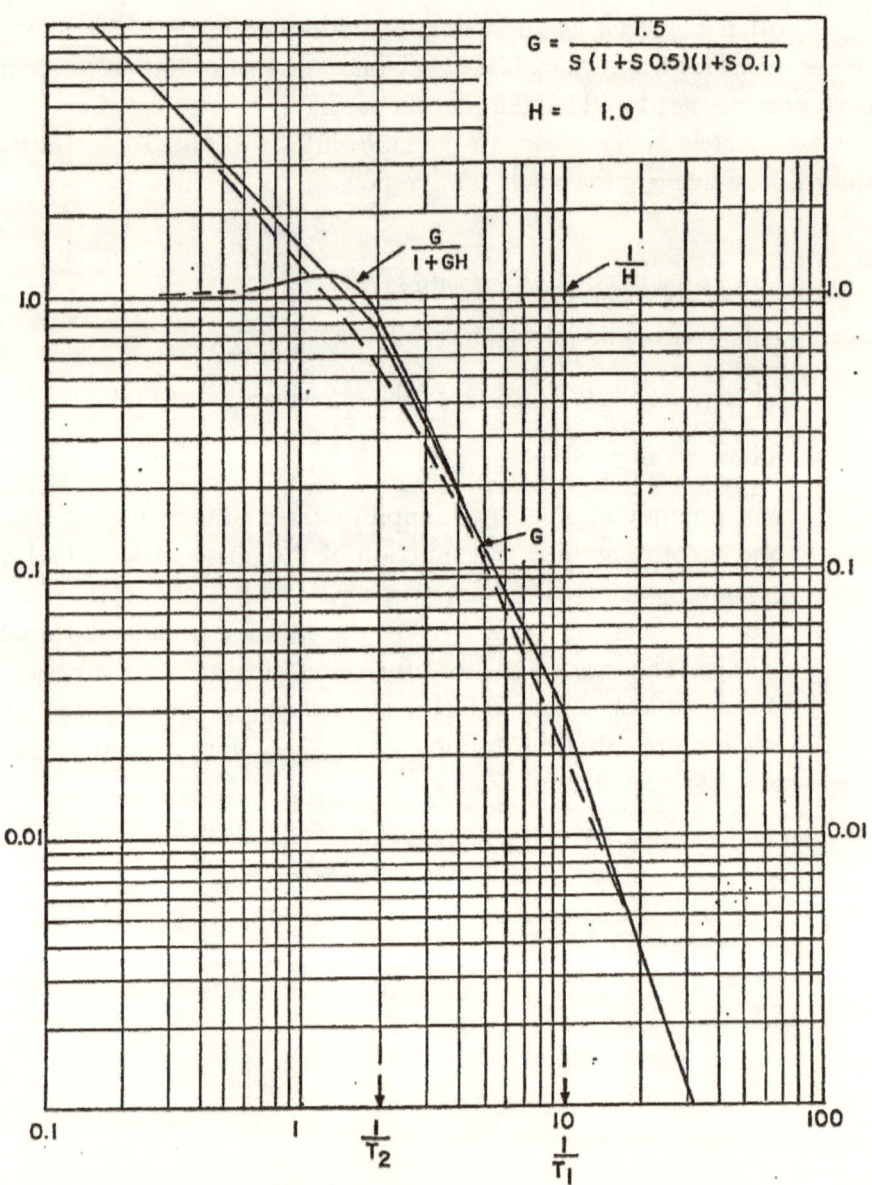

$$G = \frac{1.5}{S(1+S\,0.5)(1+S\,0.1)}$$

$$H = 1.0$$

$\frac{G}{1+GH}$

$\frac{1}{H}$

G

$\frac{1}{T_2}$

$\frac{1}{T_1}$

FIGURE E-14

APPENDIX F

NOTES ON PROCESS CONTROL

Historically automatic control technology developed along two parallel paths—one path being that taken by the instrumentation industry serving the process control field; such as control of chemical processes, steam plant controls, etc. The other path developed in the field of positioning, guidance controls, military gun turret controls, machine tool control, etc. This particular area saw developed the Bode, Nyquist, Root locus and other analytic techniques under the impetus of large outlays of Government research funds during and following World War II.

While the principles of control are universal, the usage of terms has been quite distinct among these groups. The latter talk about lead/lag functions, pole-zero locations, root locus, etc. While the former, rooted in process control terminology which precedes World War II, talks about two-mode and three-mode controls, proportional bands, reset and rate times, etc. These brief notes are to familiarize the student with process control terminology and practices which form an important field in Power Plant and Power System control.

Control Modes

A controller is a device which shapes control action by operating on the error signal. Figure F-1 is helpful in defining terms.

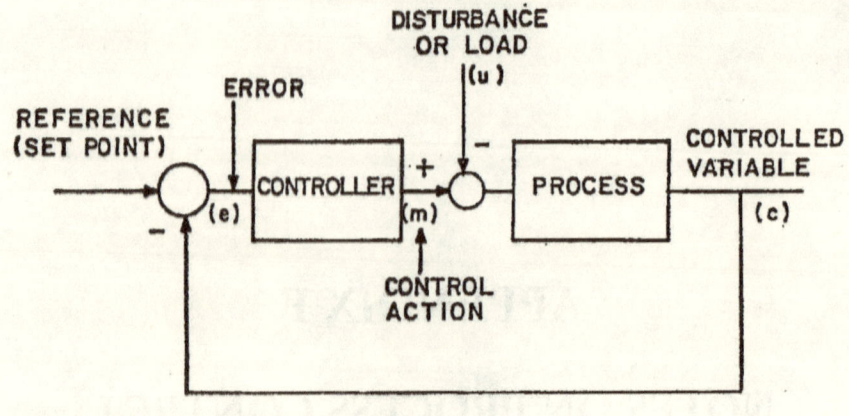

FIGURE F-1

A <u>proportional controller</u> is one where the relationship between control action and error is a simple proportional gain, K_p.

$$m = K_p \varepsilon \tag{F-1}$$

From previous discussions we note that a proportional controller cannot restore the error to zero unless there is an integration inherent in the process itself. The higher the gain the smaller the steady state error. Gain is expressed as "proportional band."

$$P.B = \frac{100}{K_p}\% \tag{F-2}$$

A gain of 1 yields a proportional band of 100%. A gain of 20 is equivalent to a 5% proportional band, i.e.; it takes an error of 5% to drive the control action 100%.

A <u>proportional plus reset controller</u> (a two mode controller) is one where integral action is added to proportional action.

The integral action <u>resets</u> the error to zero. The control equation is:

$$m = (K_p \varepsilon + K_1 \int \varepsilon dt)$$

or

$$m(s) = \left[K_p + \frac{K_I}{s} \right] \varepsilon(s) \qquad \text{(F-3)}$$

Equation F-3 can also be expressed as

$$\frac{m(s)}{\varepsilon(s)} = \frac{K_I \left(1 + \frac{K_p}{K_I} s\right)}{s} \qquad \text{(F-4)}$$

Reset action is generally expressed as the number of times that the integral action repeats the proportional action per unit of time.

i.e.; Repeats per second = K_I / K_p

$$\text{or Repeats per min (more commonly used)} = \frac{K_I \times 60}{K_p} \qquad \text{(F-5)}$$

Another way of expressing integral action is by "reset time," i.e.; the time that it takes the integral action to repeat the proportional action for a given fixed error. Obviously "reset time" in mins is the reciprocal of repeats per min.

A controller with <u>Proportional plus reset plus rate</u> action is called a 3 mode controller.

Its transfer function is generally expressed

$$K_p \left[1 + \frac{K_I}{K_p s} + T_R s \right] \qquad \text{(F-6)}$$

where T_R can be expressed as rate time which can be interpreted as the gain to rate of change of the error relative to the proportional gain.

In equation F-6 the term $T_R s$ implies an ideal differentiator. In actual practice differentiation usually has a maximum cut-off frequency and of necessity must be of the form $T_R s / (1 + T'_R s)$ where $T'_R \ll T_R$.

Practical 3 mode controllers either pneumatic, hydraulic or electronic have the form

$$\frac{K_I(1+\frac{K_p}{K_I}s)(1+T_Rs)}{s(1+\frac{T_R}{K}s)} \tag{F-7}$$

where   Prop. Band in percent $= 100/K_p$

$$\text{Reset time} = \frac{K_p}{K_I} \times \frac{1}{60} \text{ mins}$$

$$\text{Reset time} = T_R \times \frac{1}{60} \text{ mins}$$

The additional time constant T_R/K in the denominator is usually about $T_R/10$. It sets the maximum instantaneous gain for a step input at 10 times the proportional gain.

Figure F-2 shows the time response characteristics of the various controller types discussed above for a unit step input of error.

Figure F-3 shows the frequency response characteristics of these controllers

Tuning of Controllers and Process Response Curves

The important criteria in designing a process control system are:

(1) Instability or cycling of the controlled variable must be avoided (except where on-off controls are used).
(2) The steady state error should be minimized or even reduced to zero.
(3) The control should return the process variable to set point as soon as possible.

The type of controller and its adjustments that should be used to meet these requirements should be a function of the process characteristics. The essential information on these characteristics can be derived from the step response curves of the process.

Figure F-4 shows typical response curves of systems with multiple lags. One simple way of characterizing the shape of these response curves is by noting the times T_1 and T_2 which, as indicated on Fig. F-3, are the equivalent dead time and the time to maximize rate of rise of the response curve.

The relative case or difficulty of controlling a process is revealed by the shape of the response curve. In general the smaller the ratio L/D, i.e.; the sharper the "S" shape of the response curve the more difficult is the control task. In the limit, when L/D = 0 we have a process with almost pure "dead time" or "transport delay."

There are numerous articles giving guide rules on the choice of the type of controller and on the optimum values of controller parameters. Some references on this subject are included at the end of this Appendix.

In general these methods of controller tuning can be classified as "closed loop methods" or "open loop methods."

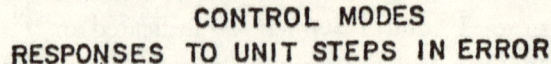

CONTROL MODES
RESPONSES TO UNIT STEPS IN ERROR

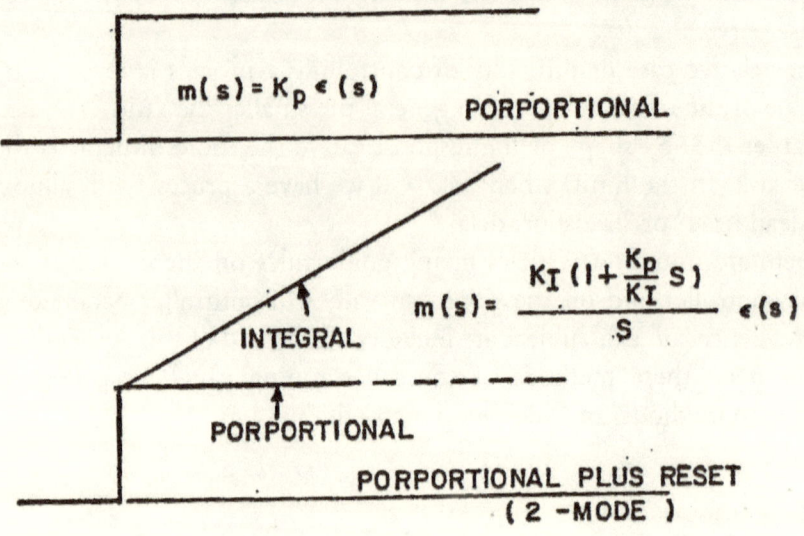

$$m(s) = K_p \, \epsilon(s)$$

PORPORTIONAL

$$m(s) = \frac{K_I \left(1 + \dfrac{K_p}{K_I} s\right)}{s} \, \epsilon(s)$$

INTEGRAL

PORPORTIONAL

PORPORTIONAL PLUS RESET
(2 -MODE)

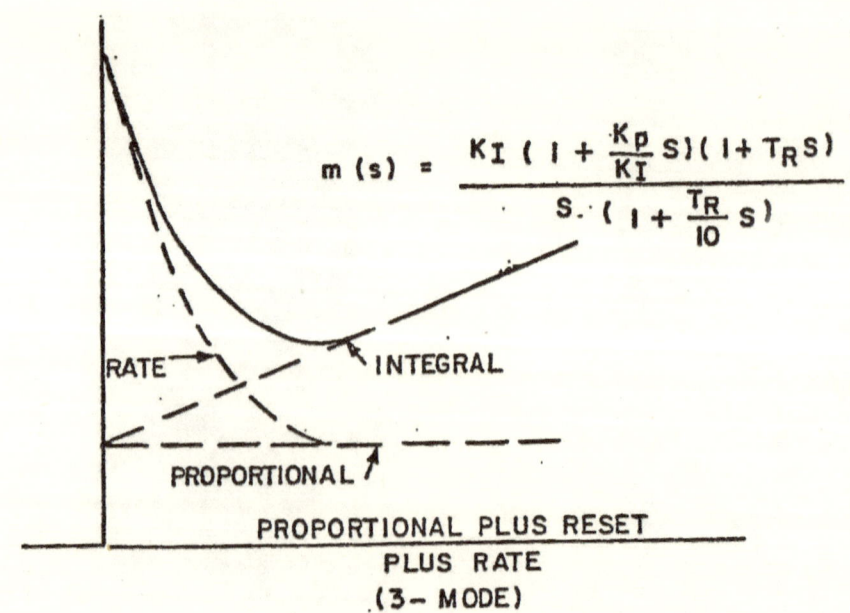

$$m(s) = \frac{K_I \left(1 + \dfrac{K_p}{K_I} s\right)\left(1 + T_R s\right)}{s \cdot \left(1 + \dfrac{T_R}{10} s\right)}$$

RATE

INTEGRAL

PROPORTIONAL

PROPORTIONAL PLUS RESET
PLUS RATE
(3- MODE)

FIGURE F-2

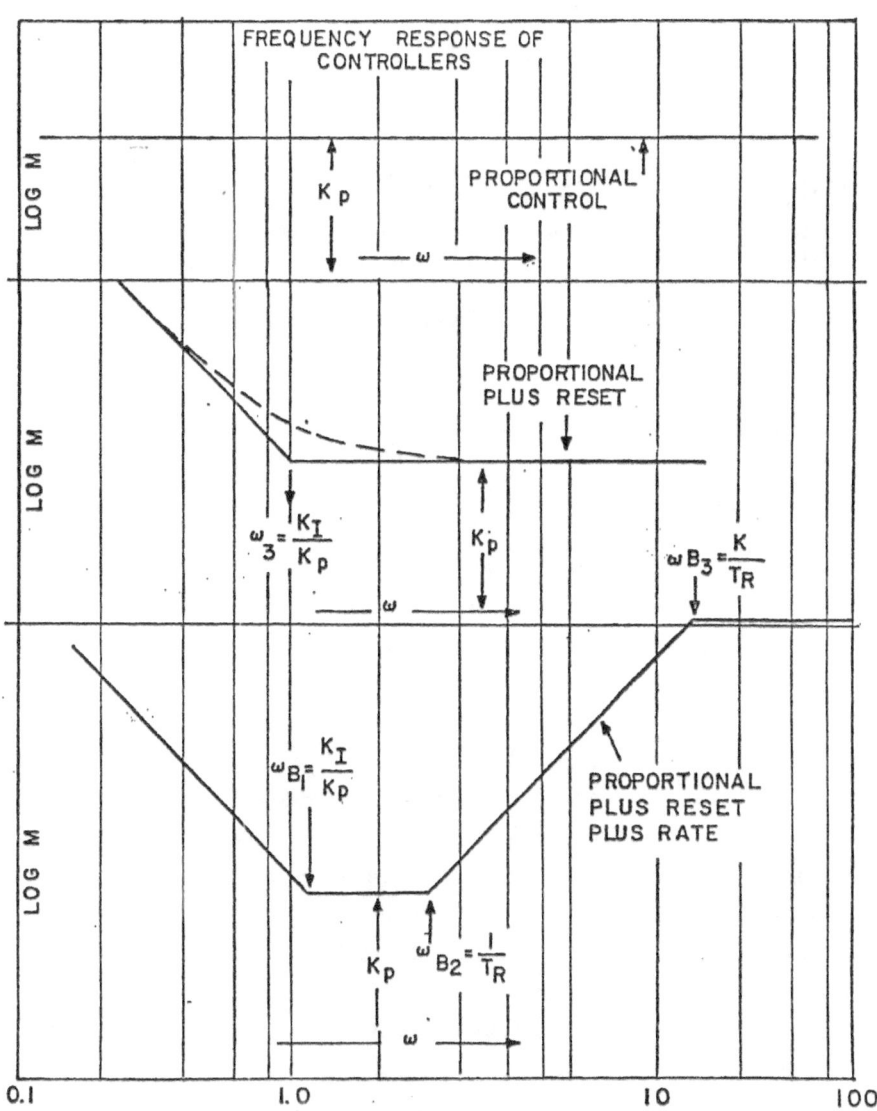

FIGURE F-3

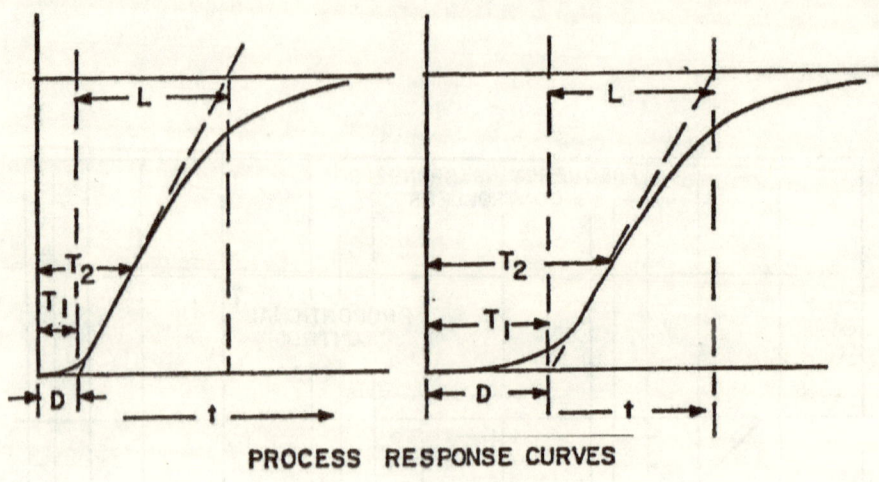

PROCESS RESPONSE CURVES

FIGURE F-4

Closed Loop Methods

Some of the oldest guide rules on controller tuning were developed by Ziegler and Nichols[34] and involved tuning by means of on line experiments with the closed loop system. The experiment involves using proportional control only and determining the so-called "ultimate proportional gain," K_u which causes the control system to be barely unstable. The period of oscillation in seconds P_u is also noted. On the basis of these two values the controller parameters are found which give a control response characterized by the so-called 1/4 decay ratio. Figure F-5 summarizes the essentials of this method of tuning.

It should be noted that these guide rules are quite empirical and do not necessarily yield the optimum parameters in cases where the process has significant dead time. Also the 1/4 decay ratio criterion often may not be acceptable since there are many control situations where the criterion should be one with no overshoot.

Open Loop Methods

The derivation of control parameters from the process open loop characteristics is probably of much wider use since it permits tuning from knowledge of the process response.

The material included here is from the work of Chien, Hrones and Reswick[32]. Many others have developed useful and more sophisticated guide rules. Nevertheless the general rules listed here have fairly wide applicability, especially when it is recognized that the exact criterion of optimality of the closed loop response is rather subjective to begin with.

Referring to Fig. F-4, the figure of merit R is defined as follows from the times L and D.

$$R = L/D$$

Also, let C be the plant gain. Based on these two figures of merit, R and C, the table below gives recommended controller parameters for two types of response, the quickest response without overshoot and quickest recovery with about 20% overshoot as shown in Fig. F-6.

CLOSED LOOP METHODS — ZIEGLER & NICHOLS

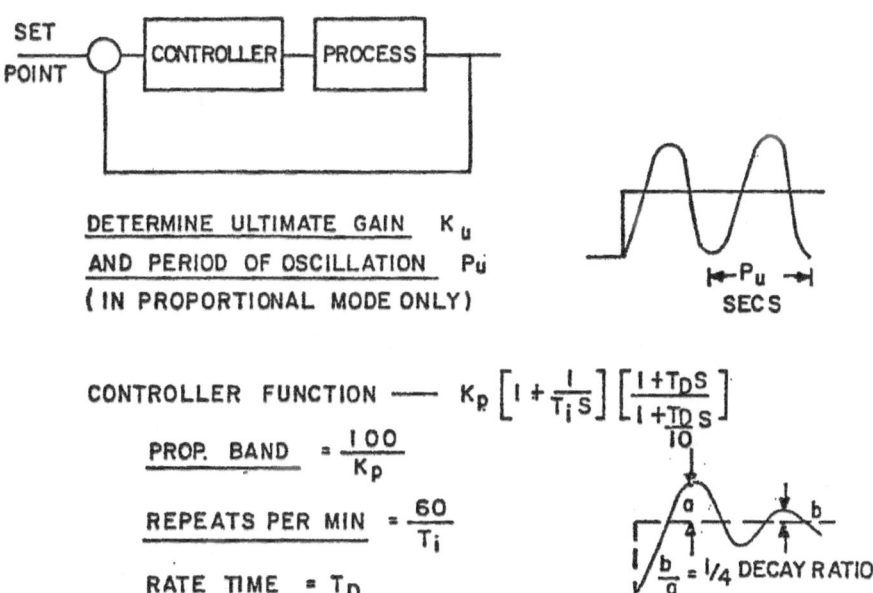

DETERMINE ULTIMATE GAIN K_u
AND PERIOD OF OSCILLATION P_u
(IN PROPORTIONAL MODE ONLY)

$\vert\!\!\leftarrow\!P_u\!\rightarrow\!\!\vert$
SECS

CONTROLLER FUNCTION —— $K_p \left[1 + \dfrac{1}{T_i S} \right] \left[\dfrac{1 + T_D S}{1 + \frac{T_D}{10} S} \right]$

PROP. BAND $= \dfrac{100}{K_p}$

REPEATS PER MIN $= \dfrac{60}{T_i}$

RATE TIME $= T_D$

$\dfrac{b}{a} = 1/4$ DECAY RATIO

CONTROLLER ADJUSTMENT FORMULAE

MODES	GAIN	PROP. BAND	REPEATS / MIN	RATE TIME-MIN
PROPORTIONAL	0.5 Ku	200/Ku	— — —	— — —
PROPORTIONAL & RESET	0.45 Ku	222/Ku	$\dfrac{1.2 \times 60}{Pu}$	— — —
PROPORTIONAL RESET & RATE	0.6 Ku	167/Ku	$\dfrac{2 \times 60}{Pu}$	$\dfrac{0.125\ Pu}{60}$

FIGURE F-5

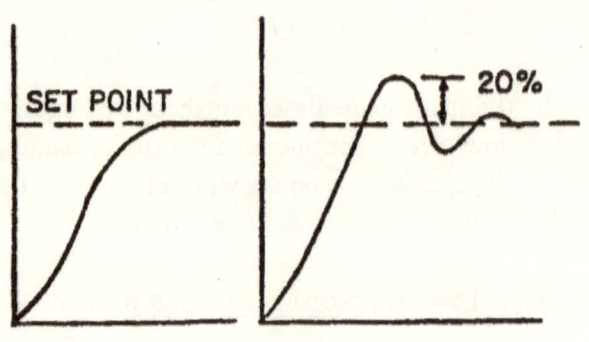

FIGURE F-6

TYPE OF CONTROL	QUICKEST RESPONSE WITHOUT OVERSHOOT		QUICKEST RESPONSE WITH 20% OVERSHOOT	
	Step Change in set point	Step Change in Load	Step Change in set point	Step Change in Load
STRAIGHT PROPORTIONAL Proportional band (percent) =	$\dfrac{333C}{R}$	$\dfrac{333C}{R}$	$\dfrac{143C}{R}$	$\dfrac{143C}{R}$
PROPORTIONAL AND RESET Proportional band (percent) =	$\dfrac{286C}{R}$	$\dfrac{167C}{R}$	$\dfrac{167C}{R}$	$\dfrac{143C}{R}$

Reset time (time units of D) =	3.33	$\dfrac{6.67DC}{R}$	1.67DC	$3.33\dfrac{D}{R}$
PROPORTIONAL AND RESET AND DERIVATIVE Proportional band (percent) =	$\dfrac{167C}{R}$	$\dfrac{105.2C}{R}$	$\dfrac{105.2C}{R}$	$\dfrac{83.3C}{R}$
Reset (time units of D) =	1.67DC	$2.5\dfrac{DC}{R}$	1.43DC	$1.67\dfrac{DC}{R}$
Rate time (time units of D) =	$0.3\dfrac{RD}{C}$	$0.4\dfrac{RD}{C}$	$0.45\dfrac{RD}{C}$	$0.5\dfrac{RD}{C}$

Note that two types of disturbance for which the table above gives recommended control parameter values are:

(1) Step change in set point or reference
(2) Step load change

Refer to the block diagram of Fig. F-1 for definition of these disturbances.

Controller Tuning Based on Open Loop Process Response Measurement

A wide range of processes exhibits responses to control action characterized by a combination of dead time and time lags. The response curve can be values of time to reach a certain fraction of the final value. An arbitrary characterization of the response curve is suggested in terms of the times to reach 10%, 63%, and 90% of the process final value as shown in Fig. F-7.

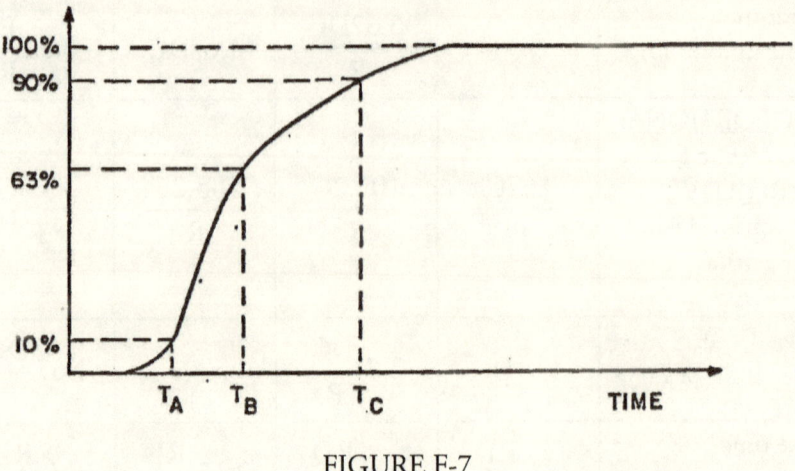

FIGURE F-7

Ratios of time T_C/T_B and T_A/T_B appear to characterize the type of response, i.e.; the relative amount of dead time to other lags.

A number of computer runs performed to optimize controller tuning, i.e.; to minimize a performance index of the form

$$I = \int_0^\infty (|\text{error}| \times \text{time}) \, \alpha t + K \times (\text{Max. overshoot})$$

has yielded nomographs for arriving at two and three mode controller settings as function of the ratios T_A/T_B and T_C/T_B.

Figures F-8 to F-12 show these nomographs for three mode and two mode controllers.

<u>Example</u>

In order to illustrate the use of this method in determining controller settings, the following sample problem is presented.

A pressure controller modulates feeder speed as follows:

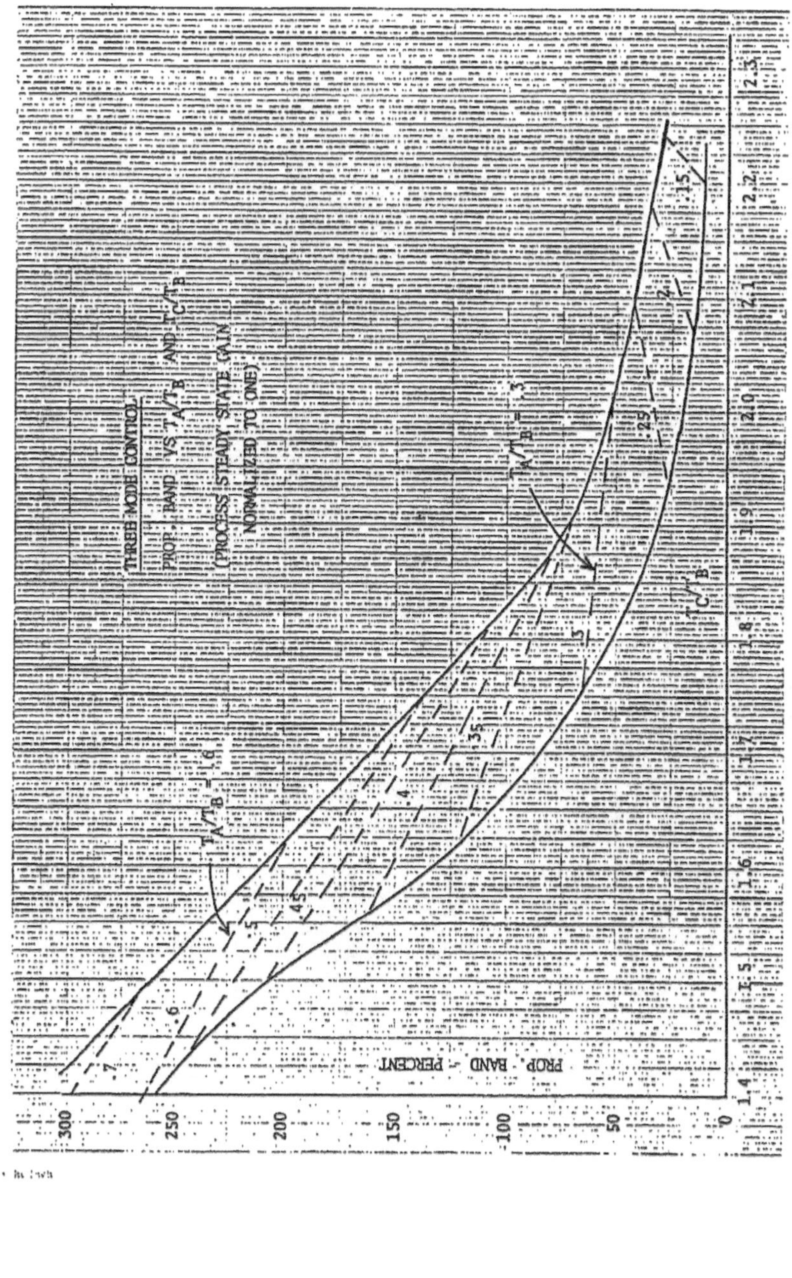
THREE MODE CONTROL
PROP. BAND VS T_A/T_B AND T_C/T_B
(PROCESS STEADY STATE GAIN NORMALIZED TO ONE)
PROP. BAND - PERCENT
T_A/T_B = .6
T_A/T_B = .3
T_C/T_B

FIGURE F-8

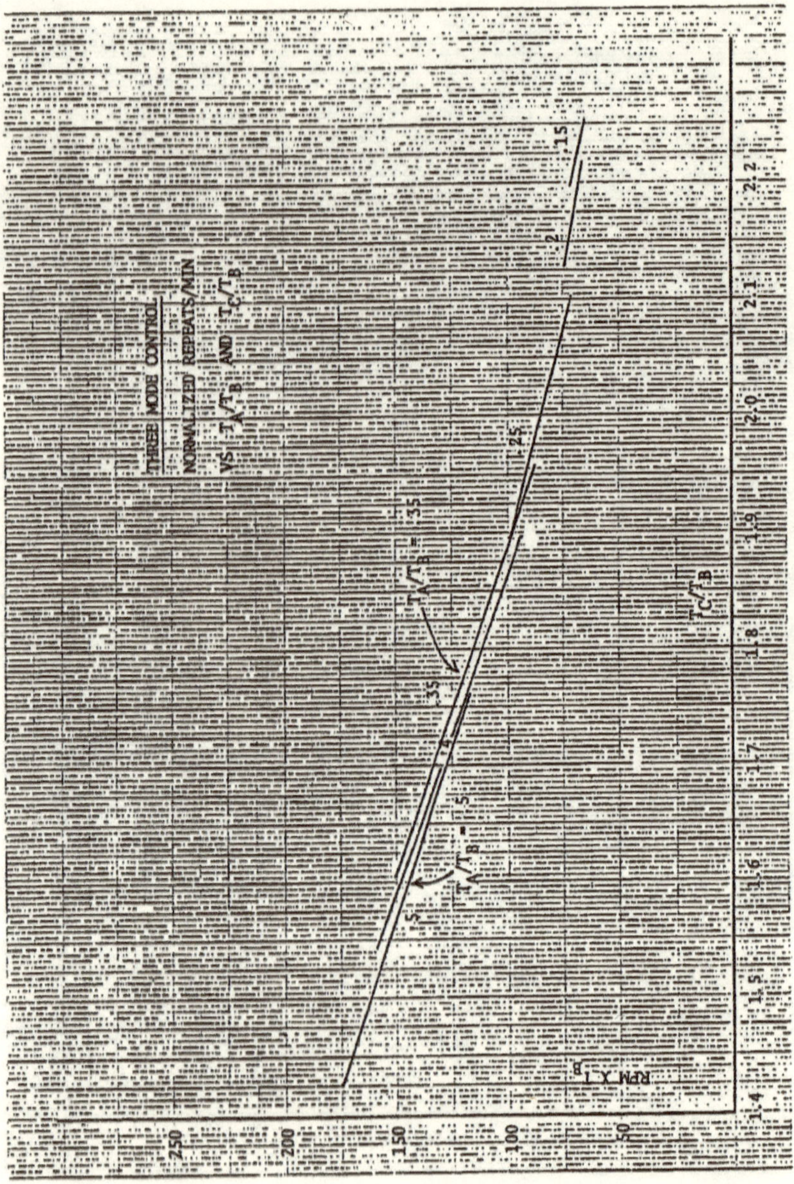
THREE MODE CONTROL
NORMALIZED REPEATS/MIN
VS T_A/T_B AND T_C/T_B
T_A/T_B = .35
T_A/T_B = .5
RPM X T_B
T_C/T_B
250
200
150
100
50
1.4
1.5
1.6
1.7
1.8
1.9
2.0
2.1
2.2
.5
.4
.35
.25
.2
.15

FIGURE F-9

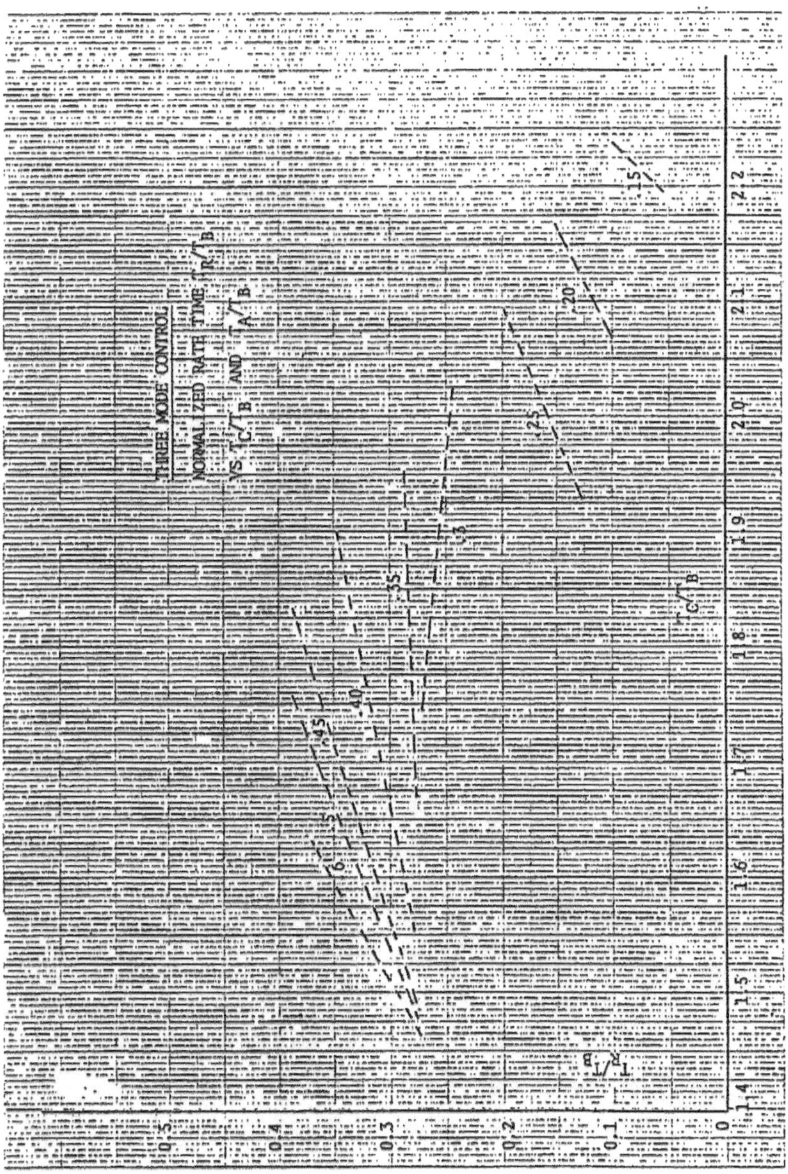
THREE MODE CONTROL
NORMALIZED RATE TIME τR/τB
VS τC/τB AND τA/τB
τR/τB
τC/τB
0.5
0.4
0.3
0.2
0.1
0
1.4
1.5
1.6
1.7
1.8
1.9
2.0
2.1
2.2

FIGURE F-10

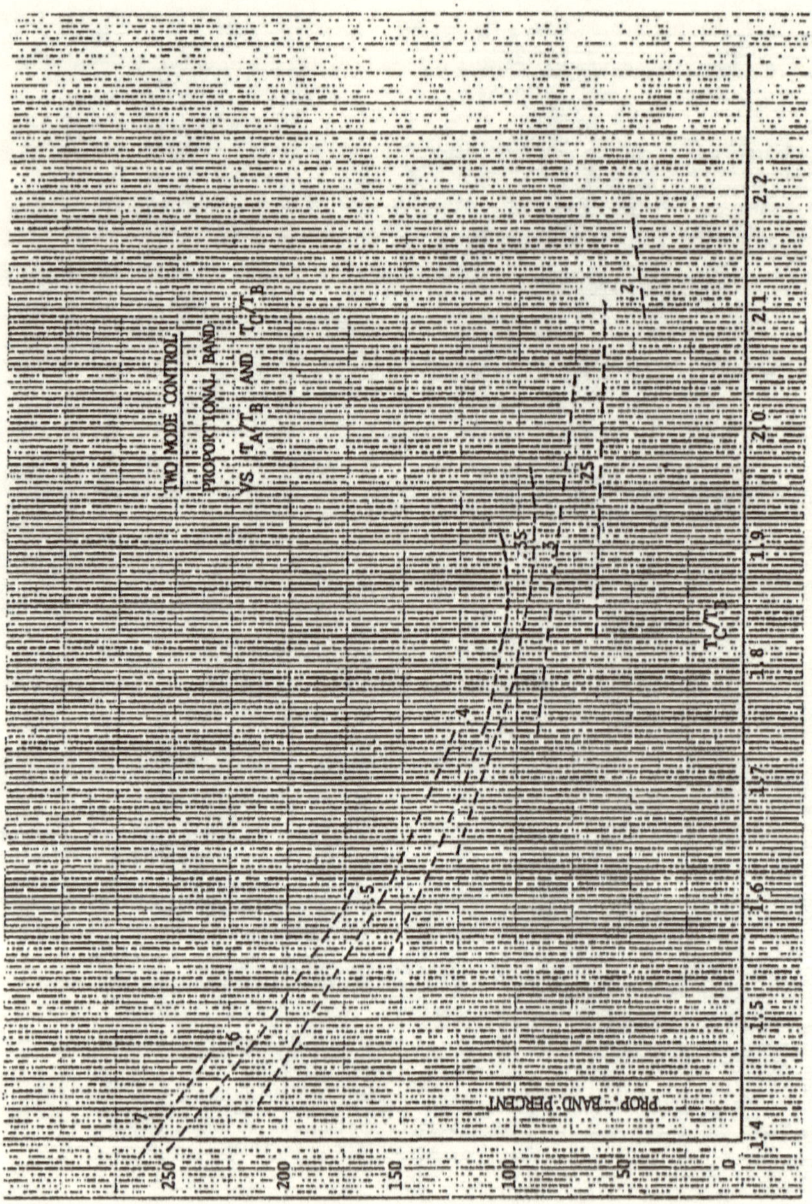
TWO MODE CONTROL
PROPORTIONAL BAND
VS T_A/T_B AND T_C/T_B
PROP. BAND PERCENT
T_C/T_B
0
50
100
150
200
250
1.4
1.5
1.6
1.7
1.8
1.9
2.0
2.1
2.2
.7
.6
.5
.4
.35
.3
.25
.2

FIGURE F-11

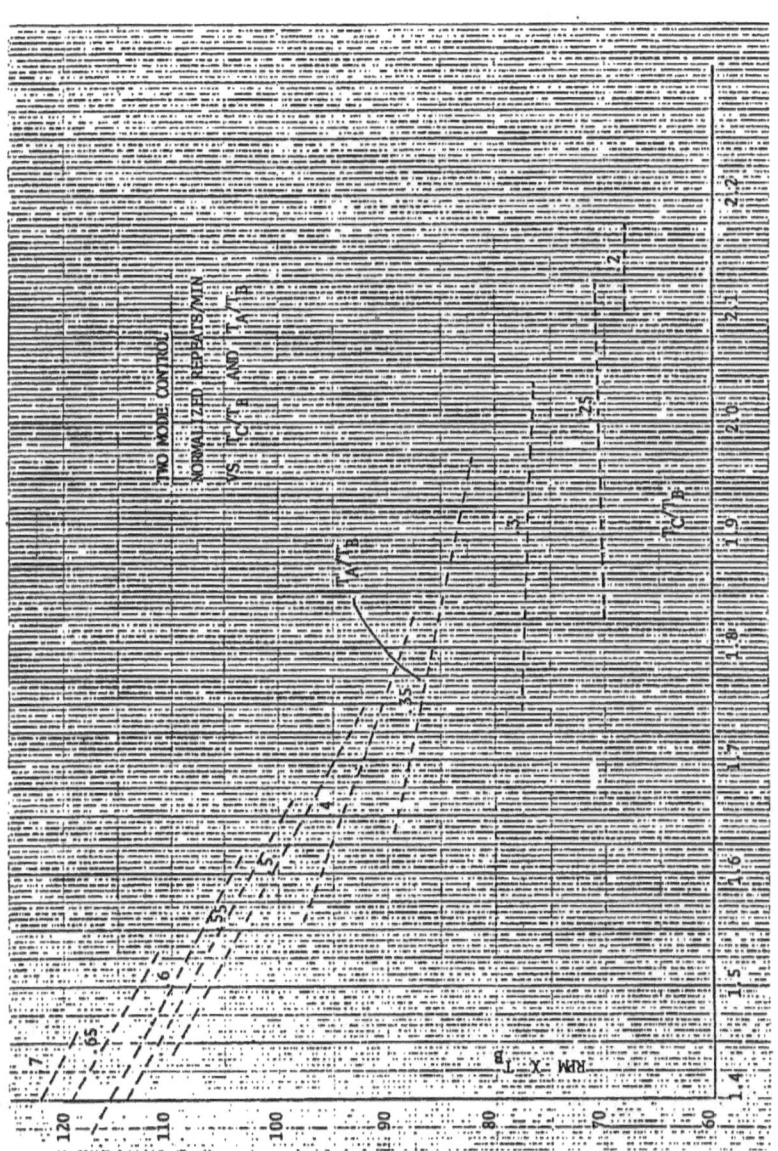

FIGURE F-12

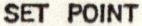

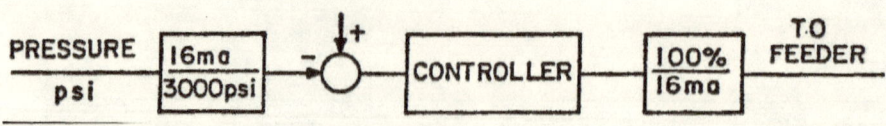

FIGURE F-13

With the controller in manual, a 2 ma step in the manual signal to feeder produces the pressure response shown in Fig. F-14. The steady state change in pressure is 1.6 ma. Thus, we may measure the time to 10%, 63%, and 90% (0.16 ma, 1.0 ma, and 1.44 ma, respectively) in order to obtain T_A, T_R and T_C.

$$
\begin{aligned}
T_A &= 15.5 \\
T_B &= 52.5 \\
T_C &= 96.25 \\
T_A/T_B &= .3 \\
T_C/T_B &= 1.83
\end{aligned}
$$

Referring to Fig. F-8 we may find controller settings as follows: Draw a vertical line through 1.83 on the $T_C/T_B = .3$ line by interpolation. Draw a horizontal line to the PB axis and read PB = 180%. This procedure may be repeated for each parameter using Figs. F-8 through F-12. For this example we obtain results:

2 Mode	3 Mode
PB = 85%	PB = 62%
RPM × T_B = 77	RPM × T_B = 105
	T_R/T_B = .265

We may now apply our gain correction factors due to time scaling (multiply RPM/T_B and T_R/T_B by T_B) and due to loop gain correction.
(Multiply P.B. by 1.6 ma pressure/2 ma feeder)

2 Mode	3 Mode
PB = 85% × 1.6/2 = 68%	PB = 62% × 1.6/2 = 49.6%
RPM = 77/52.5 = 1.467	RPM = 105/52.5 = 2.00
	T_R = .265 × 52.5 = 13.9

The controlled responses for these controller settings are shown in Fig. F-15 for two and three mode controllers.

REFERENCES

G.1 Chien, Hrones & Reswick, "On The Automatic Control of Generalized Passive Systems," ASME Transactions, Vo. 74, No. 2, Feb. 1952, p. 175.

G.2 Ziegler, J.G., and Nichols, N.B., "Optimum Settings for Automatic Controllers," ASME Transactions, Vo. 64, 1942, pp. 769-775.

G.3 Smith, C.L., and Murrill, P.W., "A More Precise Method for Tuning Controllers," ISA Journal, May 1966, pp. 50-58.

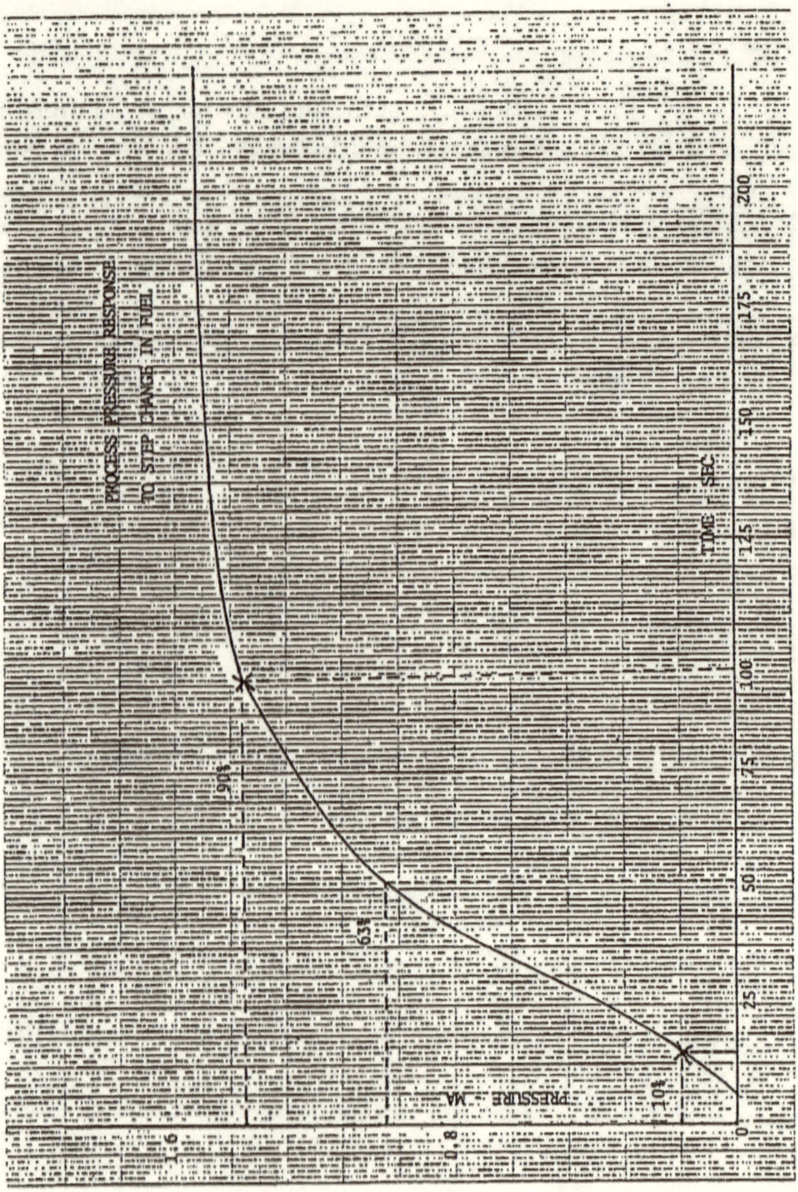
PROCESS PRESSURE RESPONSE
TO STEP CHANGE IN FUEL
PRESSURE - MA
1.6
0.8
0
10%
63%
90%
TIME - SEC
25
50
75
100
125
150
175
200

FIGURE F-14

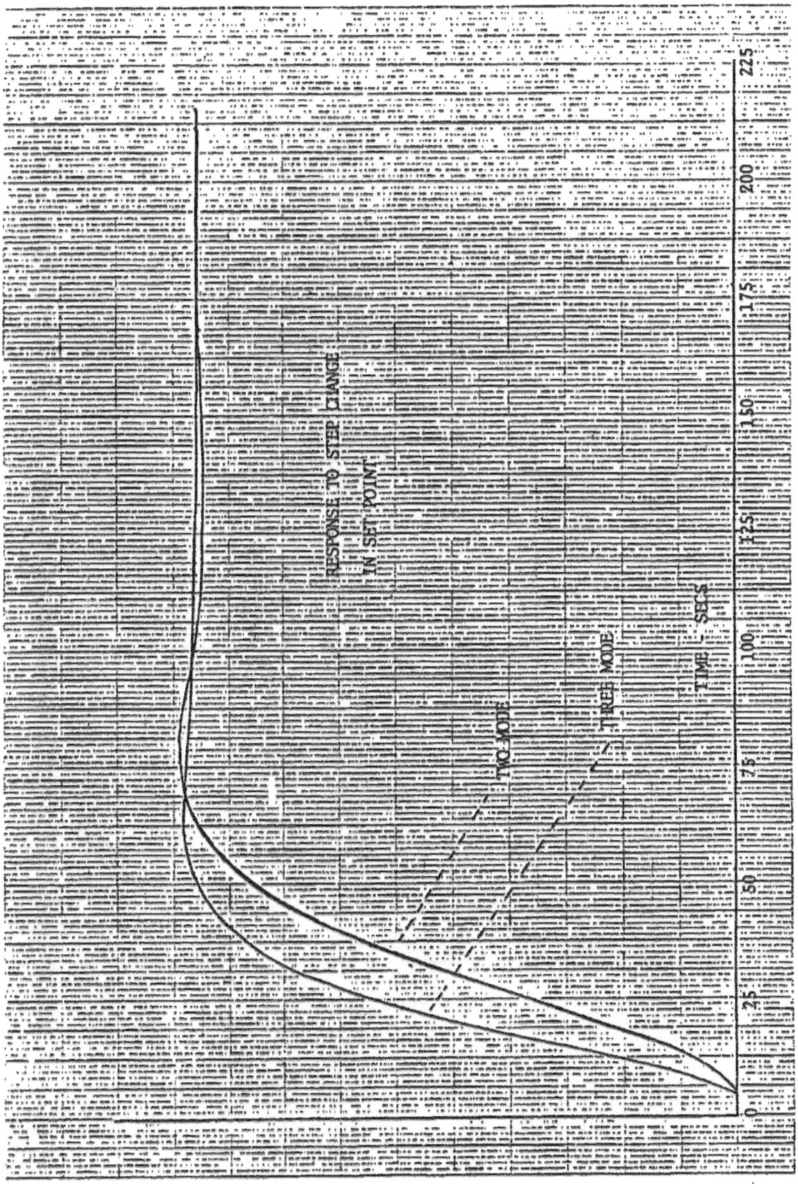
RESPONSE TO STEP CHANGE
IN SET POINT
TWO MODE
THREE MODE
TIME - SECS
0
25
50
75
100
125
150
175
200
225

FIGURE F-15

ADDENDUM

DIGITAL CONTROL ALGORITHMS AND CONTROL TUNING

Implementation of control laws with a digital computer can be done in an infinite number of ways. Many of these give essentially equivalent performance, and one should be careful not to spend a career analyzing trivial or marginal differences between algorithms. One of the great advantages of the digital computer is the flexibility with which all types of control and logic calculations can be incorporated to suit the particular need. Such different control modes as the use of limits and dead-bands that may be fixed or functions of some variable, plus switching on or off of integral or other control action depending on input conditions are purely a function of the process requirements and the ingenuity of the control engineer.

Control know-how is nevertheless well rooted in analog practices, and it is logical to evolve from these practices developing digital equivalents to well-known analog control modes, limiting enhancements to those easily implementable with the digital approach. Such enhancements, with particular reference to dead-time compensation, are proposed in this paper as are methods of parameter tuning from knowledge of the process open loop response.

As a preamble and for reference purposes, some terms frequently used in process control are described below.

THREE MODE CONTROL

Two most commonly used forms of three mode control are shown in Figures F-16 and F-17. The parallel configuration of Figure F-16 has non-interacting control modes in the time domain, i.e., proportional, integral or derivative gains can be individually set without interacting with each other.

When the overall control is expressed as one transfer function made up of zeros and poles, we note that the adjustment of individual terms of the time-domain function results in changes of the poles and zeros of the control function. For this reason, this form is known as interacting in the frequency domain. One should note that the parallel form of Figure F-16, when expressed into series form, can give rise to complex zeros which at times may be advantageous.

The form of Figure F-17. called the series form, is that normally encountered with analog three mode controllers using one operational amplifier. The form of the transfer function for this type of three mode controller is:

$$\frac{K_I(1+T_i s)(1+T_D s)}{s(1+\frac{T_D}{K} s)} \tag{F-8}$$

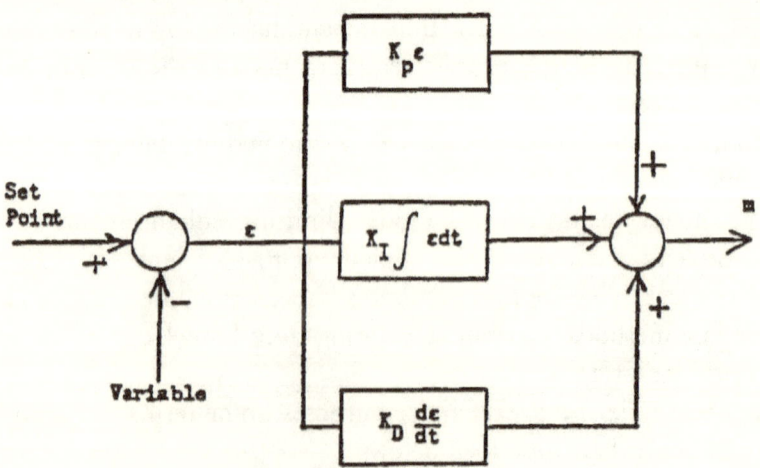

Parallel Configuration of 3-mode Controller

Figure F-16

Set Point

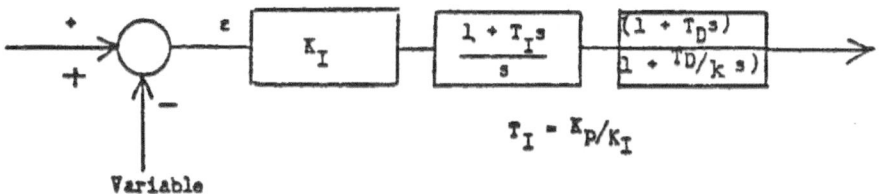

$$T_I - K_P/K_I$$

Series Configuration of 3-Mode Controller

Figure F-17

and, the adjustments normally available are non-interacting in the frequency domain, $= \dfrac{100}{K_p}$ i.e. proportional band adjusts the gain of the overall function. Reset time T_i is generally expressed in reciprocal form as repeats per $= \dfrac{60}{T_i}$ and can be adjusted without affecting the other parameters of the transfer function, and likewise rate time T_D can be adjusted independent of the other parameters. Note that in this form there is no possibility of complex zeros. K is often fixed at some value such as 10.

The considerable amount of frequency response thinking, guide rules. etc., that have evolved over the years make it desirable to have control algorithms that are non-interacting in the frequency domain. In other words, the control engineer is more at home with locating such things as lead breaks (T_D and T_I) or corner frequencies $\left(\dfrac{1}{T_D}\right), T\dfrac{1}{I}$ and gain K_p as independent parameters than to work with the basic proportional reset and rate gains of the parallel system. As more control experience and guide rules are developed in the time domain, these preferences may change. Digital techniques permit, of course, a wide range of control laws which can be implemented with algorithms and algebraic Z form manipulations, as explained in many tests and technical papers.

Some of the variations that will find application are control modes written such that derivative and/or proportional action is bypassed on set point changes, provision for limiting the change in output per computation step and provision for automatic resetting of accumulators to prevent integral windup.

A major opportunity for digital control is the relative ease of accounting for process dead-time and the implementation of linear, nonlinear, adaptive etc. algorithms not necessarily constrained by the linear differential equations of electrical circuits.

In addition to ensuring a control algorithm that is capable of coping with a wide range of process response characteristics, it would be very desirable to be able to derive parameter settings directly from knowledge of process response and desired closed loop response characteristics. The derivation and examples below show a logical approach to achieve both objectives of having a flexible control algorithm whose parameters are directly relatable to the process response and desired control response characteristics.

DERIVATION OF DESIRED CONTROLLER TRANSFER FUNCTION

The method outlined below is essentially similar to the ideas first introduced by Guillemin many years ago. The control design has been adapted for processes with dead-time and using the control logic capabilities of digital computers. Figure F-18 shows the traditional closed loop control structure with the controller function g(s) and the process function $\varepsilon^{-st} G(s)$.

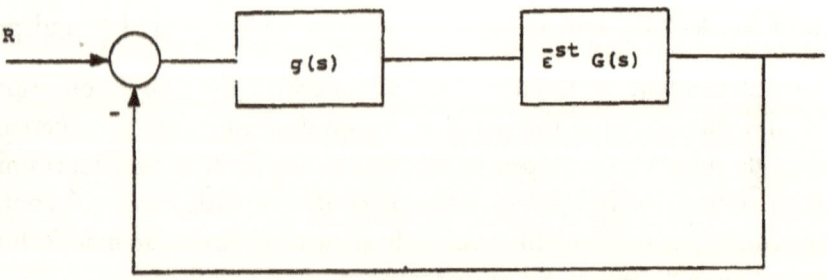

Figure F-18

The process response characteristic is described by the transfer function $\varepsilon^{-st}G(s)$ where ε^{-st} represents a dead-time (transport lag) and $G(s)$ is a ratio of polynomials in s.

Let the desired controlled response (closed loop) be $\varepsilon^{-st}D(s)$. To be physically realizable the desired controlled response cannot avoid exhibiting a dead-time at least equal to the dead-time in the process response.

The control action to achieve the desired response would evidently be given by the impulse response function:

$$\frac{\varepsilon^{-st}D(s)}{\varepsilon^{-st}G(s)} = \frac{D(s)}{G(s)} \tag{F-9}$$

This same control action expressed in terms of the elements of the closed loop in Figure F-18 is:

$$\frac{g(s)}{1+g(s)\varepsilon^{-st}G(s)} \tag{F-10}$$

Equating (F-8) and (F-9) and solving for g(s):

$$g(s) = \frac{D(s)}{G(s)\,[1-\varepsilon^{-st}D(s)]} \tag{F-11}$$

Expression (F-10) is the desired transfer function of the controller.

Application To Typical Processes

A wide range of process response characteristics can be described approximately by the function:

$$\varepsilon^{-st}G(s) = \frac{\varepsilon^{-st}\,\omega\dfrac{2}{p}}{s^2 + 2\zeta_p\omega_p s + \omega\dfrac{2}{p}} \tag{F-12}$$

normalized to yield a steady state gain of one. This function exhibits a dead-time T and a second order response with a natural frequency ω_p and a damping ratio ζ_p.

Figures F-19 and F-19a show the step response of this function beyond the dead-time T for a range of values of ζ_p. Addition of dead-time results in a translation of the response by the amount of dead-time.

The desired closed loop response of the process can also be expressed by a similar function with symbols subscripted "d" for desired versus "p" for process in (F-12).

$$\varepsilon^{-st} D(s) = \frac{\varepsilon^{-st} \omega\frac{2}{d}}{s^2 + 2\zeta_d \omega_d s + \omega\frac{2}{d}} \tag{F-13}$$

The speed of response and overshoot of the desired function are controlled by the parameter ω_d and ζ_d. Typically $\omega_d > \omega_p$ and ζ_d would be about 0.8.

Using (F-12) and (F-13) in (F-11), the controller function becomes

$$g(s) = \frac{\omega\frac{2}{d}(s^2 + 2\zeta_p \omega_p s + \omega\frac{2}{p})}{\omega\frac{2}{p}(s^2 + 2\zeta_d \omega_d s + \omega\frac{2}{d})(1 - \varepsilon^{-st})} \tag{F-14}$$

A block diagram implementation of (F-14) in terms of integrations, time delays and other arithmetic operations is shown in Figure F-20.

The controller characteristics are thus completely specified if the process response can be expressed as a time delay followed by a second order response as in (F-12).

Examples

Figures F-21 to F-24 show examples of closed loop controlled performance for different process open loop response characteristics described by the parameters ω_p, ζ_p and T. As expected the control adjustments yield the same response shape whether or not the process has dead-time, the difference being that the output response is delayed by the process dead-time. Figures F-21 and F-22 show the controlled response for the desired damping ratio ζ_d = 0.8. Figures F-23 and F-24 are for a desired damping ratio ζ_d = 1.0.

The sensitivity of the controlled performance to errors in estimates of the process dead-time is shown in Figures F-25 to F-28. Figures F-25 and F-27 show controlled performance for the case where the controller compensation is for a dead-time larger than the actual process dead-time. Figures F-26 and F-28 are for cases where the controller compensation is for less than the actual process dead-time.

 PROCESS RESPONSE TO UNIT STEP INPUT FOR VARIOUS
VALUES OF ZETA NORMALIZED FREQUENCY (W = 1)

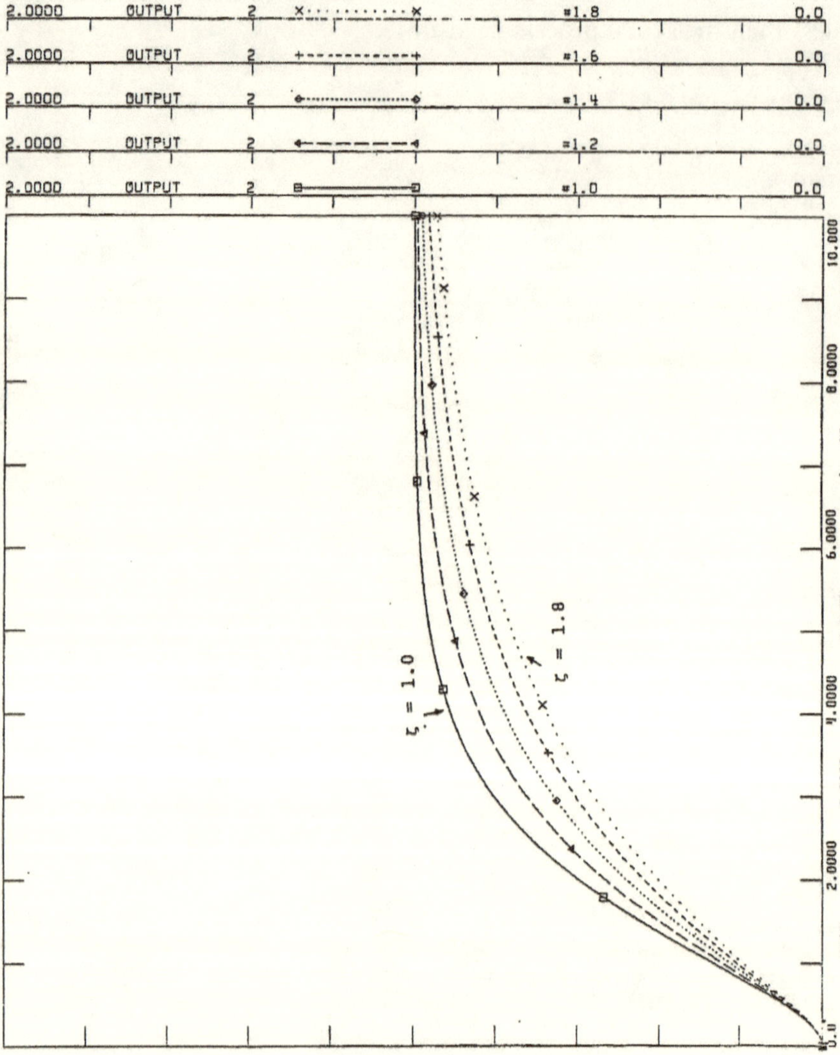

Figure F-19

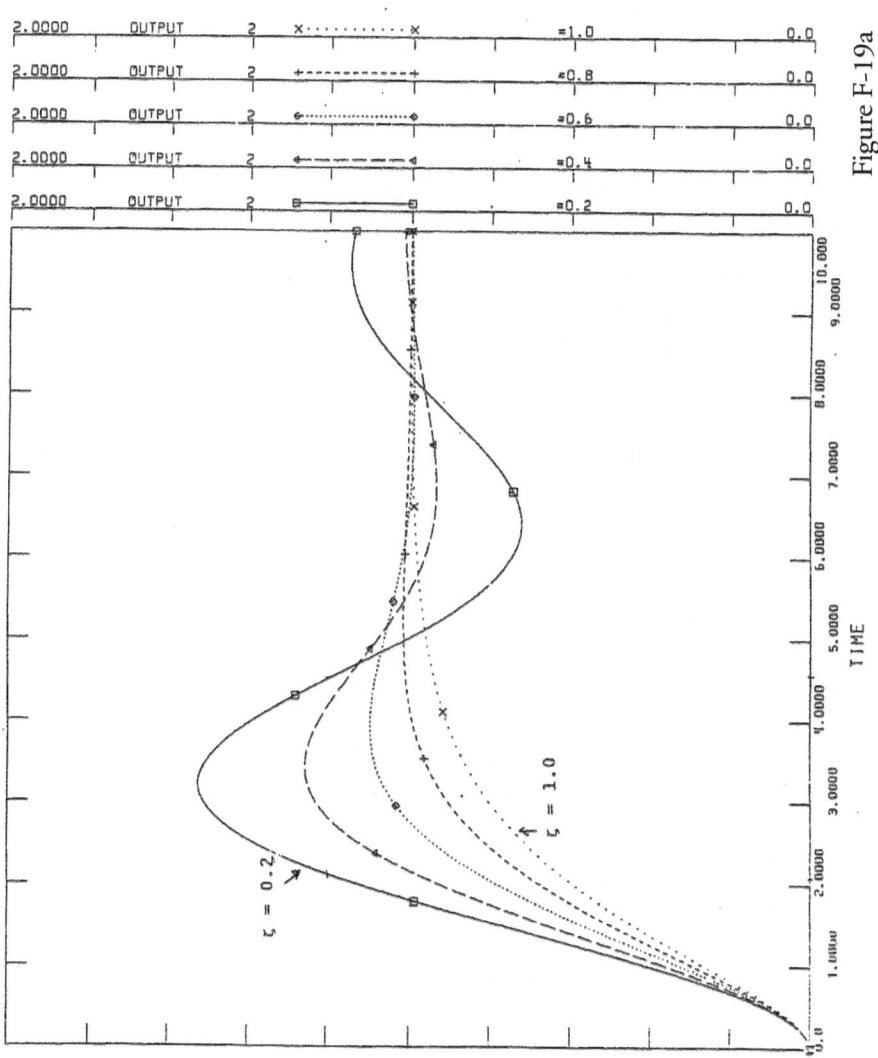

Figure F-19a

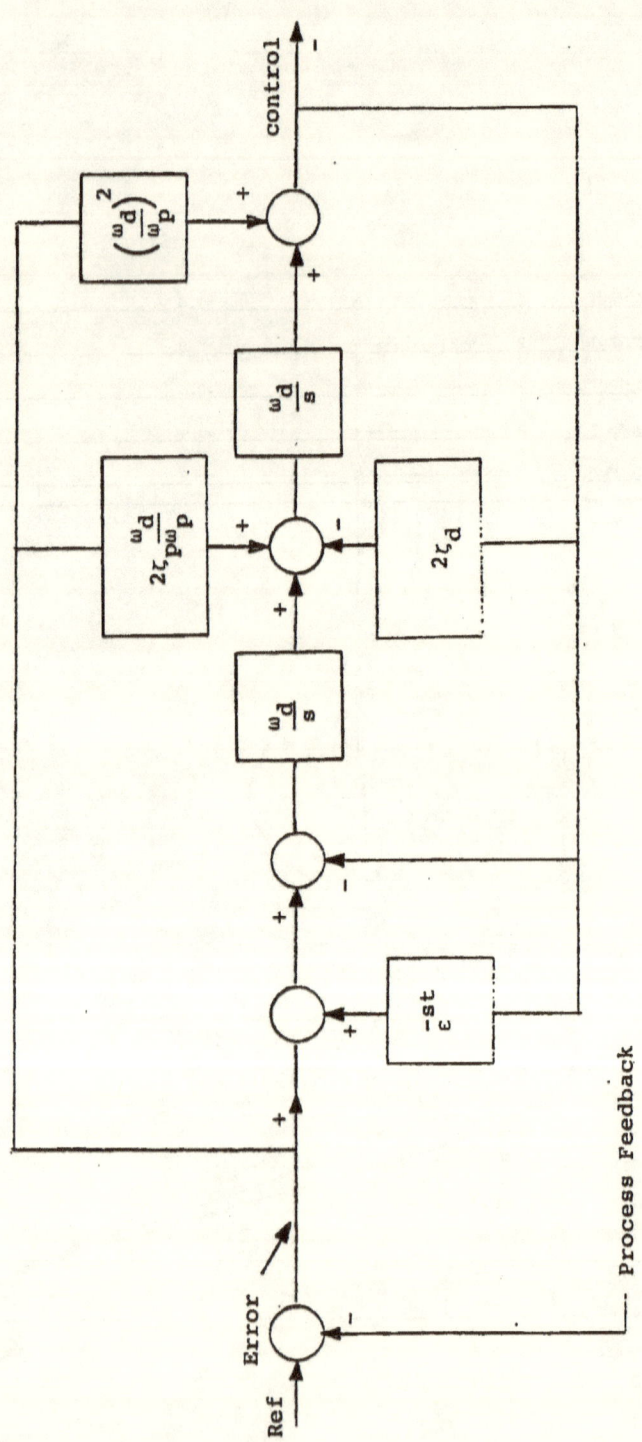

Figure F-20

DIGITAL CONTROLLER MODEL SECOND ORDER MODEL
OF FROCESS AND OF DESIRED OUTPUT WD = 5.0;
ZETRD = 0.8; WP = 1.0; ZETAP = 0.3 T = .25; T1 = .25

FILE: = P01

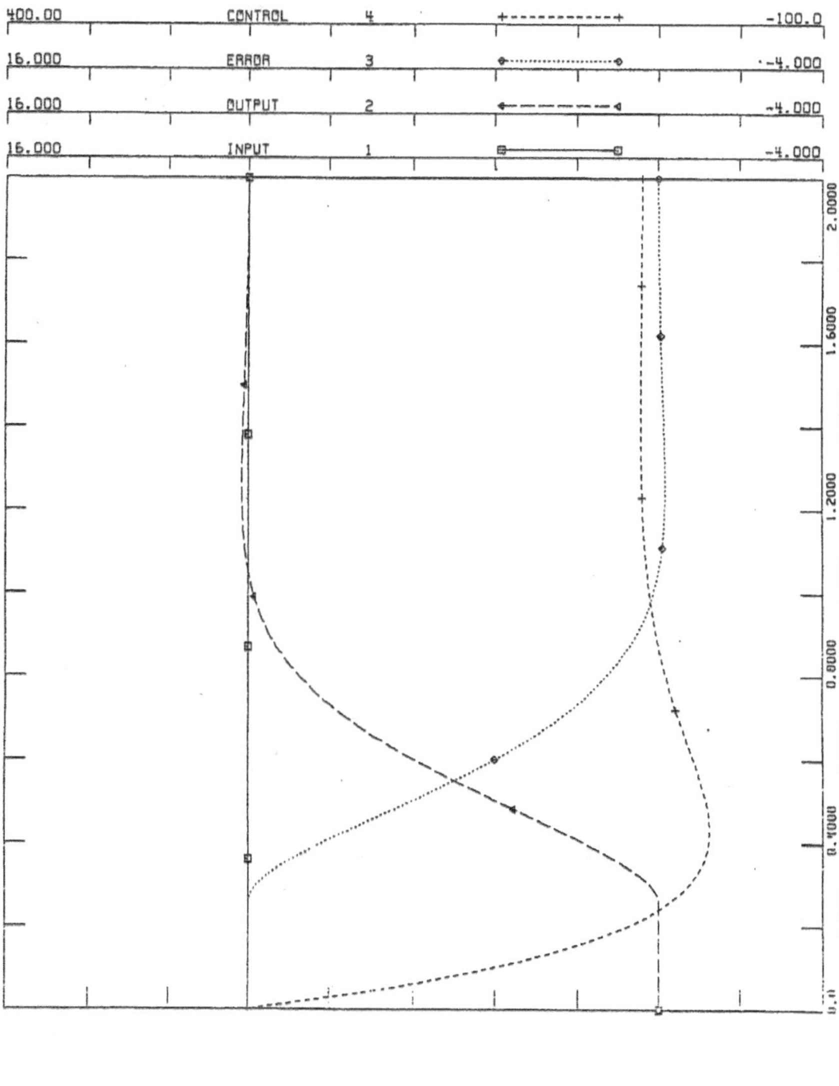

Figure F-21

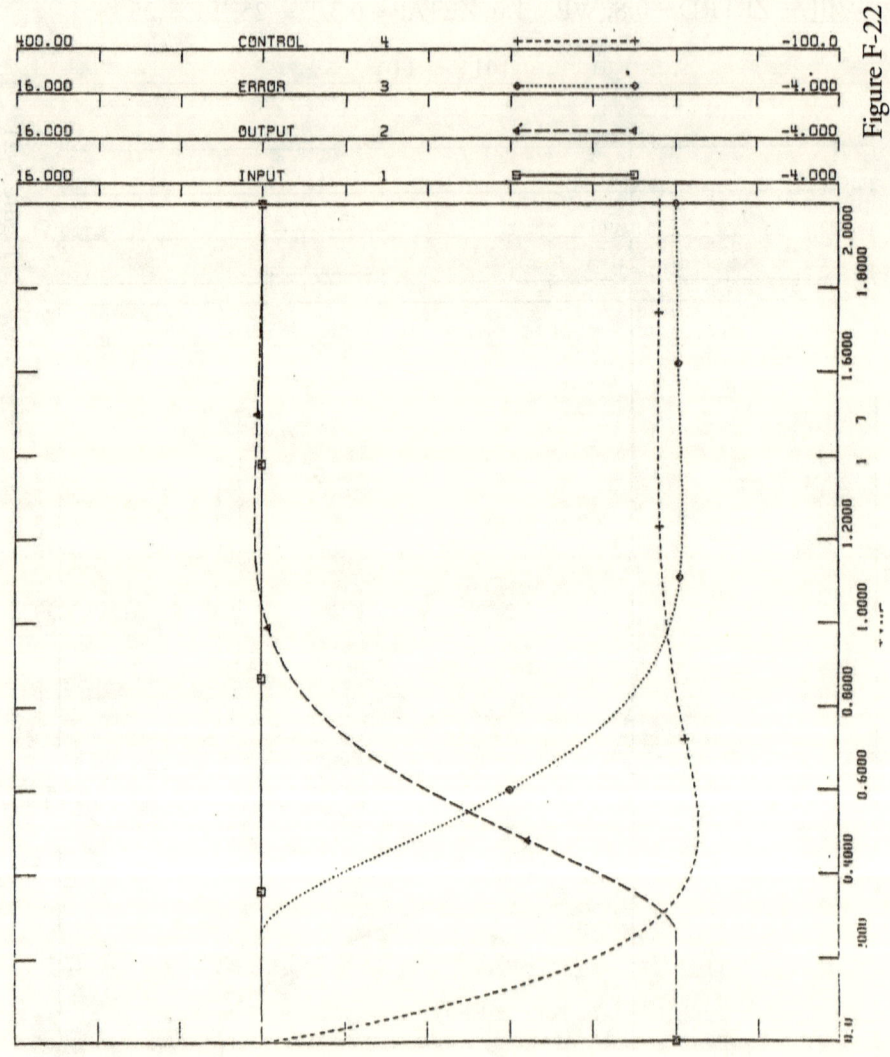

Figure F-22

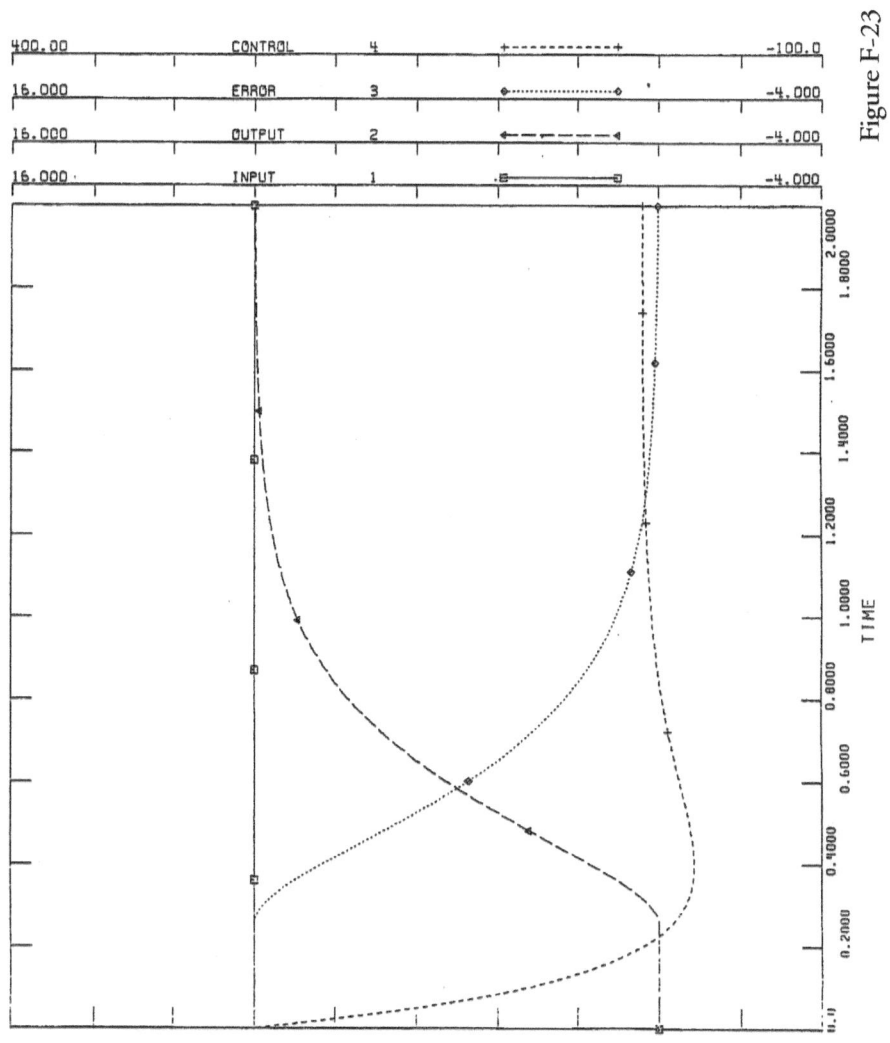

400.00	CONTROL	4	+------------+	-100.0
16.000	ERROR	3	◆·············◆	-4.000
16.000	OUTPUT	2	◄------◄	-4.000
16.000	INPUT	1	▣——————▣	-4.000

Figure F-23

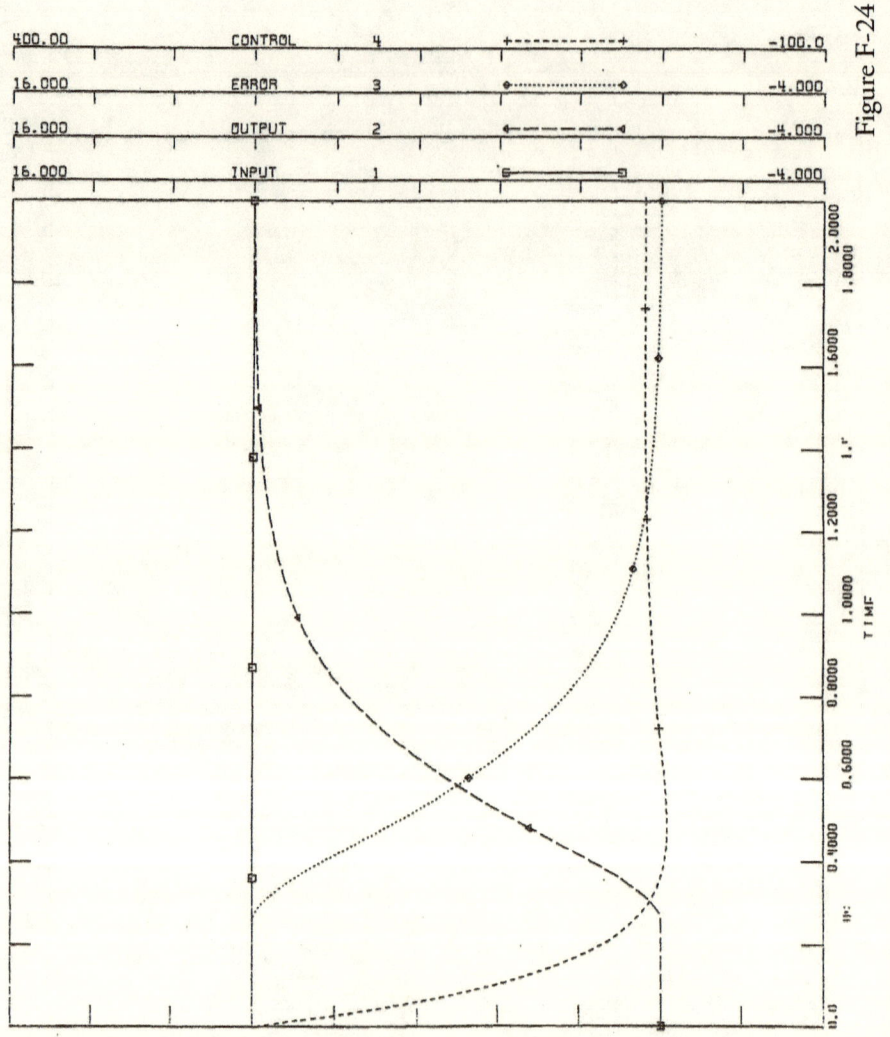

Figure F-24

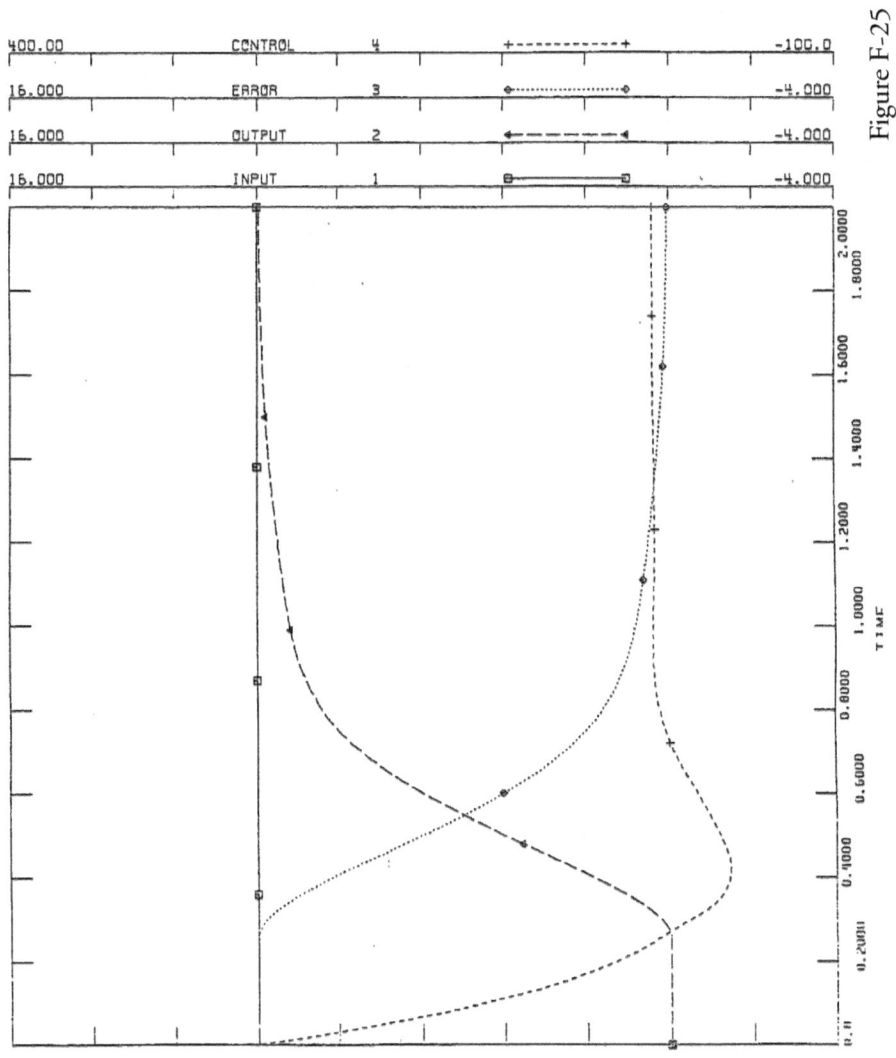

Figure F-25

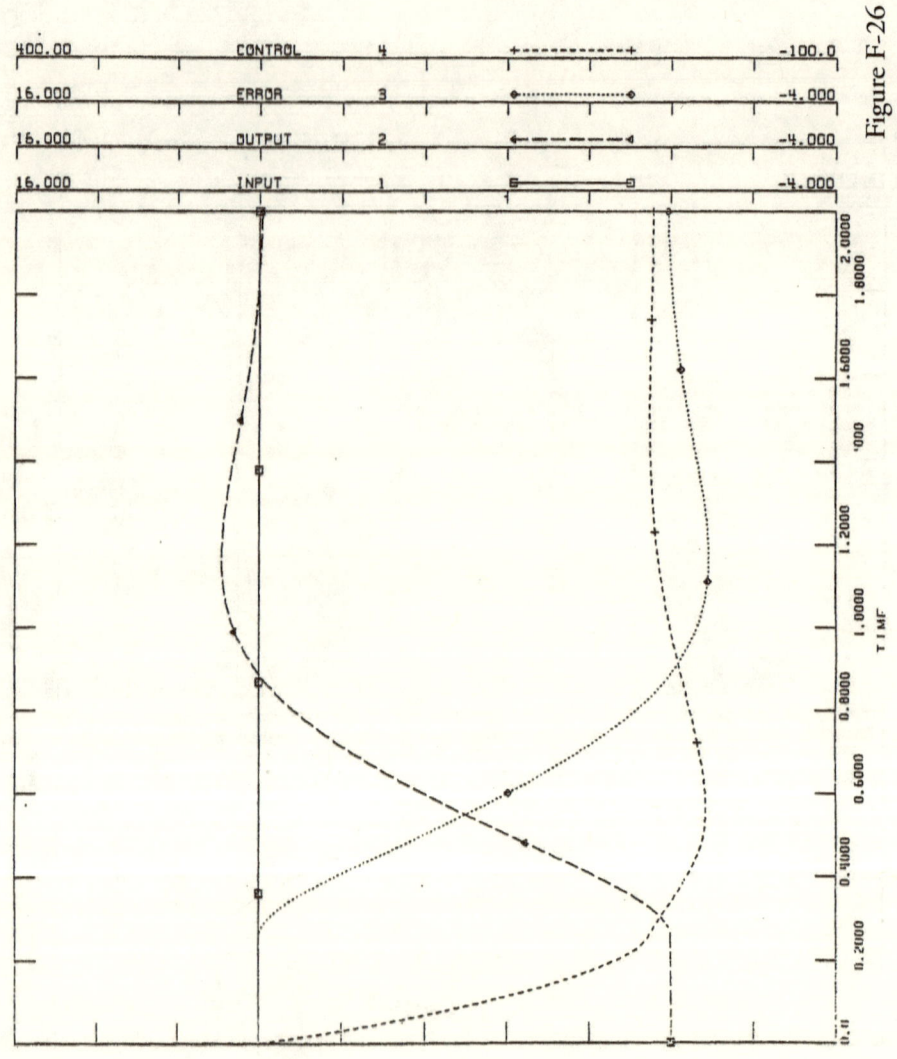

Figure F-26

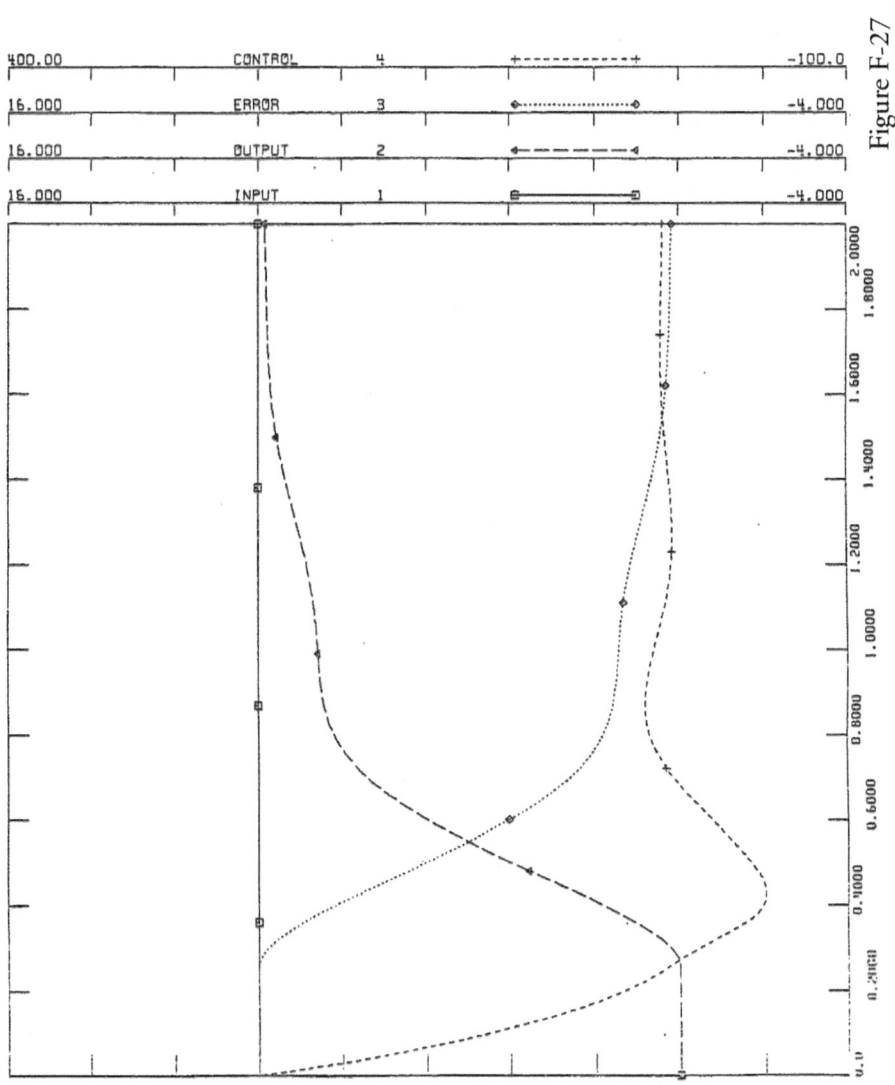

Figure F-27

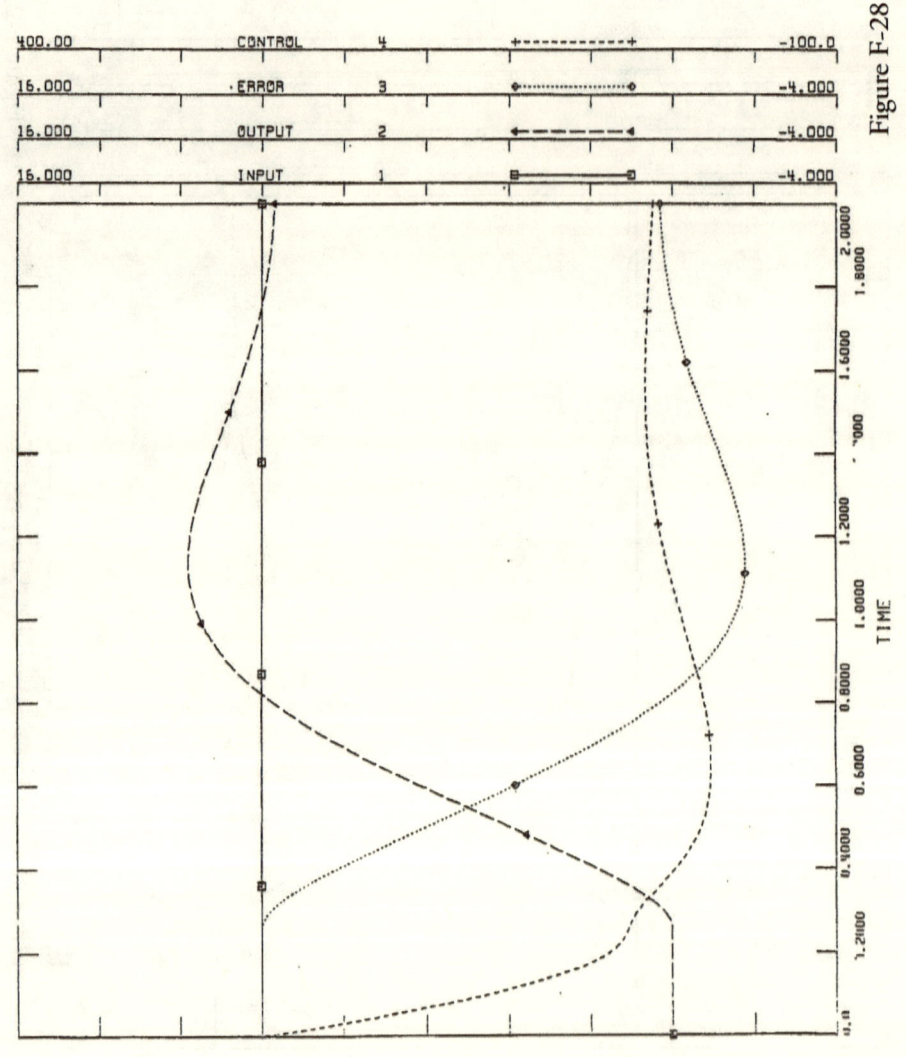

Figure F-28

Tuning Procedures

The previous examples show control performance obtainable by following the analytical rules starting from knowledge of the exact process response which turns out to fit the postulated second order function plus dead-time. In the more general problem, one does not know the process parameters or that the process necessarily fits exactly the second order plus dead-time form postulated. Rather, the process response function is likely to be known in terms of a response curve and thus the first step would be to approximate it by a second order plus dead-time function characterized by parameters T, ω and ζ.

The technique of picking the best fit of the approximation is simple on a CRT screen with an X-Y time response display of the process response. By having normalized response curves of the second order process for various damping ratios also displayed, the scale factor of the time axis of the process response curve can be adjusted until a fit is obtained with one of the normalized curves. Thus the damping ratio of the second order approximation to the process is established and the natural frequency can be obtained from the ratio of time scale factors of the process response curve and that of the normalized curve.

Another approach is to express the normalized response curves ($\omega=1.0$) in terms of points characterized by specific ratios relative to the final steady state response and ratio of time to reach the particular point to the time to reach 50% of the final value. With the actual response curve also expressed by such ratios, it is easy to search for matches between the response curve to be fitted and a particular normalized second order curve.

Table F-1 gives the ratios of times to reach particular response values to the time to read 50% of final value for normalized second order responses ($\omega=1.0$) with various damping ratios ζ. Also listed in the row corresponding to response ratio of 0.5 is the actual time for the normalized response ($\omega=1.0$) to reach 50% of the final value. For low values of damping ratio where the process has oscillatory responses, certain points are characterized by two time ratios, one corresponding to the response in the increasing direction and the other in the descending direction.

A third practical possibility is to have the normalized curves in transparencies with a variety of scale factors for the time axis. A fit is soon obtained by superposing the actual response curve on to one of the transparencies that produces the closest match. A set of second order system response curves plotted for a range of time scales is contained in Figures F-45 to F-54.

TABLE F-1

RATIO OF TIME TO REACH MAGNITUDE TO TIME TO REACH 0.5 (NORMALIZED: $\omega = 1.0$)

Magnitude	0.2	0.4	0.6	0.8	1.0	1.2	1.4	1.6	1.8
0.2	0.60	0.57	0.54	0.52	0.49	0.46	0.44	0.42	0.40
(time)*	(1.13)	(1.24)	(1.36)	(1.51)	(1.68)	(1.88)	(2.11)	(2.35)	(2.60)
0.5	1.00	1.00	1.00	1.00	1.00	1.00	1.00	1.00	1.00
0.8	1.36	1.41	1.51	1.63	1.79	1.93	2.04	2.11	2.17
1.0	1.60	1.75	2.04	2.76					
1.2	1.88	2.27							
	4.05	3.35							
1.4	2.25								
	3.45								

*Time to reach 0.5 of final value

 FOURTH ORDER PROCESS RESPONSE TO UNIT STEP
WP1 = 1.0; ZETAP1 = 0.8; WP2 = 0.3; ZETAP2 = 1.5

FILE: =P01

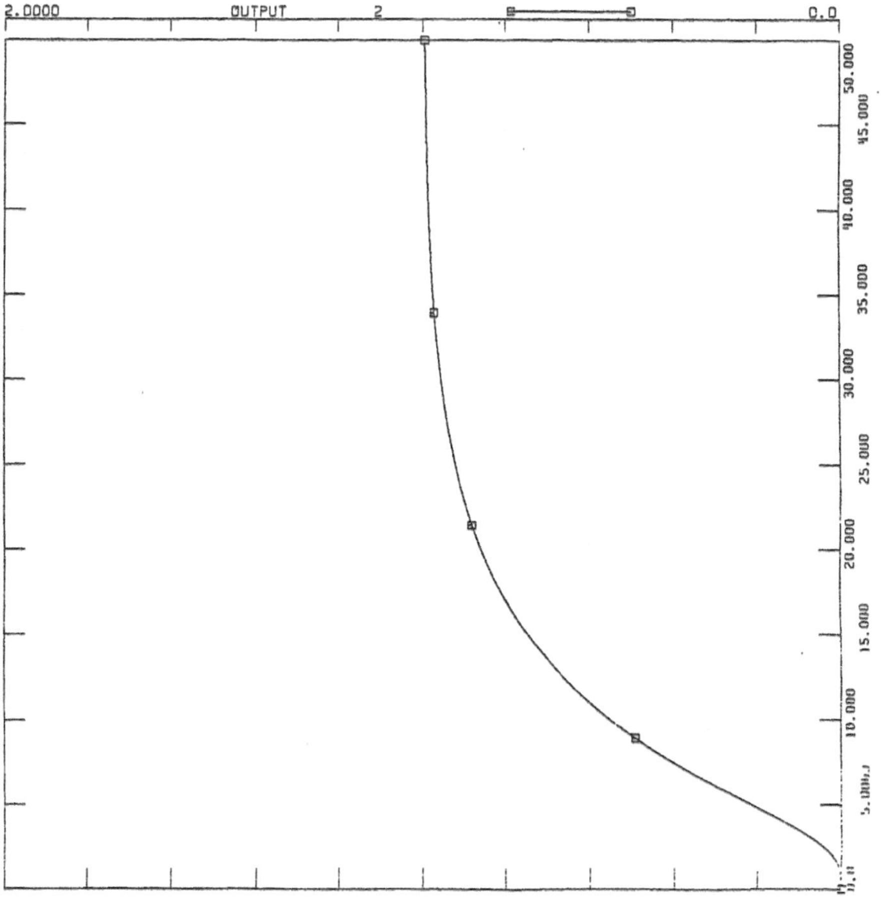

Figure F-29

2.0000
OUTPUT
0.0
TIME
0.0
8.0000
12.000
16.000
20.000
24.000
32.000
36.000
40.000

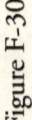

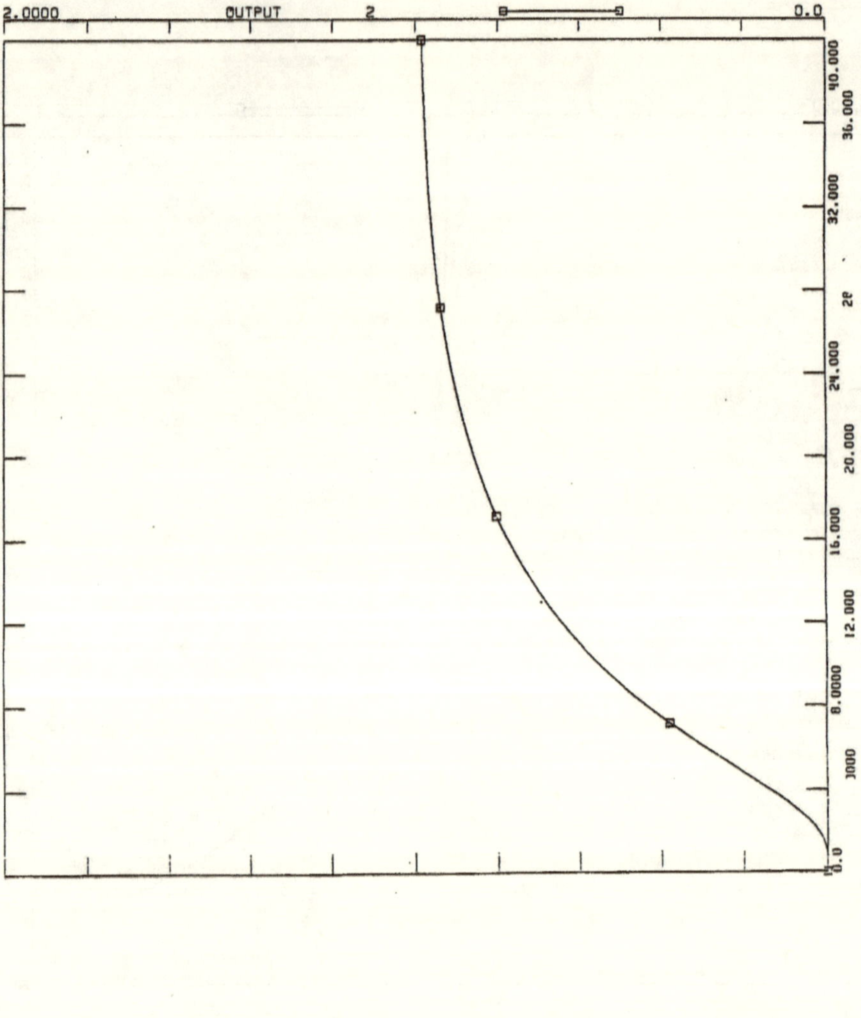

Figure F-30

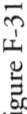

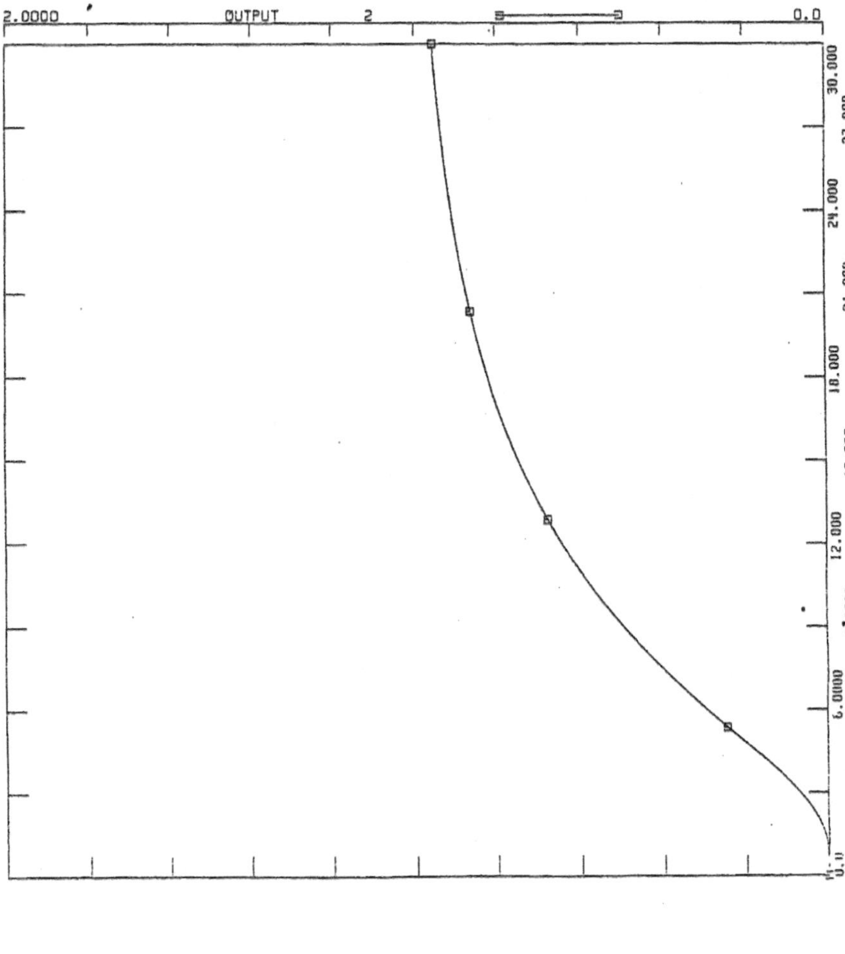
2.0000
OUTPUT
2
0.0
30.000
27.000
24.000
21.000
18.000
15.000
TIMF
12.000
9.0000
6.0000

Figure F-31

4TH ORDER PROCESS: WP1=1.0; ZETAP1=0.8; WP2=0.3: ZETAP2=1.5
APPROX BY 2ND ORDER PROCESS: WP=0.22; ZETAP=1.1; TDELAY=1.00

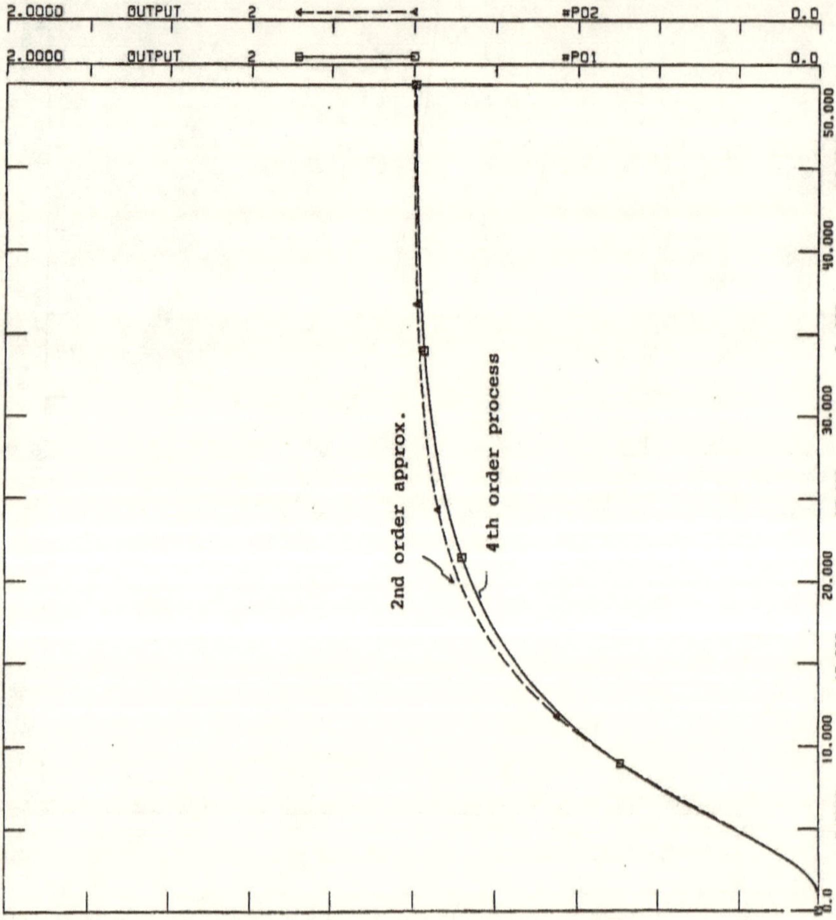

Figure F-32

4TH ORDER PROCESS: WP1=1.0; ZETAP1=0.8; WP2=0.3;
ZETAP2=1.5
APPROX BY 2ND ORDER PROCESS: WP=0.22; ZETAP=1.1;
TDELAY=1.00
DESIRED OUTPUT: WD=0.66; ZETAD=0.8
T = 0.00; T1 = 1.00

FILE: #P4TH4

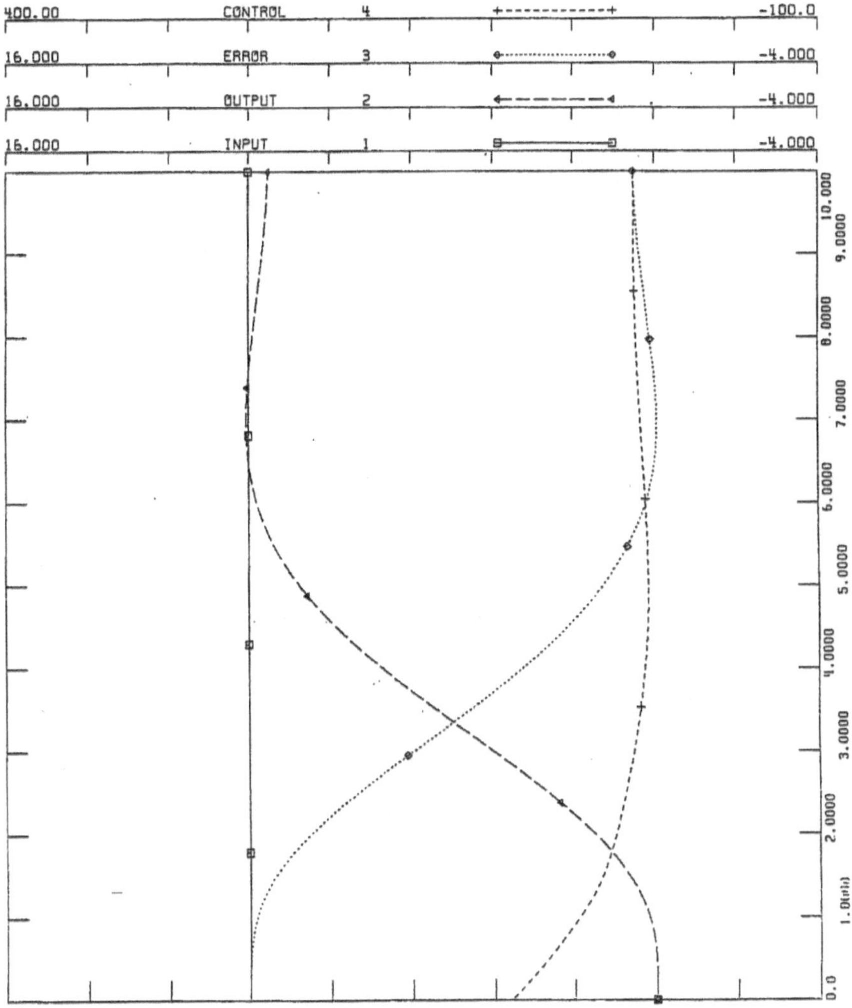

Figure F-33

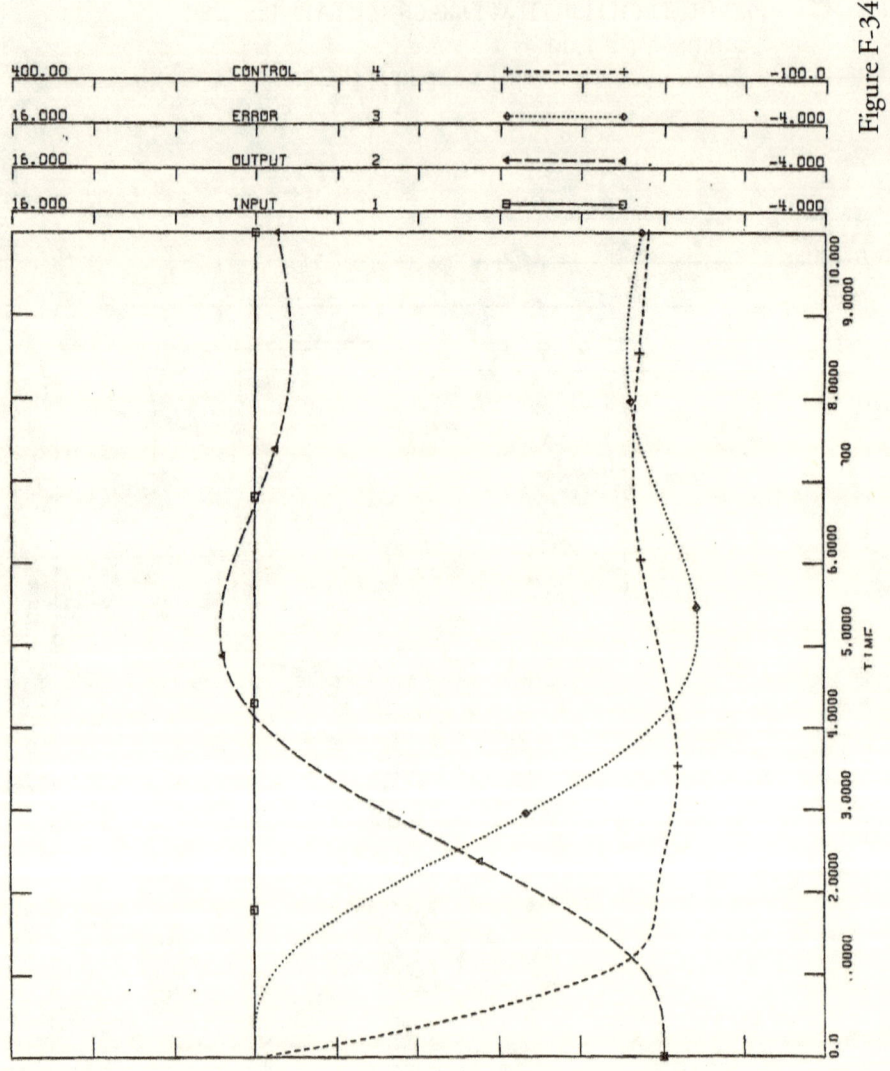

Figure F-34

Example of Tuning Technique

A process characterized by a dead-time and fourth order function (two second order functions in series) is to be used as an example of the tuning process. The actual process function is

$$\frac{\varepsilon^{-st}\,\omega_1^2\omega_2^2}{(\omega_1^2 + 2\zeta_1\omega_1 s + s^2)\,(\omega_2^2 + 2\zeta_2\omega_2 s + s^2)} \qquad \text{(F-15)}$$

where
$$T = 0 \qquad \omega_2 = 0.3$$
$$\omega_1 = 1.0 \qquad \zeta_2 = 1.5$$
$$\zeta_1 = 0.8$$

Figures F-29, F-30 and F-31 show the actual process response to a step plotted to different time scales. One can clearly discern an equivalent dead-time of 1 sec followed by a response which we wish to approximate with a second order function of natural frequency ω_p and damping ratio ζ_p.

Using any of the methods outlined above, we note that the actual process can be approximated by the function

$$\frac{\varepsilon^{-st}\,\omega_p^2}{(\omega_p^2 + 2\zeta_p\omega_p s + s^2)} \qquad \text{(F-16)}$$

where
$$\omega_p = 0.22$$
$$\zeta_p = 1.1$$
$$T_p = 1.0$$

Figure F-32 shows the response of the actual 4th order process and the 2nd order approximation.

If we now specify a desired ratio of closed loop to open loop response frequencies $\omega_d/\omega_p = 3.0$ and a desired closed loop damping ratio of 0.8, the resulting controlled response is shown on Figure F-33.

The closed loop response for a desired ω_d/ω_p ratio of 5 is shown on Figure F-34.

Relationship to Three Mode Control

The algorithm described in Figure F-20 has a similarity to that which describes a traditional three mode controller, except that it compensates for dead-time permitting faster controlled response for processes exhibiting dead-time.

If dead-time compensation is set equal to zero the diagram of Figure F-20 reduces to that in Figure F-35.

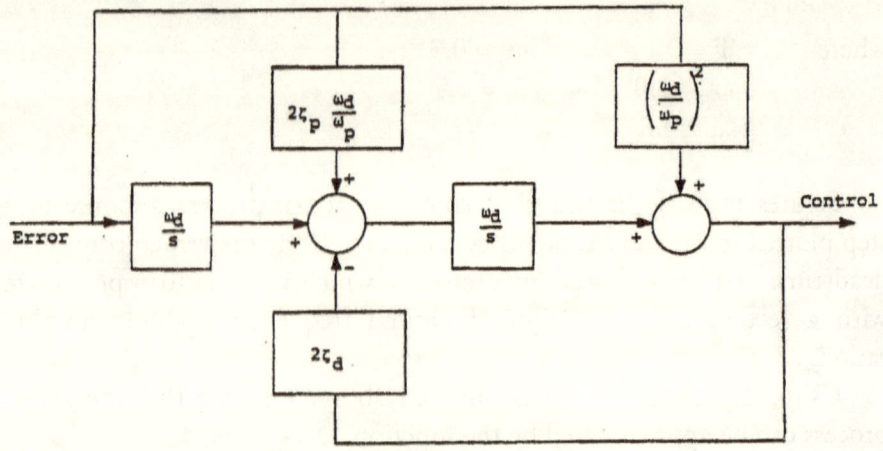

Figure F-35

and the corresponding transfer function can be expressed as

$$\frac{\omega_d^2}{\omega_p^2} \frac{(s^2 + 2\zeta_p\omega_p s + \omega_p^2)}{s(s + 2\zeta_d\omega_d)}$$

or

(F-17)

$$\frac{\omega_d}{2\zeta_d} \frac{\left(1 + \dfrac{2\zeta_p}{\omega_p}s + \dfrac{s^2}{\omega_p^2}\right)}{s\left(1 + \dfrac{s}{2\zeta_d\omega_d}\right)}$$

Comparing with the expression for one form of a three mode controller described in (F-8), one notes that the form is similar except that the numerator zeros are not constrained to be real as is the case in the three mode controller.

The correspondence between parameters of the three mode controller function

$$\frac{K_I(1+sT_I)(1+sT_R)}{s(1+sT_{R/K}}$$

(F-18)

in Figure F-17 is

$$K_I = \frac{\omega_d}{2\zeta_d}$$

$$(T_I+T_R) = 2\zeta_p / \omega_p$$

(F-19)

$$T_IT_R = \frac{1}{\omega_p^2}$$

$$\frac{K}{T_R} = 2\zeta_d\omega_d$$

T_I and T_R would be real values for critically damped and over-damped processes i.e. those with $\zeta_p \geq 1$.

For such processes (second order, no dead-time and $\zeta_p > 1.0$). The adjustments in terms of ω_p, ω_d, ζ_d and ζ_p are

$$T_I = \frac{K_p}{K_I} = \frac{\zeta_p}{\omega_p}\left[1+\sqrt{1-\frac{1}{\zeta_p^2}}\right]$$

$$T_R = \frac{\zeta_p}{\omega p}\left[1-\sqrt{1-\frac{1}{\zeta_p^2}}\right]$$

(F-20)

$$K_I = \frac{\omega_d}{2\zeta_d}$$

The controller zeros $1/T_I$ and $1/T_R$ are essentially positioned to cancel the process poles or roots. An over-damped second order process has two real roots s_1 and s_2. The ratio of the roots to the natural frequency as function of damping ratio is shown in Table F-2.

TABLE F-2

ζ	$s_{1/\omega}$	$s_{2/\omega}$
1.0	1.0	1.0
1.2	0.5367	1.863
1.5	0.382	2.618
2.0	0.268	3.732

When three mode controllers are used for processes with dead-time, the specification of a second order desired response cannot be made in the same way as was possible using the controller with dead-time compensation.

For second order processes with dead-time, i.e. where the process function is

$$\frac{\omega_p^2 \, \bar{\varepsilon}^{st}}{(\omega_p^2 + 2\zeta_p \omega_p s + s^2)} \tag{F-21}$$

and $\zeta_p = 1.0$ or greater, two of the three mode controller adjustments would logically be:

$$T_I = K_p / K_I = \frac{\zeta_p}{\omega_p}\left[1 + \sqrt{1 - \frac{1}{\zeta_p^2}}\right] \tag{F-22}$$

$$T_R = \frac{\zeta_p}{\omega_p}\left[1 - \sqrt{1 - \frac{1}{\zeta_p^2}}\right]$$

2ND ORDER PROCESS: WP=1.0; ZETAP=1.5; TDELAY=1.00 3
MODE CONTROL

FILE: #P2ND

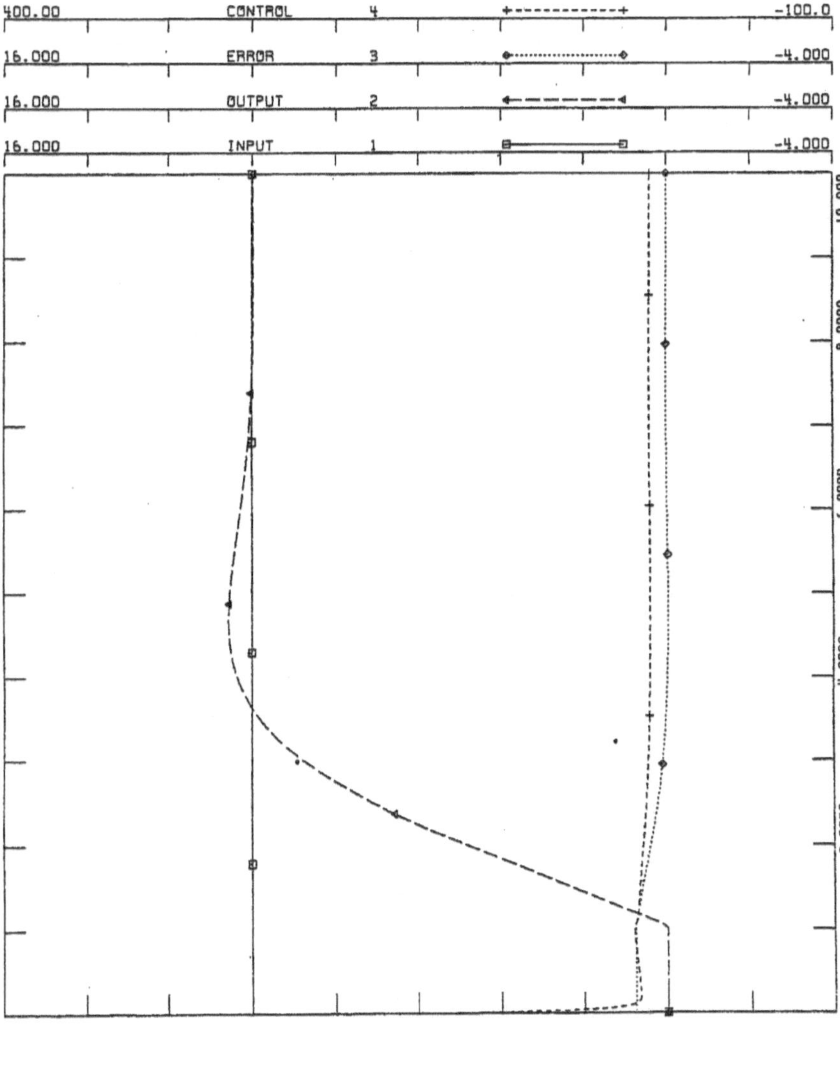

Figure F-36

400.00	CONTROL	4	+---------+	-100.0
16.000	ERROR	3	●·········●	-4.000
16.000	OUTPUT	2	◄-------◄	-4.000
16.000	INPUT	1	□——————□	-4.000

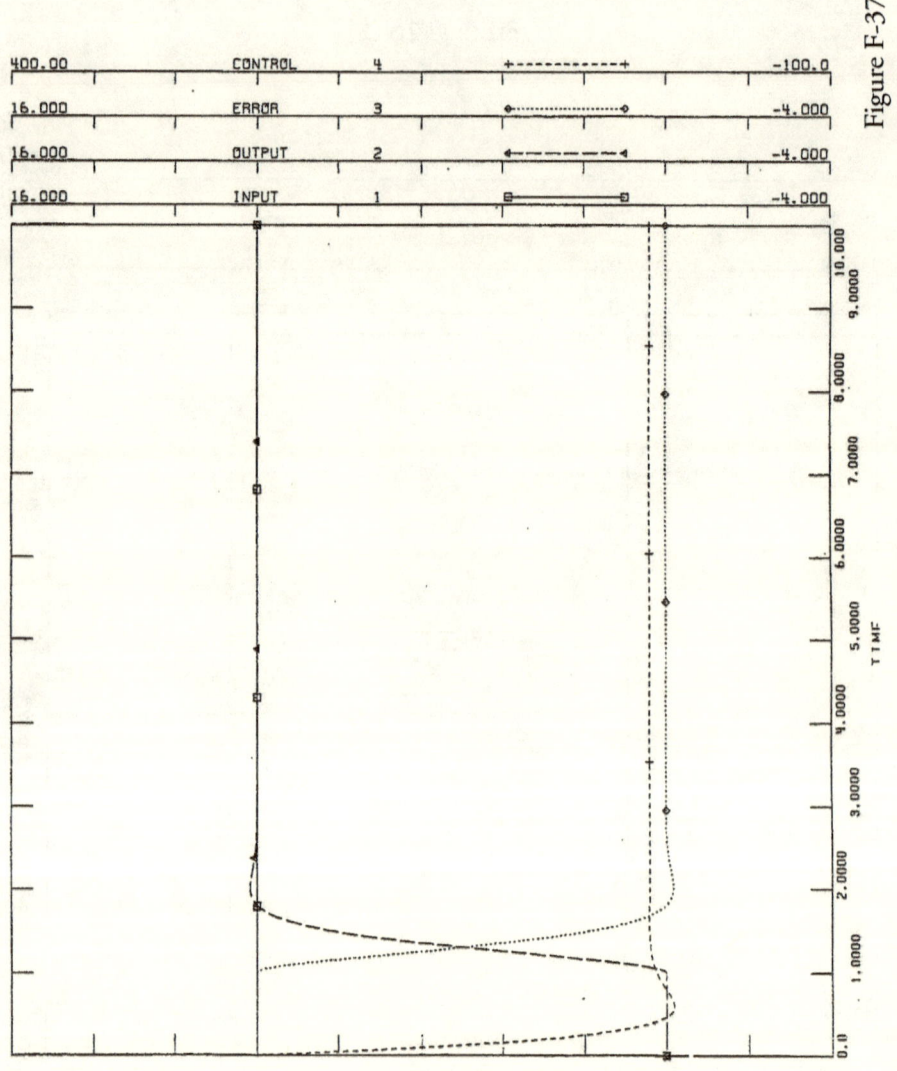

Figure F-37

The denominator pole would normally be $T_R/10$ and thus the only remaining parameter to be established is K_I.

With the process poles cancelled by the controller zeros, the overall open loop function now becomes

$$\frac{K_I \bar{\varepsilon}^{st}}{s(1+T_R/10^s)} \tag{F-23}$$

One method of determining K_I would be to establish the frequency $S=j\omega$ for which the function (F-19) has a phase lag of about 117.5°, and then calculate the gain K_I which would make the magnitude of the function unity at this frequency. The choice of a higher phase lag than 117.5° would give more oscillatory performance.

ω is thus calculated from the relation

$$\tan^{-1}\left(\frac{T_R\omega}{10}\right)+\frac{\omega Tx180}{\pi}=27.5° \tag{F-24}$$

or

$$\frac{T_R\omega}{10}=\tan\left[27.5-\frac{\omega Tx180}{\pi}\right]^\circ$$

making the approximation $\tan\theta=\theta$ for small θ,

$$\frac{T_R\omega}{10}=[.48-\omega T]$$

or

$$\omega\left[T+\frac{T_R}{10}\right]=0.48$$

i.e.

$$\omega=\frac{0.48}{\left[T+\frac{T_R}{10}\right]} \tag{F-25}$$

Hence

$$K_I=\frac{0.48}{\left(T+\frac{T_R}{10}\right)}\left[1+\left(\frac{T_R}{10}\right)^2\frac{(.48)^2}{\left(T+\frac{T_R}{10}\right)^2}\right]^{1/2}$$

$$= \frac{0.48}{\left(T+\dfrac{T_R}{10}\right)}\left[1+\frac{.0023\,T_R^{\,2}}{\left(T+\dfrac{T_R}{10^2}\right)^2}\right]^{1/2}$$

An example of three mode control tuning is illustrated for a process characterized by $\omega_p=1.0$, $\zeta_p=1.5$ and $T=1.0$.

Using the above formulas

$$T_I = \frac{K_p}{K_I} = 1.5\left[1+\sqrt{1-\frac{1}{1.5^2}}\right] = 2.618$$

$$T_R = 1.5\left[1-\sqrt{1-\frac{1}{1.5^2}}\right] = 0.382 \qquad\qquad \text{(F-26)}$$

$$K_I = \frac{0.48}{[1+.038]}\left[1+\frac{.0023\times.382^2}{(1.0382)^2}\right]^{1/2}$$

The corresponding adjustments would be: proportional band = 100/K_p = 100/2.618x.46 = 83%. Repeats per minute = 60/2.618 = 22.9 and rate time = 0.38.

The closed loop response of the process with the above three mode controller settings is shown in Figure F-36.

For comparison, using the controller of Figure F-20, with dead-time compensation, and ω_d = 5, ζ_d, T = 1.0, one obtains the closed loop response also plotted on Figure F-37. The striking improvement in response is evident by noting that the process is essentially at set point within 1.8 sec whereas it reaches set point and overshoots at 3.5 sec in Figure F-36.

Application To Process With Integration

For processes characterized by a dead-time, single time constant and an integration (typical of level control)

$$\frac{\overline{\varepsilon}^{sT}}{s(1+sT_1)} \tag{F-27}$$

a similar derivation yields the control algorithm described in Figure F-38.

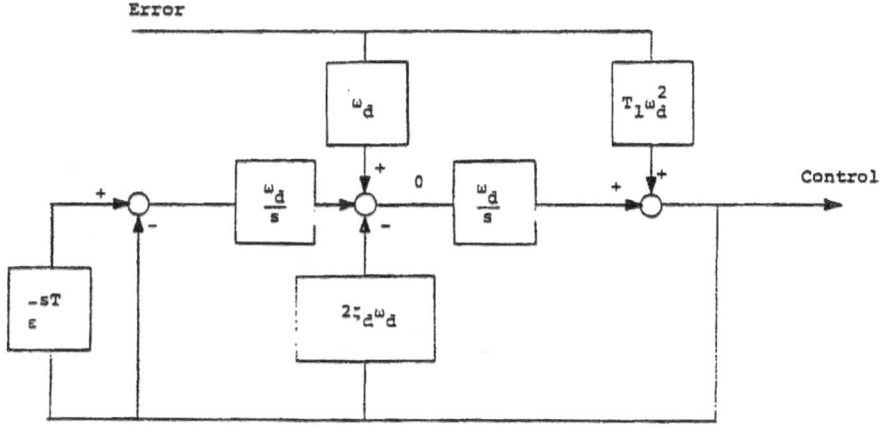

Figure F-38

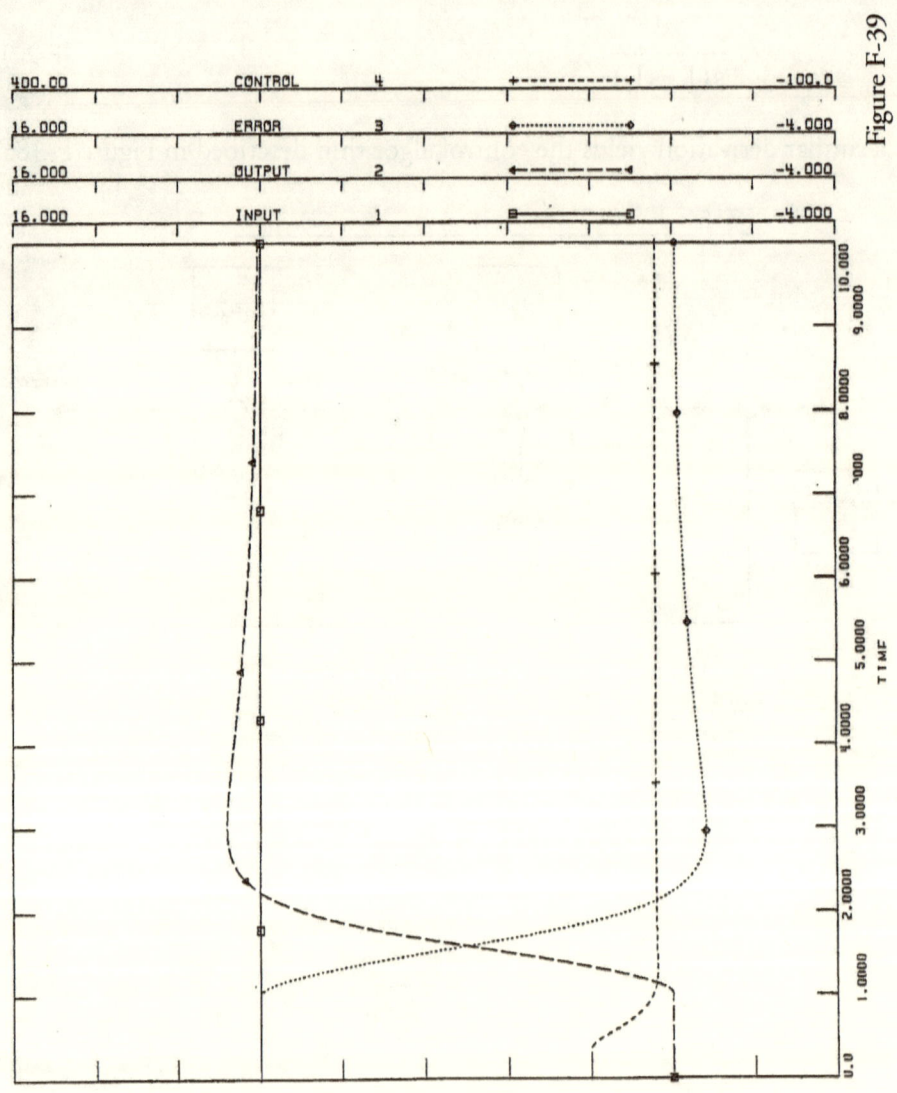

If one wishes to preserve integral action on the error, the same algorithm as was applied for a process with finite gain could be used by assuming the process with very large rather than infinite steady state gain. e.g. The process

$$\frac{\overline{\varepsilon}^{sT}}{s(1+sT_1)} \tag{F-28}$$

could be approximated by

$$\frac{\overline{\varepsilon}^{sT} T_2}{(1+sT_2)(1+sT_1)} \quad \text{where } T_2 \text{ is very large.} \tag{F-29}$$

The equivalent natural frequency and damping ratio formulation is

$$\frac{T_2 \overline{\varepsilon}^{st} \omega_p^2}{[\omega_p^2 + 2\zeta_p\omega_p s + s^2)} \tag{F-30}$$

Picking $T_2 >> T_1$ or T say $T_2 = 100\,T_1$ or $100T$ whichever is larger.

$$T_1 + T_2 = T_2 = 2\zeta_p/\omega_p$$

and

$$1/T_1 T_2 = \omega_p^2 = \frac{1}{100_1^2}$$

i.e.

$$\omega_p = \frac{1}{10 T_1} \tag{F-31}$$

$$\therefore \frac{2\zeta_p}{\omega_p} = 2\zeta \times 10 T_1 = 100 T_1$$

or

$$\zeta_p = 5$$

The process approximation is thus

$$\frac{100 T_1 \overline{\varepsilon}^{st} \omega_p^2}{[\omega_p^2 + 2\zeta\omega_p s + s^2]} \tag{F-32}$$

where $\omega_p = \dfrac{1}{100T_1}$

$\zeta = 5$

Gain $= 100_1$

The controller of Figure F-20 would be adjusted using the desired parameters ω_d and ζ_d and process parameters ω_p and ζ_p and the overall control attenuated (multiplied) by $1/100T_1$ or $\omega_p/10$.

Operation Under Limits

When control output is limited, the output of the right hand side integrator ω_d/s in Figure F-20 is continuously adjusted so that its output plus the error signal multiplied by $(\omega_d/\omega_p)^2$ is equal to the limit. This avoids reset windup and provides predictive characteristics to control action.

Figure F-39 shows performance for a step change in set point with limits on control action.

Digital Implementation

The power of modern computers and the bandwidth of a/d and d/a converters is so large that for most process applications the problem of sampling rate is no longer germane and suffice it to state that sampling will be executed at a rate compatible with the process response characteristics.

If ω_d is indicative of the desired response bandwidth it is natural to pick a sampling time smaller than $1/4\omega_d$.

Again if dead-time is significant, an appropriate submultiple of the dead-time could be the criterion for choice of sampling rate.

In any case there is no reason to have control response suffer because of sampling rate.

The integration function in the controls denoted as $1/s$ in the continuous domain, is accomplished with a simple accumulation algorithm.

Examples

The response of a process is characterized within the form of expression (F-12) by

$$\omega_p = 1 \text{ rad/sec}$$
$$\zeta_p = 0.3$$
$$T = 2 \text{ sec}$$

The desired closed loop response is described by the parameters $T = 2$, $\omega_d = 2$ rad/sec, $\zeta_d = 0.8$. Using a sampling time of 0.1 sec and the control algorithm of Figure F-20, the closed loop response is shown in Figure F-40.

Another example for the same type of process except that the dead-time is increased to 4 sec is shown on Figure F-41.

The effect of faster sampling rate $T_{samp} = 0.05$ sec. for the process with the 2 sec dead-time is shown on Figure F-42. (Compare with Figure F-41.)

The performance with an ideal continuous controller which performs with dead-time compensation as in Figure F-20 is shown for the process with a 2 sec dead-time in Figure F-44. (Compare with Figures F-39 and F-41.)

CONCLUSIONS

Digital control permits implementation of complex control logic with relative ease. Compensation for process dead-time, in particular, makes it possible to obtain practically the same controlled response (excluding the inevitable dead-time delay) independent of process dead-time.

The control structure and technique of arriving at control parameter adjustments have been developed using simple analytical operational methods that apply to continuous systems, it being recognized that translation into the discrete domain is not a problem provided adequate sampling rates are used.

The method is based on the observation that most process open loop response characteristics can be approximated by a dead-time in series with a second order transfer function whose parameters can be expressed in terms of a natural frequency ω_p and damping ratio ζ_p. Likewise the desired closed loop control response can be expressed in a similar form, where the achievable, hence desired, dead-time must be equal to the process deadtime, the natural frequency of the closed loop response ω_d selected as some reasonable multiple of the process natural frequency, and the damping ratio of the desired response ζ_d selected for acceptable overshoot (usually $\zeta_d = 0.8$).

The controller parameters are then directly related to these constants characterizing the process and closed loop control.

The relationship of the usual analog three mode controller (proportional with reset and rate) to the controller structure with dead-time compensation is demonstrated and rules established for parameter adjustment of three mode controllers from knowledge of the process as characterized by dead-time, natural frequency and damping ratio.

2ND ORDER PROCESS: WP=1.0; ZETAP=0.3; TDELRY=2.00
CONTROLLER WD=2.0 ZETAD=0.8 T1=2.0
DIGITAL CONTROL: T_{samp} = .10 sec

FILE: =P2ND

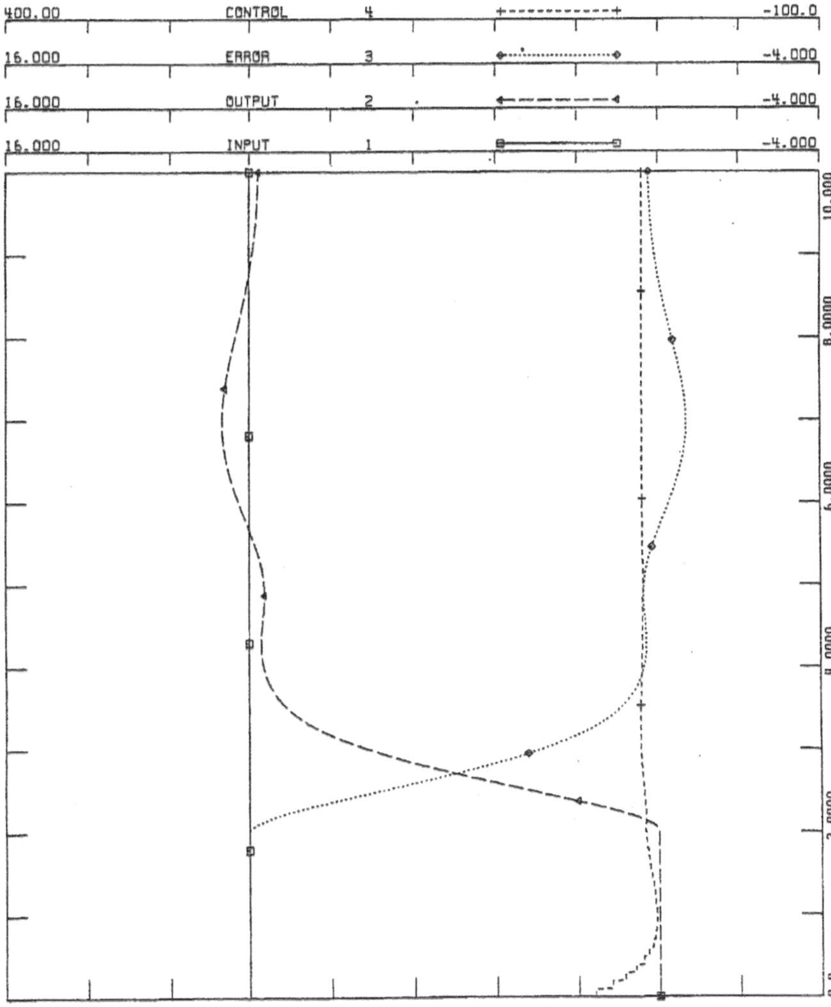

Figure F-40

400.00	CONTROL	4	+---------+	-100.0
16.000	ERROR	3	⬥·········⬥	-4.000
16.000	OUTPUT	2	◄--- ---◄	-4.000
16.000	INPUT	1	▣——▣	-4.000

TIME

10.000
9.0000
8.0000
?0
6.0000
5.0000
4.0000
3.0000
2.0000
0000
0.0

 2ND ORDER PROCESS: WP=1.0; ZETAP=0.3; TDELAY=4.00
CONTROLLER WD=2.0 ZETAD=0.8 T1=4.0

FILE: =P2ND

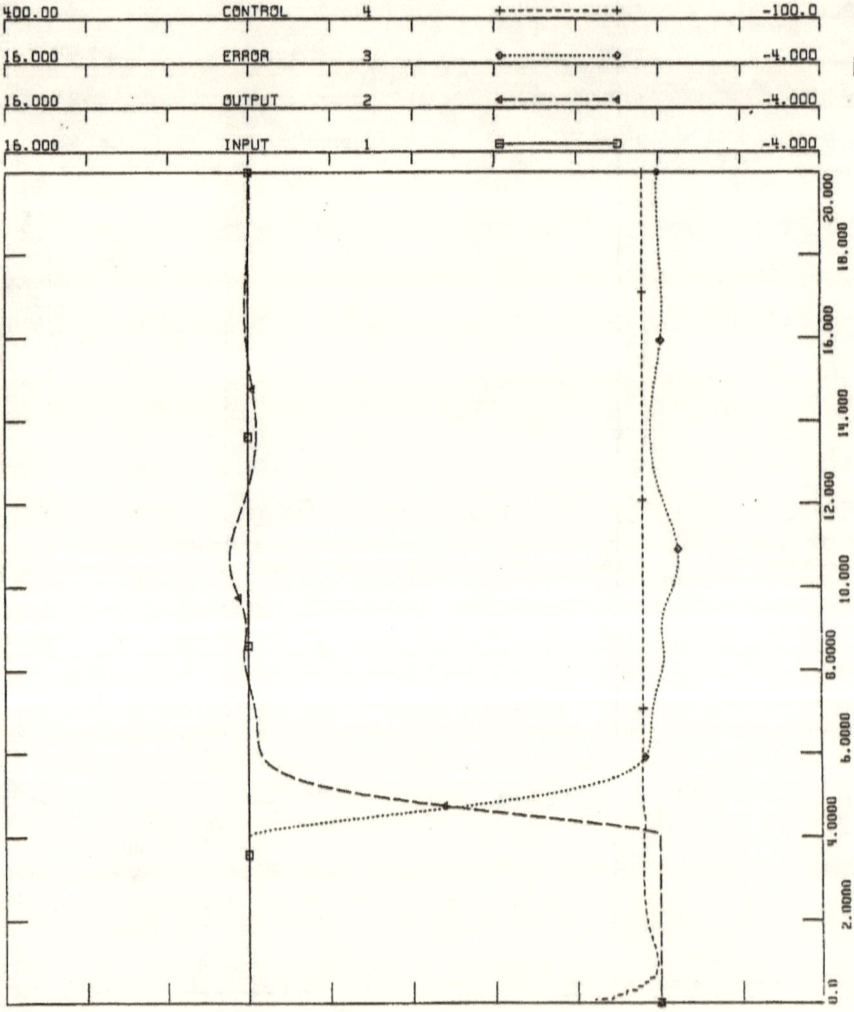

Figure F-41

 2ND ORDER PROCESS: WP=1.0; ZETAP=0.3; TDELAY=4.00
CONTROLLER WD=2.0 ZETAD=0.8 T1=2.0
DIGITAL CONTROL—TSAMP = .05
TME IN SEC.

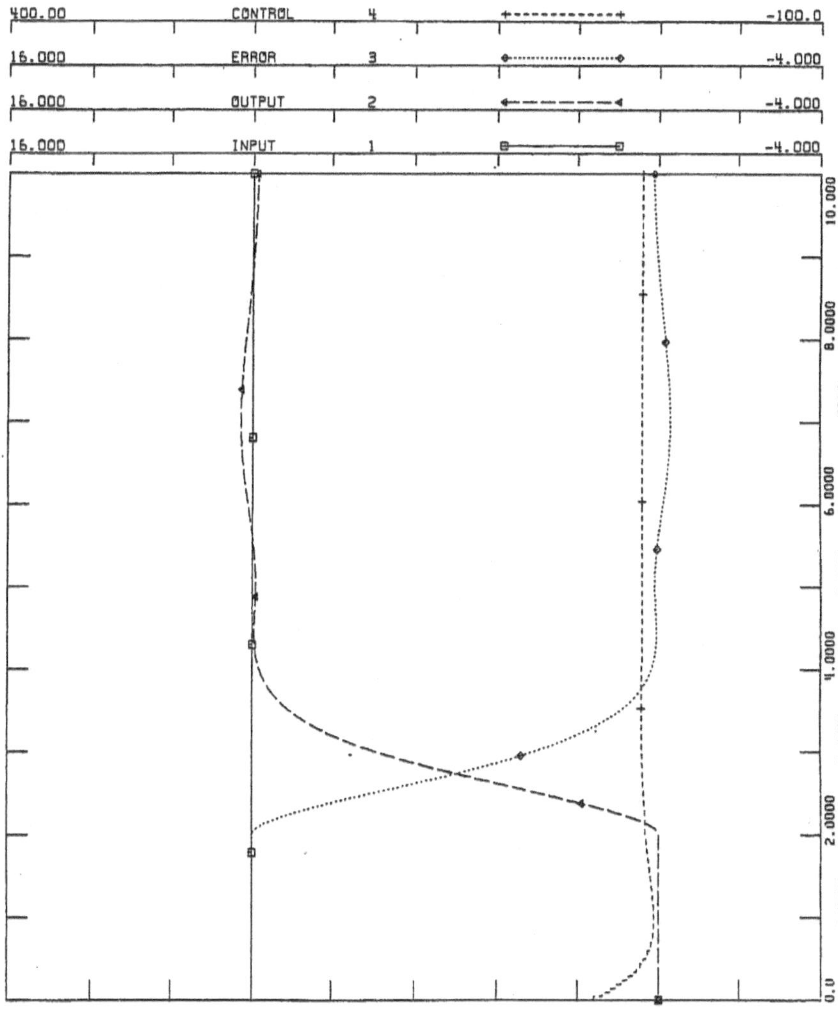

Figure F-42

400.00	CONTROL	4	+----------+	-100.0
16.000	ERROR	3	◆·········◆	-4.000
16.000	OUTPUT	2	◀— — — —◀	-4.000
16.000	INPUT	1	▣————▣	-4.000

2ND ORDER PROCESS: WP=1.0; ZETAP=0.3; TDELAY=2.00
CONTROLLER WD=2.0 ZETAD=0.8 T1=2.0. TIME IN SEC.

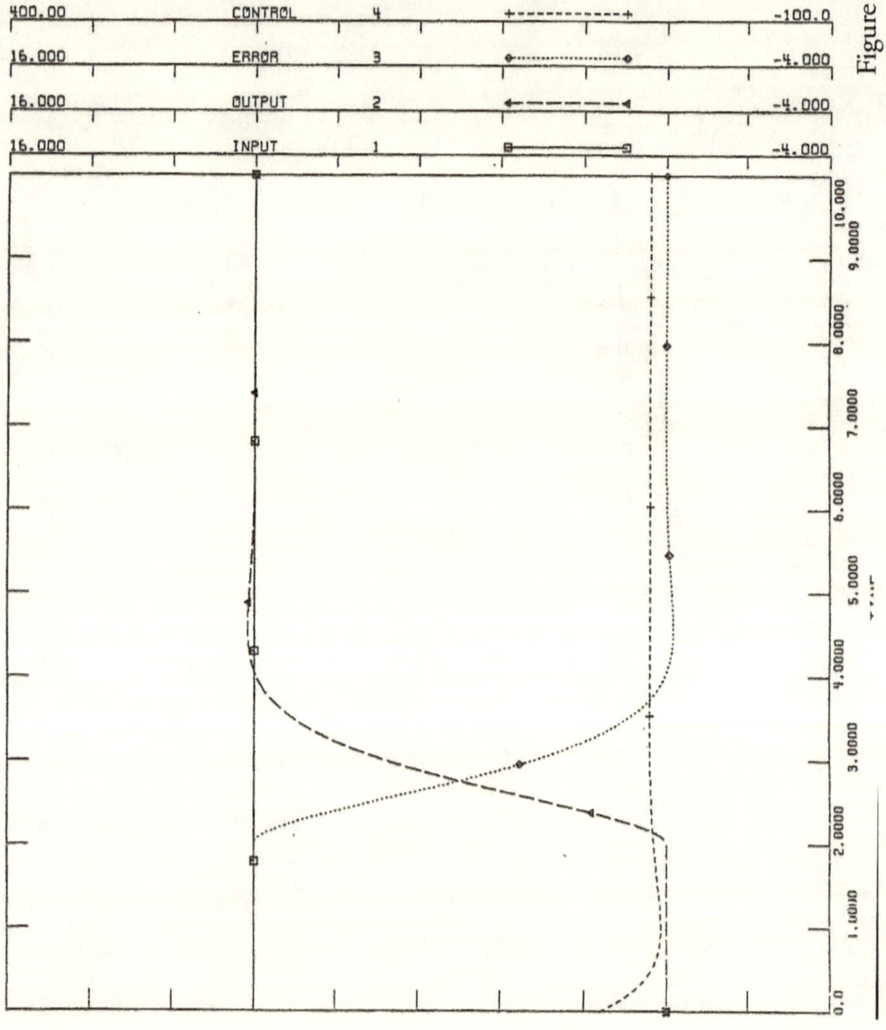

Figure F-43

PROCESS RESPONSE TO UNIT STEP INPUT FOR VARIOUS
VALUES OF ZETA NORMALIZED FREQUENCY (W = 1)

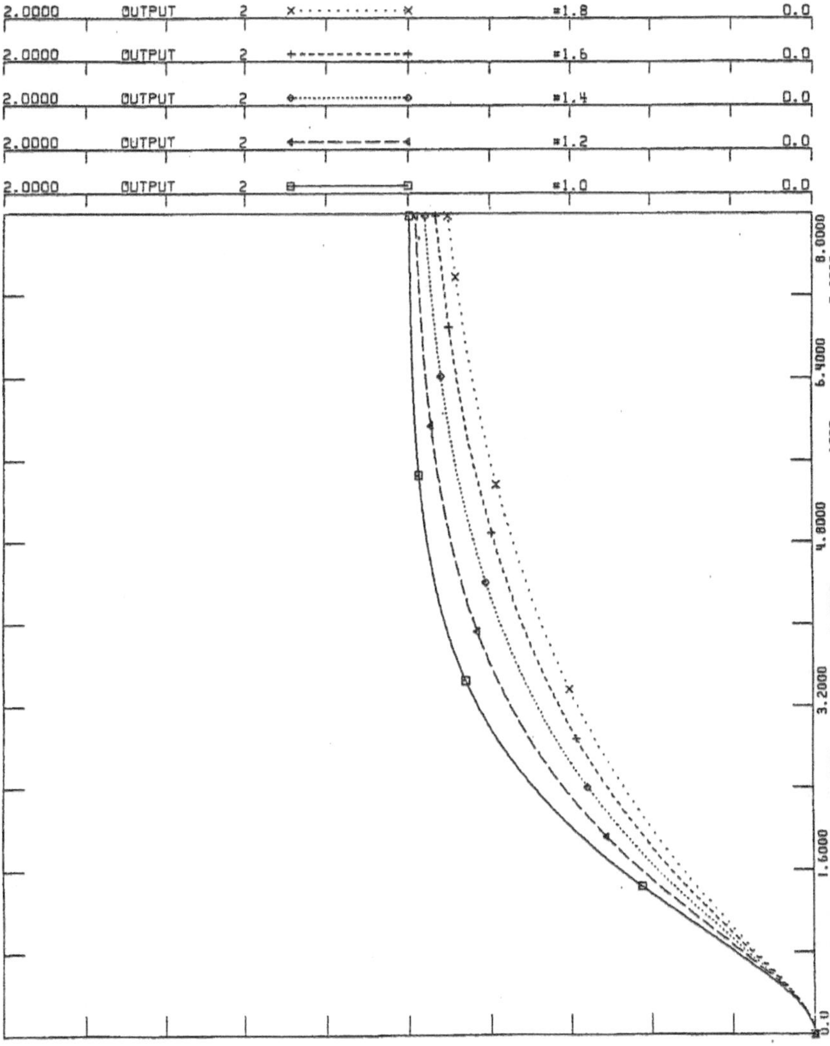

Figure F-44

370   F. PAUL DE MELLO

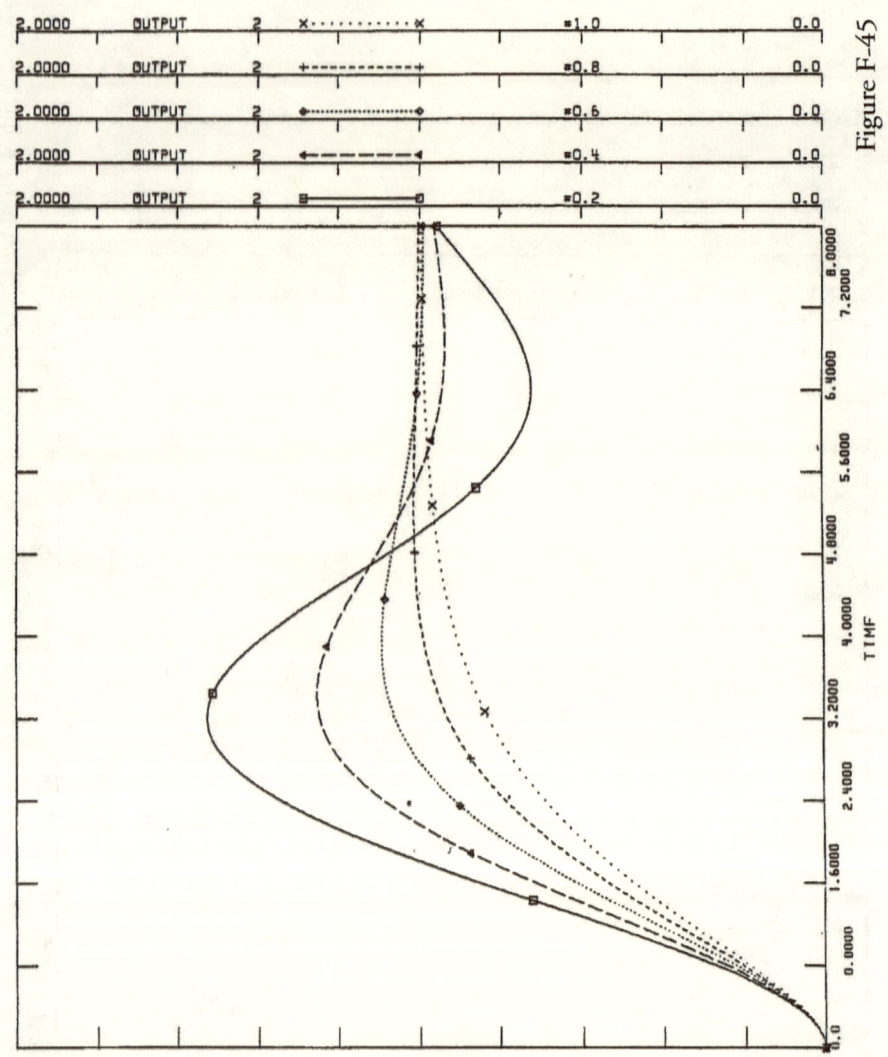

Figure F-45

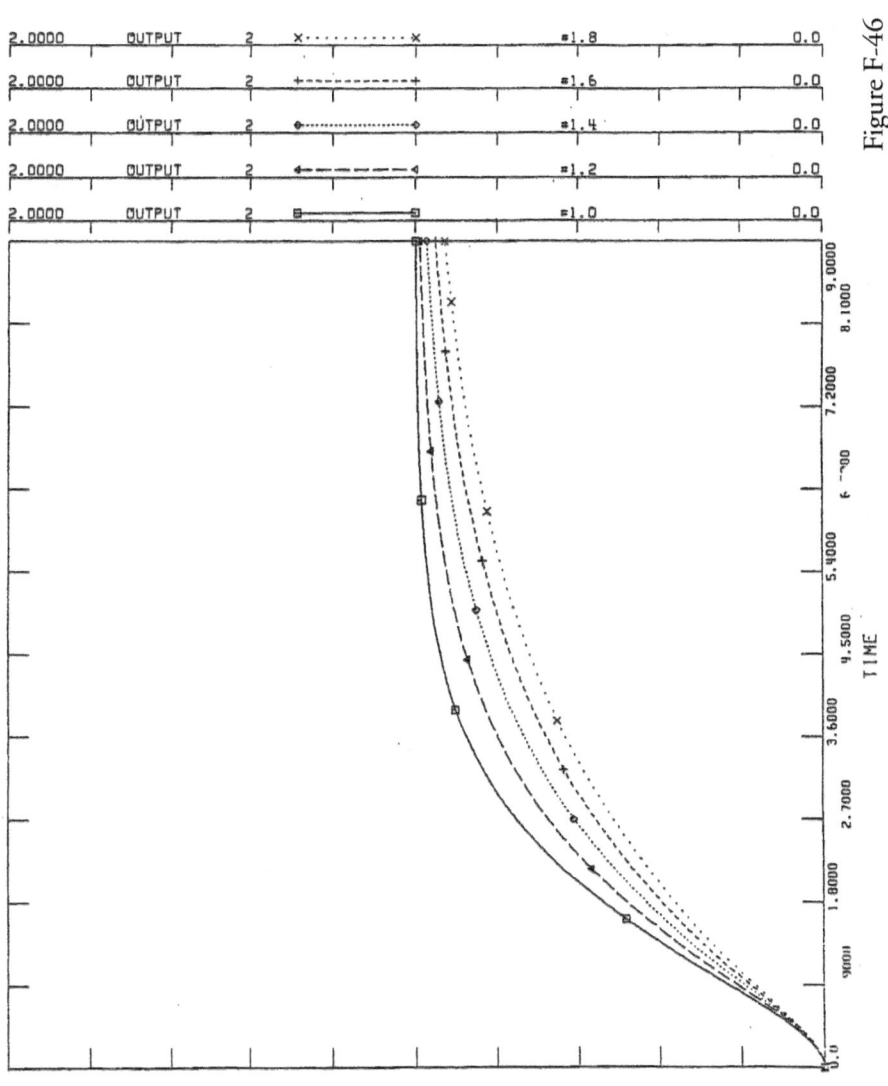

Figure F-46

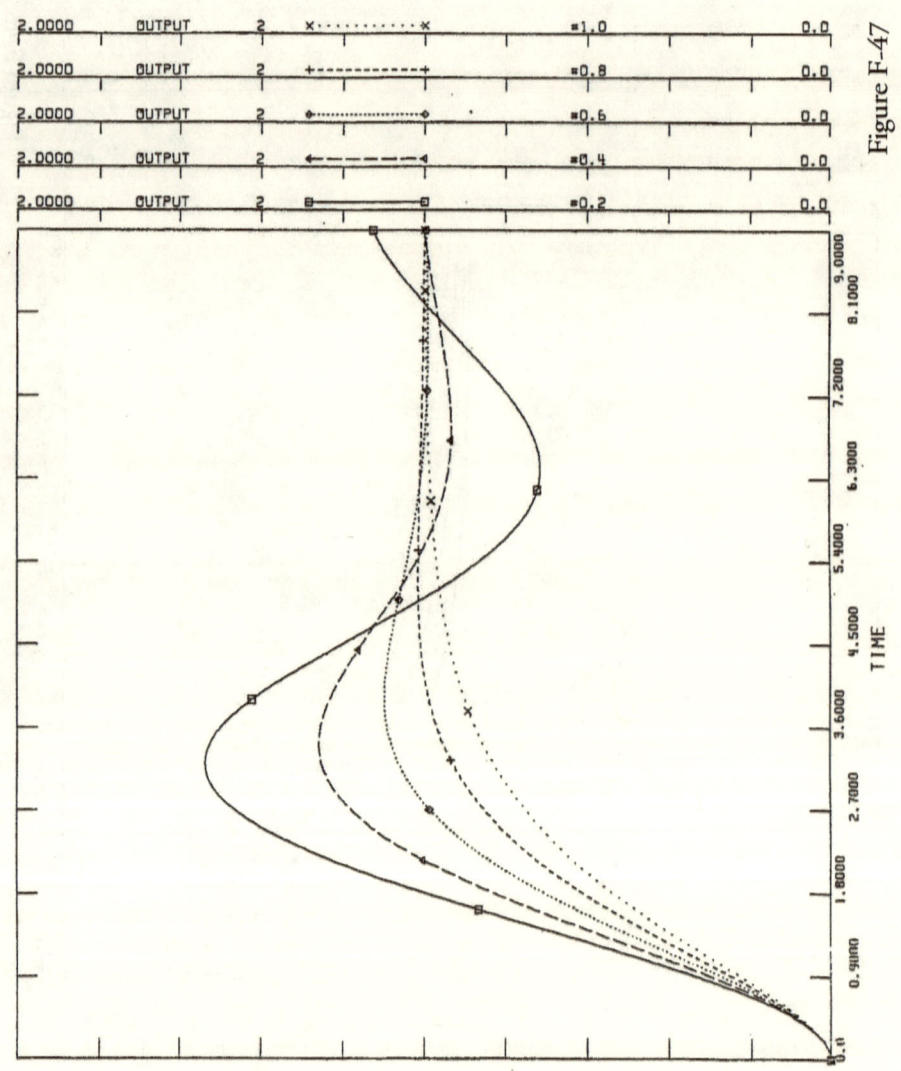

Figure F-47

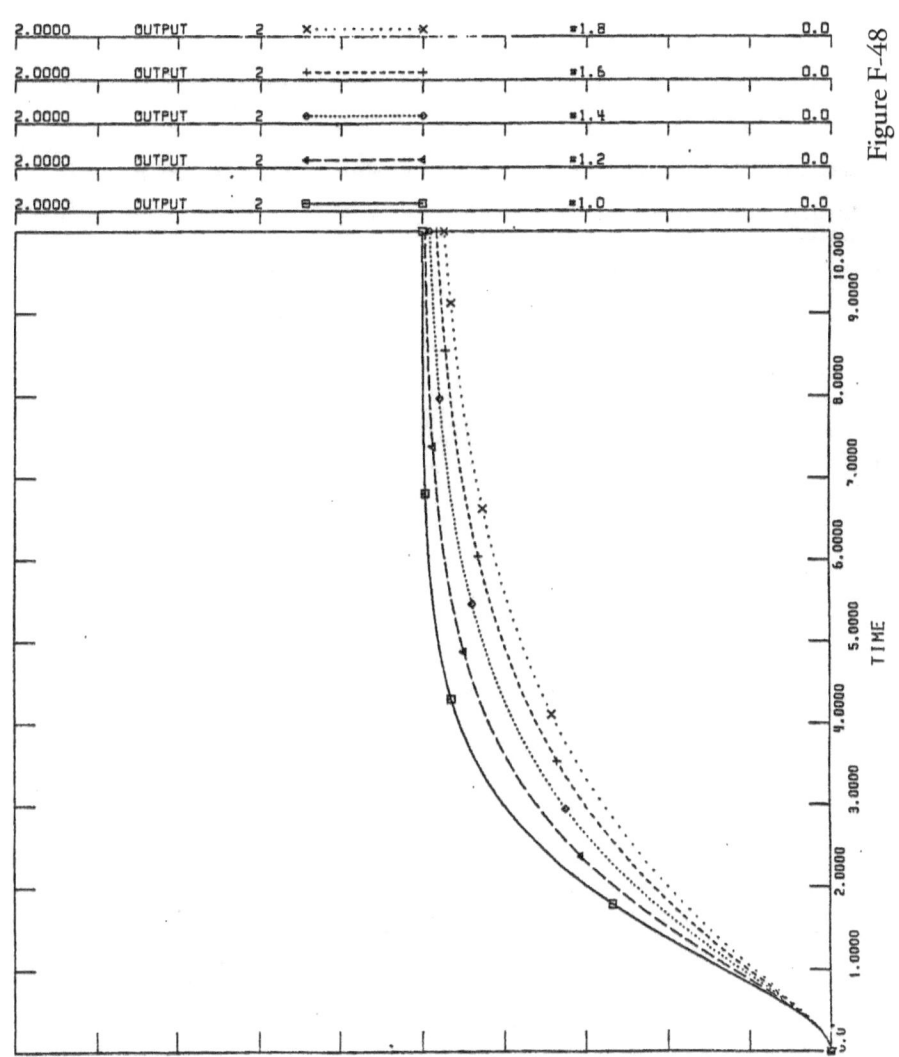

Figure F-48

F. Paul de Mello

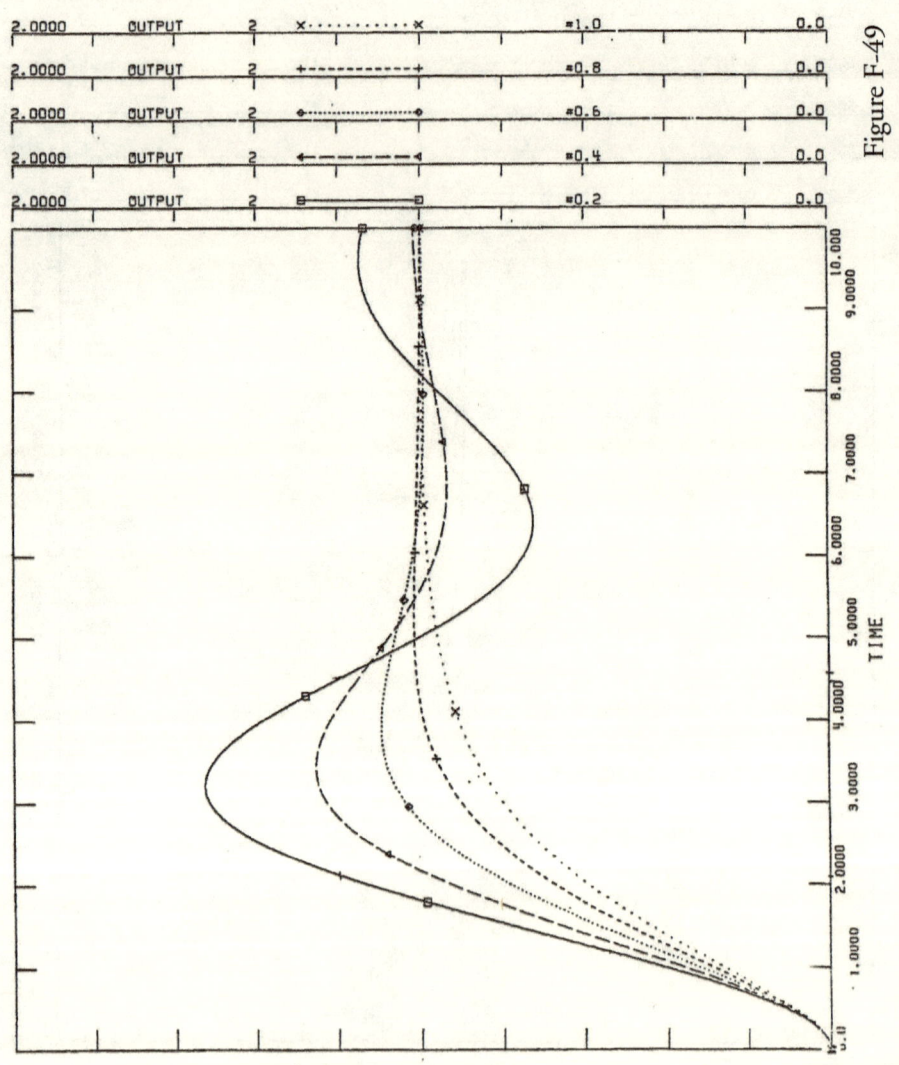

Figure F-49

PROCESS RESPONSE TO UNIT STEP INPUT FOR VARIOUS
VALUES OF ZETA NORMALIZED FREQUENCY (W = 1)

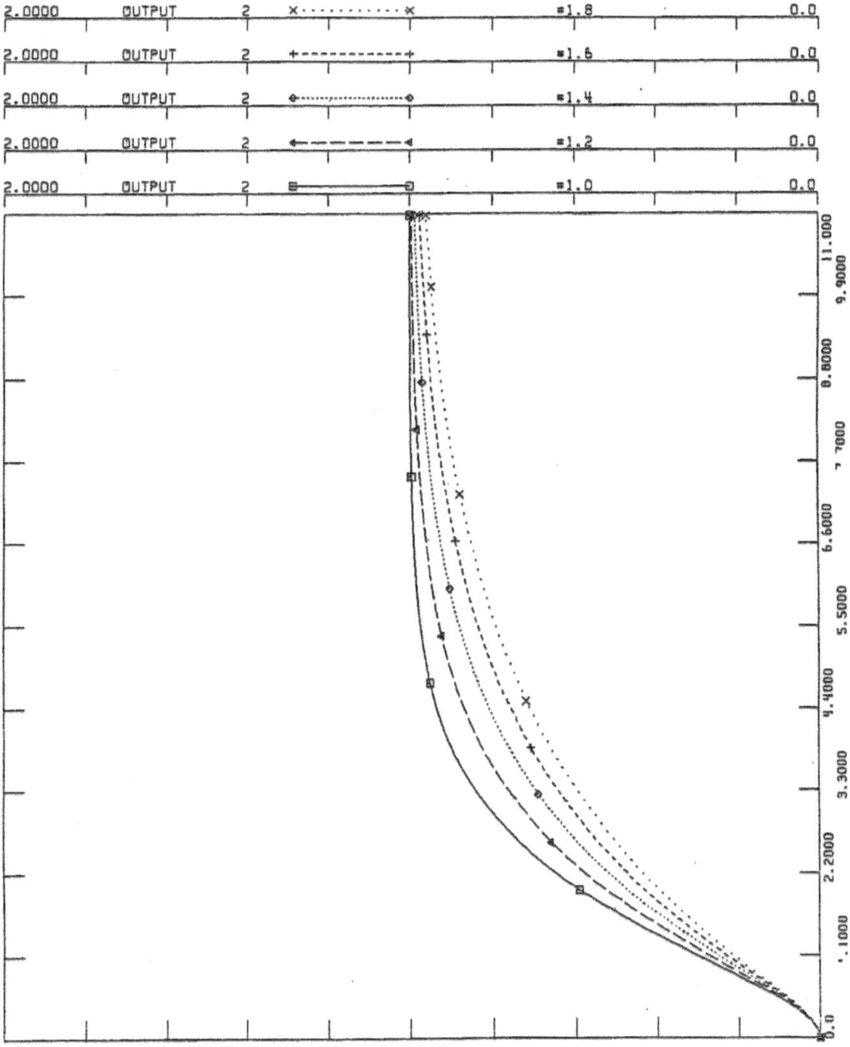

2.0000	OUTPUT	2	x·········x	=1.8	0.0
2.0000	OUTPUT	2	+----------+	=1.6	0.0
2.0000	OUTPUT	2	◆·········◆	=1.4	0.0
2.0000	OUTPUT	2	◀------◀	=1.2	0.0
2.0000	OUTPUT	2	▣——————▣	=1.0	0.0

Figure F-50

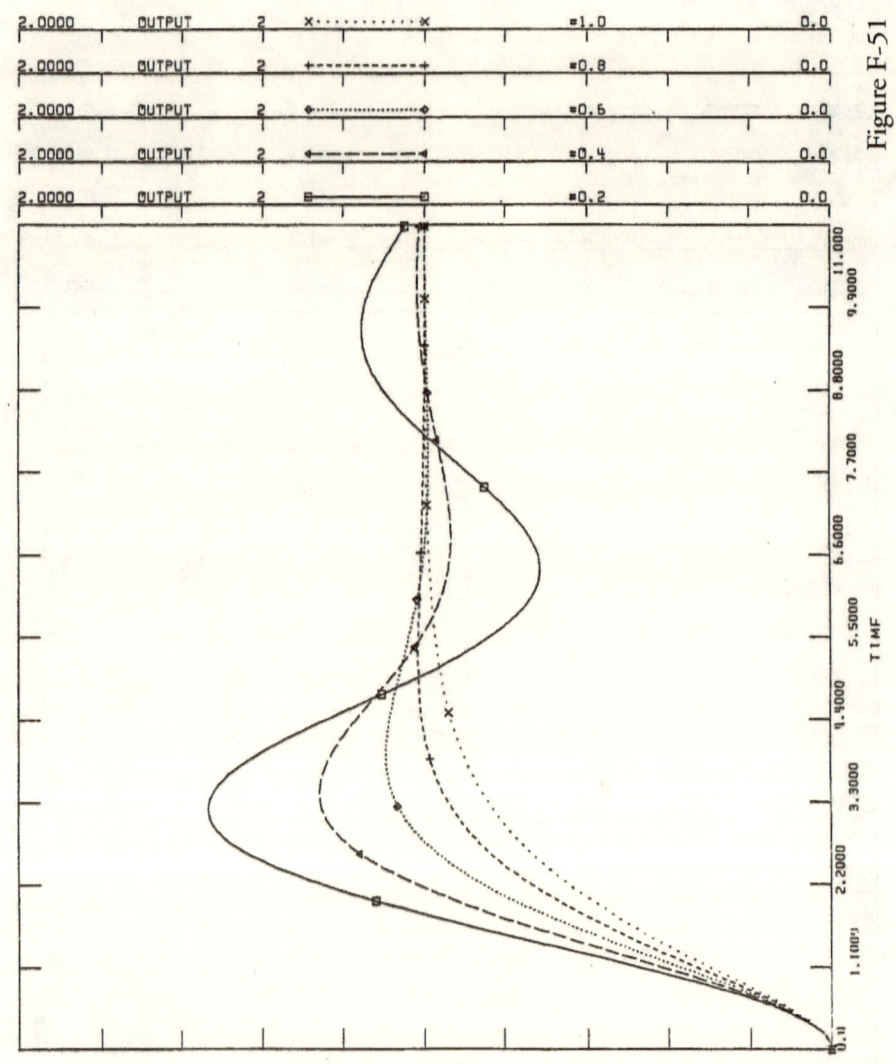

Figure F-51

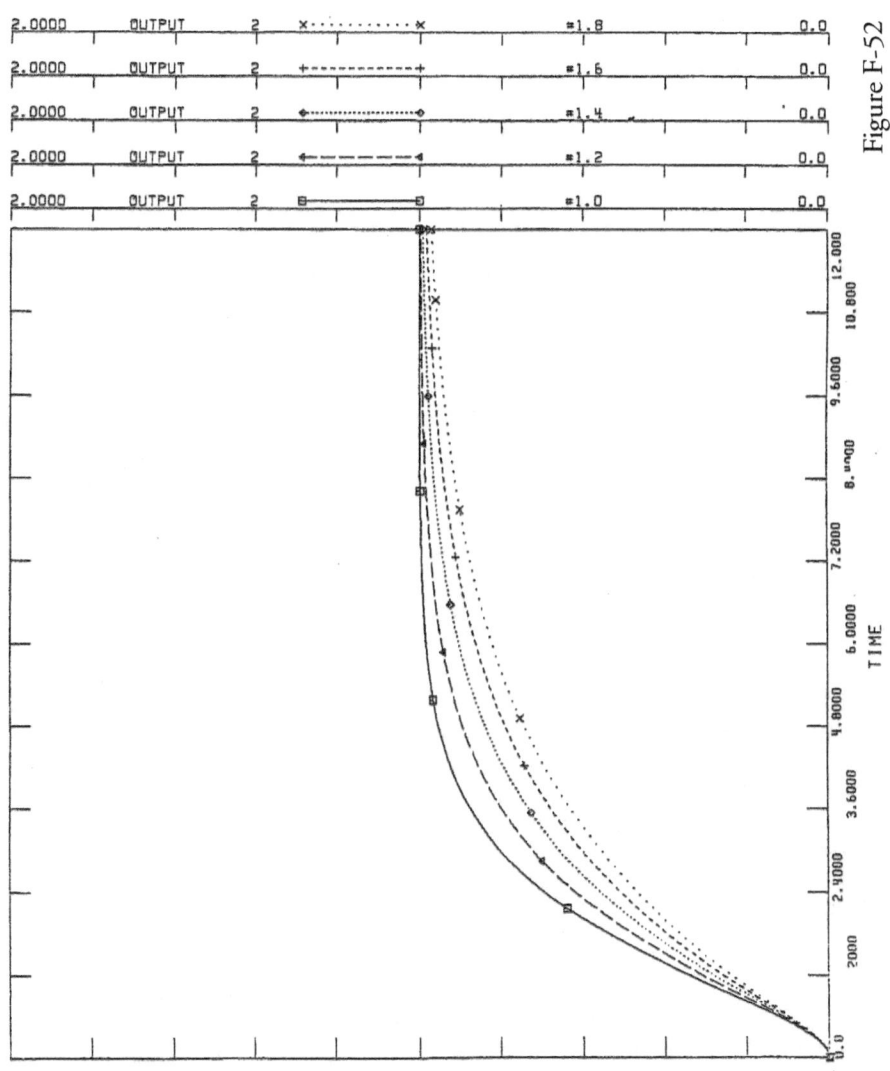

Figure F-52

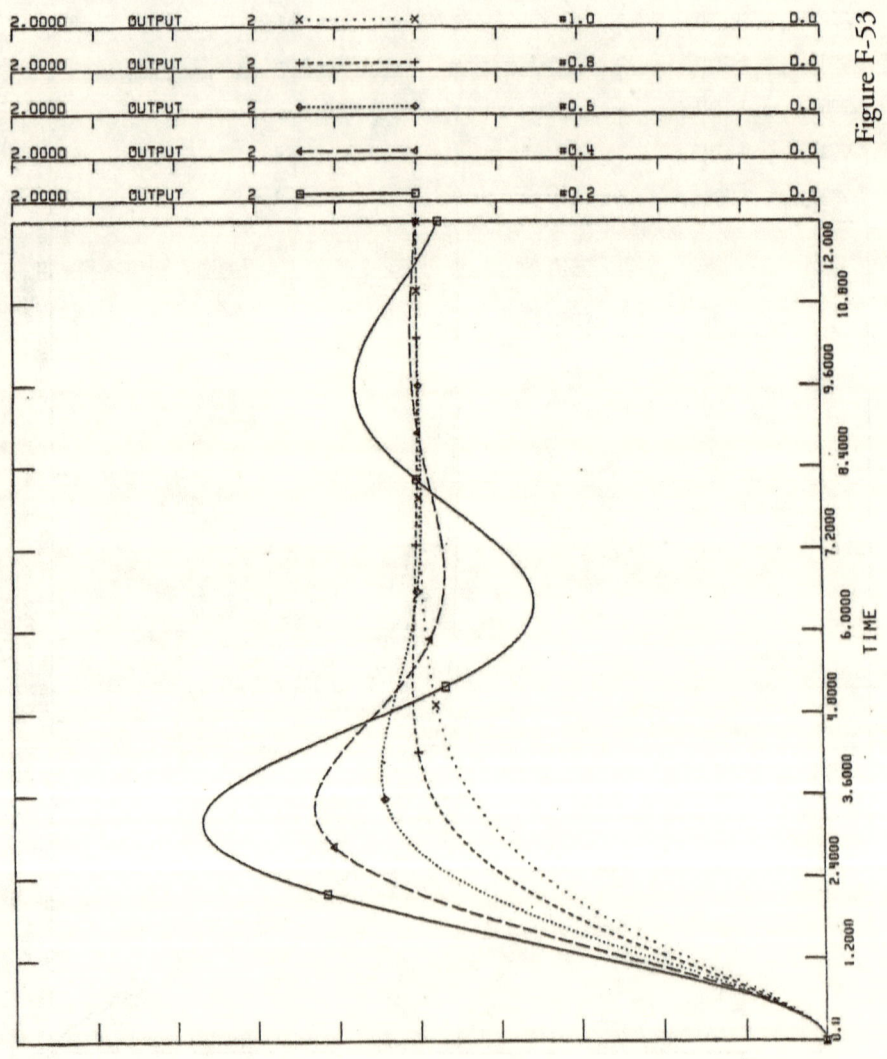

Figure F-53

REFERENCES

1. "Optimum Settings for Automatic Controllers", J. G. Ziegler and N. B. Nichols, Taylor Data Sheet No. TD5 10A 100, Taylor Instrument Companies. Rochester, NY.

2. "A Comparison of Controller Tuning Techniques", J. A. Miller, A. M. Lopez, C. L. Smith and P. W. Murrill, Control Engineering Dec. 1967, pp. 72-76.

3. "Analytical Tuning of Under-damped Systems", C. L. Smith and P. W. Murrill, ISA Journal Sept. 1966, pp. 48-53.

4. "A More Precise Method for Tuning Controllers", C. L. Smith and P. W. Murrill, May 1966, pp. 50-58.

5. "Tuning Controllers with Error-Integral Criteria", A. M. Lopez, J. A. Miller, C. L. Smith and P. W. Murrill, Instrumentation Technology, Nov. 1967. pp. 57-62.

6. "A Controller to Overcome Dead-time", O. J. Smith, ISA Journal, Feb. 1959, Vol. 6, No. 2, pp. 28-33.

7. "How Modeling Accuracy Affects Control Response", W. M. Whicater, Control Engineering. Oct. 1966, pp. 85-87.

8. "Interactive Dynamic Simulation by Digital Computer", J. M. Undrill and T. F. Laskowski, ISA Transactions. Vol. 14, 1975, pp. 33-40.

www.ingramcontent.com/pod-product-compliance
Lightning Source LLC
Chambersburg PA
CBHW020723180526
45163CB00001B/87